CONCEPT AND APPLICATION OF GROUP THEORY

CONCEPT AND APPLICATION OF GROUP THEORY

By

Dr. Kishor Arora

ANMOL PUBLICATIONS PVT. LTD.
NEW DELHI - 110 002 (INDIA)

ANMOL PUBLICATIONS PVT. LTD.
4374/4B, Ansari Road, Daryaganj
New Delhi - 110 002
Ph.: 3261597, 3278000
Visit us at: www.anmolbooks.com

Concept and Application of Group Theory

First Edition, 2003

ISBN 81-261-1431-2

PRINTED IN INDIA

Published by J.L. Kumar for Anmol Publications Pvt. Ltd New Delhi - 110 002 and Printed at Mehra Offset Press, Delhi.

Contents

Preface

Quantum mechanics provides the basis for solution of molecular structure and group theory is the mathematics which serves as a link between the molecular symmetry and quantum mechanics. Therefore, the knowledge of group theory is essential to understand the structural aspects of molecules.

The basic purpose of the present book is to describe the fundamental concepts related to symmetry and group theory and to discuss their applications, in lucid manner with optimum mathematical part which is required to understand the chemical problems. The book is divided into ten chapters emphasizing on the concepts and applications of group theory. It is tried to include enormous and sufficient number of examples to explain the concepts. Hope that this book will definitely serve the purpose and prove to be useful for the students who are pursuing their studies in this subject at graduation or post-graduation level.

I am highly thankful to Dr. R.K. Agarwal, Editor-in-Chief, Asian Journal Chemistry, who encouraged and guided me while the preparation of the manuscript of the book. My thanks are also due to Prof. (Dr.) G.D. Arora and Dr. D. Kumar, my fellow colleagues for critical suggestions during the preparation of this manuscript. I will like to record my special thanks to Prof. (Dr.) A.S. Tomar, Prof. (Dr.) V. Thakkar and my all departmental colleagues for their useful suggestions during this work. I am thankful to Ms. Sonal Agnihotri, my student who helped me during the preparation of the manuscript of this book.

Finally, I would like to express my regards and thanks to my parents, Mr. S.P. Arora and Mrs. Swadesh Arora, my wife, Mrs. Reena Arora my daughter Ms. Yashaswina and my son Master Chittin for their encouragement and sacrifices.

I shall be grateful to the readers of the book who may care to send their critical and healthy suggestions for further improvement in this book.

Author

1

Introduction to Symmetry, Symmetry Elements and Symmetry Operations

The word symmetry may be defined as a property of an object or *of a body* by virtue of which that object or that body can be divided into two equal and identical halves. The symmetrical structure is the most beautiful or proportionate structure. In order to understand, let us consider some different structures.

In Figure 1.1 some structures are drawn. Consider them one by one. Structure (i) is irregular or in other words it is not proportionate. Try to cut this structure along any line that passes through this structure, no two equal and identical halves can be obtained. So, it is not a symmetrical structure or this structure does not possess any symmetry. Consider the (ii) and (iii) structures, these are regular geometrical structures. In (ii) structure which is a square two equal halves can be obtained on cutting this structure along any dotted lines drawn on it. Structure (iii) which is equilateral triangle, two equal and identical halves can be obtained along the dotted lines drawn on it. So, these two structures i.e., (ii) and (iii) are symmetrical or they possess symmetry. Consider the (iv) structure. It is human hand. It is also irregular, unsymmetrical structure, but human body (v), possess bilateral symmetry as shown by the dotted line.

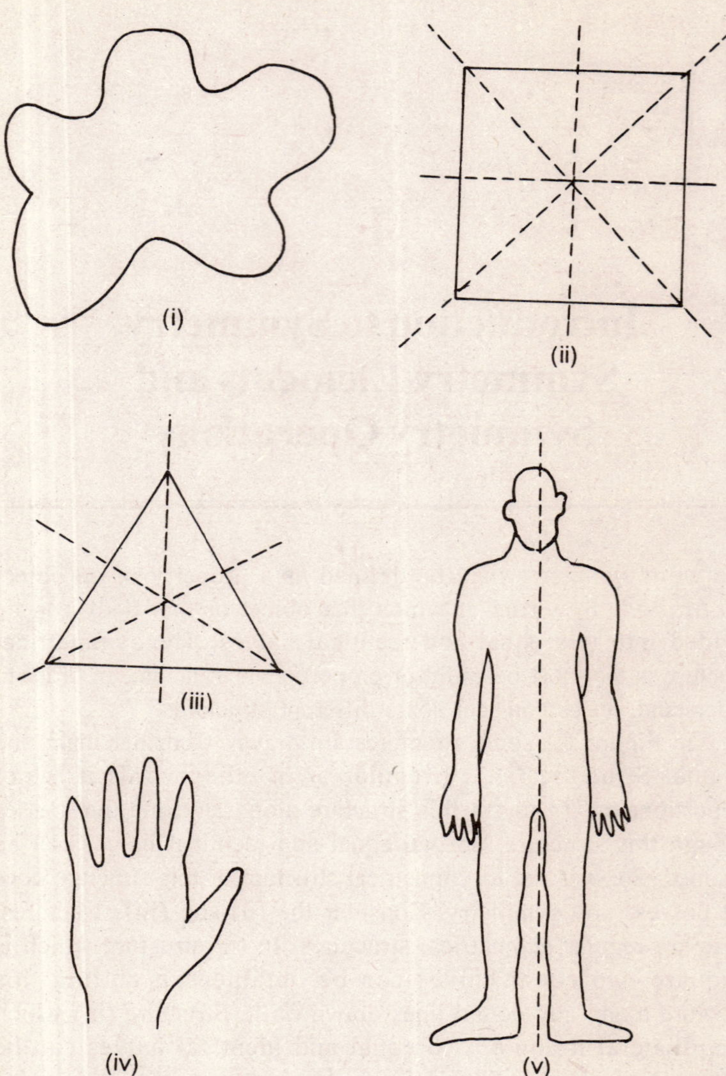

Fig. 1.1

So, it can be concluded that symmetry means regularity or proportionality in the object or in any structure because of which beauty is retained by that object. Apart from these examples there are enormous objects (or structures) *which can be seen in the nature around us which are symmetrical.* Furthermore, in different equivalent, indistinguishable orientations, the symmetry of any symmetrical object is retained i.e. for an object possessing symmetry more than one equivalent, indistinguishable three dimensional orientations are possible. Consider a cubic box (figure 1.2). In structure (ii) this box is placed inverted as in the structure (i) but in both the structures its symmetry is maintained so it is symmetrical. Both the structures are equivalent, indistinguishable orientations *of this box in space (for the sake of convenience its corners are numbered)*. After having the concepts of symmetry in general, let us study the symmetry in the molecules. Consider a homo-nuclear diatomic molecule (figure 1.3 (i)). On performing rotation along the dotted line or on placing a mirror along this dotted line, after reflection structure (ii) can be obtained which is equivalent, indistinguishable structure of this molecule. So, this molecule is said to possess symmetry (for the sake of convenience one atom A is labeled by prime (') mark). These two structures are equivalent structures for this homo-nuclear diatomic molecule. H_2, N_2, Cl_2, Br_2, I_2 etc., are some examples of these type of molecules.

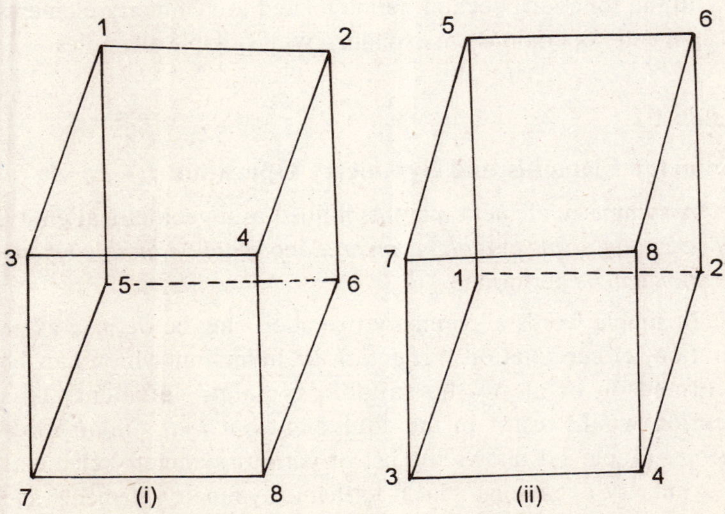

Fig. 1.2

In contrast, on considering hetero-nuclear diatomic molecules such as HCl, CO (AB type) no such symmetry is seen.

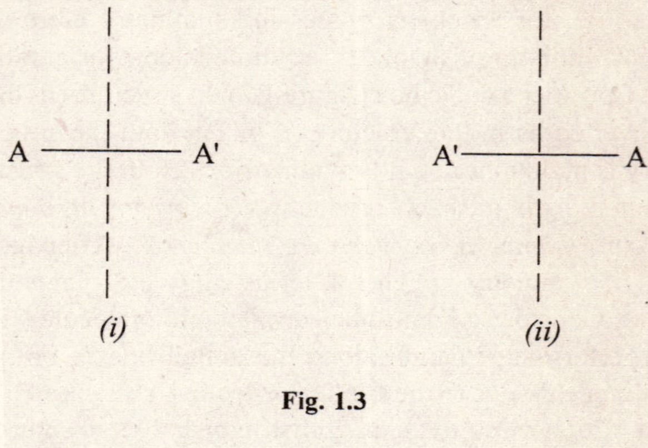

Fig. 1.3

A———B

Fig. 1.3

In the following section, terms related to symmetry elements and symmetry operations are explained with suitable examples.

Section 1.1

Symmetry Elements and Symmetry Operation :

A symmetry element may be defined as a geometrical entity, may be a *point, a line or a plane, on or along which related symmetry operation can be* performed.

In simple words a symmetry operation may be defined as an operation, either rotation, reflection or inversion which can be performed on or along the suitable symmetry element. This operation would result in the equivalent but *indistinguishable structure*. Table 1.1 shows the list of various symmetry elements and symmetry operations related to them. Symmetry elements and their related operations are so closely interrelated that operation of an element can be defined only with respect to that symmetry element.

Table 1.1 : Various symmetry elements and symmetry operations associated with them.

Symmetry element	*Symbol*	*Related Symmetry Operation*
1. Identity	E	....
2. Proper axis of symmetry	C_n	Rotation by an angle θ ($\theta = 360°/n$) along this axis.
3. Plane of symmetry	σ	Reflection through this plane.
4. Inversion centre or *centre of symmetry*	i	Inversion through this *centre (point)*
5. Improper axis of symmetry	S_n	Rotation by an angle θ ($\theta = 360°/n$) along this axis followed by reflection through a plane perpendicular to this axis.

Section 1.2

In this section let us deal with these symmetry elements and operations related to them one by one.

1. Identity (E): As identity is well-defined individuality of a person or of an object, like-vise this may be defined as well-defined structural configuration of a molecule. It is designated by the symbol 'E'. To make it clear, let us define the structure of H_2O molecule with positions of two hydrogen atoms as H_1 and H_2. This is the identity of H_2O molecule. Though both the hydrogen atoms are equivalent but for the sake of convenience these are labeled as H_1 and H_2. If after any of the operations (i.e., rotation or reflection) (ii) structure is obtained which may seems similar to the structure (i) but will not be termed as identical structure to the structure (i). It will be termed as equivalent structure. So, it is clear that equivalent and identical structures are two different terms. If after any of the operations on structure (ii) structure (i) is obtained back, then it is termed as identical structure i.e., the structure identical to the (i) well defined structure or identity.

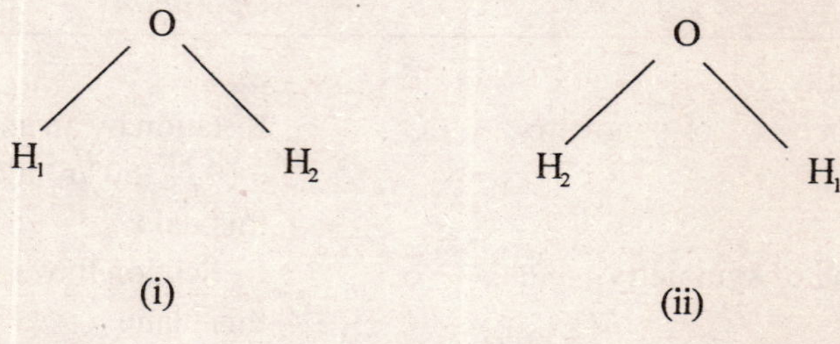

Fig. 1.4

Similarly each and every molecule has a well defined identity. Identity means doing nothing to the molecule. This is the only symmetry element which is possessed by all the molecules and there is no symmetry operation related to this symmetry element. Therefore, identity means the original orientation of the molecule. Similarly, identity can be defined for all the molecules as it is defined for water molecule above (figure 1.4). Figure 1.5 shows some examples.

2. Proper Axis of Rotation (C_n): It is defined as an imaginary line which passes through the molecular body, may be containing atom/s or through space, through which if rotation in clockwise direction by an angle θ (where $\theta = 360°/n$ or $2\pi/n$) is carried out, would result in the equivalent structure. It is designated by the symbol C_n. Here n is any integer and is known as order of the axis. There may be more than one axis present in a molecule but the axis with the highest order is considered as principal axis of rotation, and all other axes are subsidiary axes of rotation. Operation related to this symmetry element is rotation and the geometrical entity is a line.

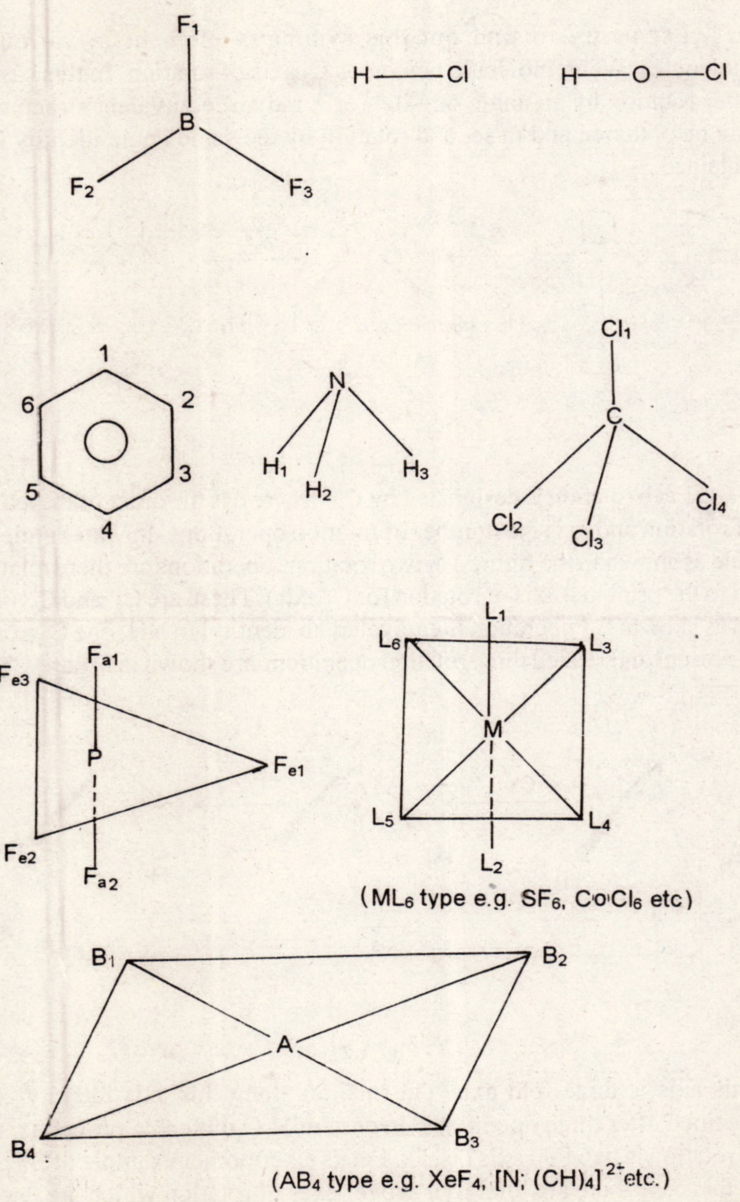
F1
B
F2
F3
H—Cl
H—O—Cl
1
2
3
4
5
6
N
H1
H2
H3
Cl1
C
Cl2
Cl3
Cl4
Fa1
Fe3
P
Fe1
Fe2
Fa2
L1
L6
L3
M
L5
L4
L2
(ML6 type e.g. SF6, CoCl6 etc)
B1
B2
A
B4
B3
(AB4 type e.g. XeF4, [N; (CH)4]2+ etc.)

Fig. 1.5

Let us try to find out this symmetry element in various molecules. Water molecule possesses C_2 axis of rotation. In this case after rotation by an angle of 180° (or π radian) equivalent structure can be obtained and in second rotation by the same angle identity is obtained.

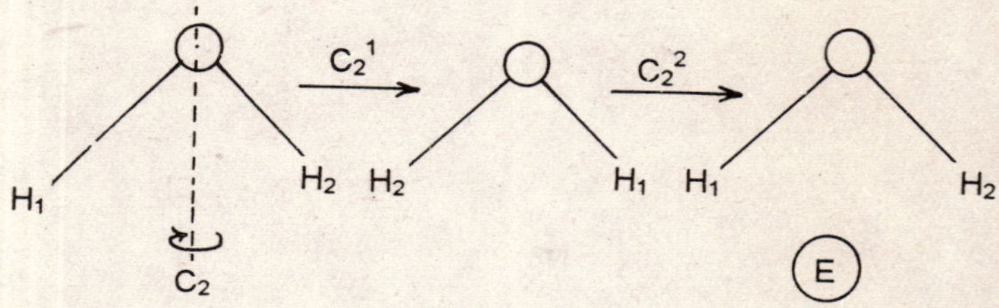

Fig. 1.6

Each rotation is designated by C_n^m where n is the order of the axis of rotation and m is the number of rotation operations. In water molecule as shown in the figure 1.6 two rotations operations are there related to the principal axis of rotation (or C_2 axis). These are C_2^1 and C_2^2 (= E) as shown. C_n^n operation is equivalent to identity. In NH_3 one C_3 axis is present and related three rotation operations are shown in Figure 1.7.

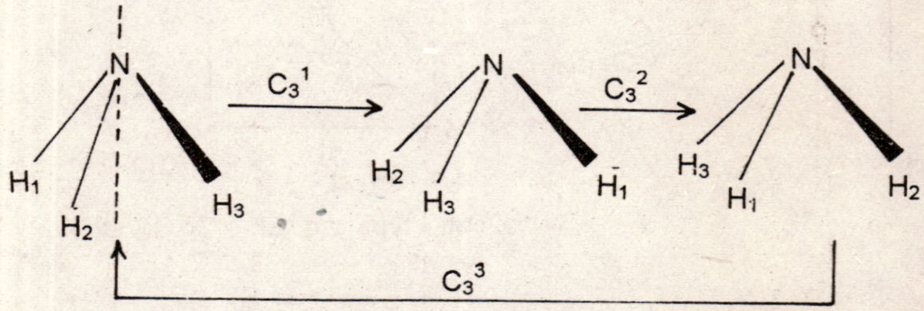

Fig. 1.7

This axis is three fold axis. On rotation along this axis identity is obtained after three operations. In case of H_2O molecule proper axis of rotation is two fold axis i.e., C_2. Let us take another example of BF_3. In this case there are total four proper axes of rotation which are one C_3 axis which passes through B-atom and which is the principal axis and three C_2 axes which are perpendicular to this C_3 principal axis.

These are known as subsidiaries axes and each one of this axis passes through B-F bond. Figure 1.8 shows the four axes of BF_3 molecule along with the related operation.

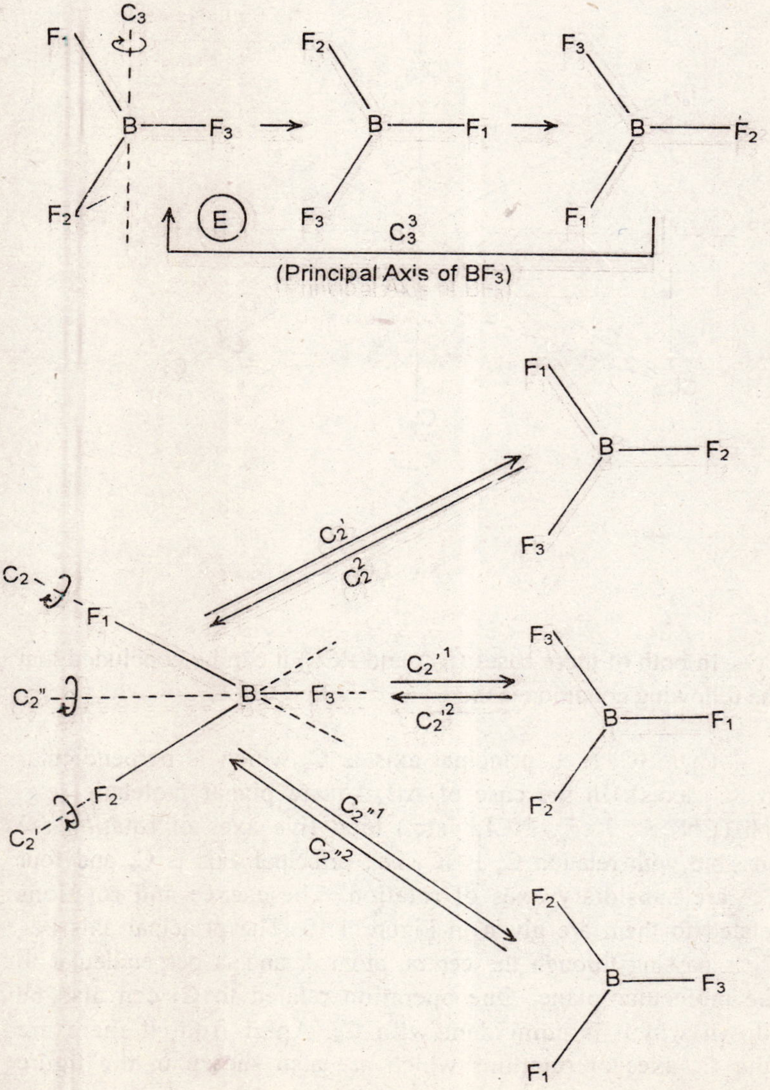

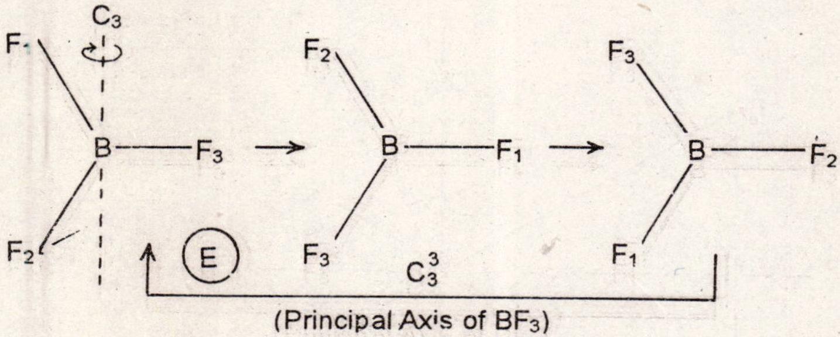

Fig,. 1.8

Similarly PCl_5 also has four proper axes of rotation i.e. one C_3, the principal axis and three C_2, the subsidiary axes. These are shown in Figure. 1.9.

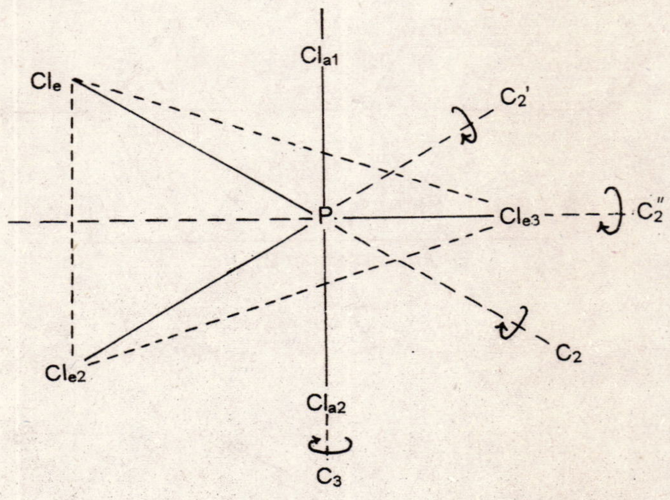

Fig. 1.9

In both of these cases (BF_3 and PCl_5) it can be concluded that the following condition is there

$C_3 \perp 3C_2$ (i.e., principal axis is C_3 which is perpendicular to $3C_2$ axes). In the case of AB_4 square planar molecule (e.g. $[Ni\ (CN)_4]^{2-}$, XeF_4, $PtCl_4^{2-}$ etc.) total five axes of rotation are possible with relation $C_4 \perp 4C_2$ i.e., principal axis is C_4 and four C_2's are subsidiary axes of rotation. These axes and rotations related to them are given in Figure 1.10. The principal axis i.e., C_4 is passing through the central atom A and is perpendicular to the molecular plane. One operation related to C_2 can also be shown which is coincident with C_4. Apart from it there are four C_2 axes or rotations which are also shown in the figure 1.10.

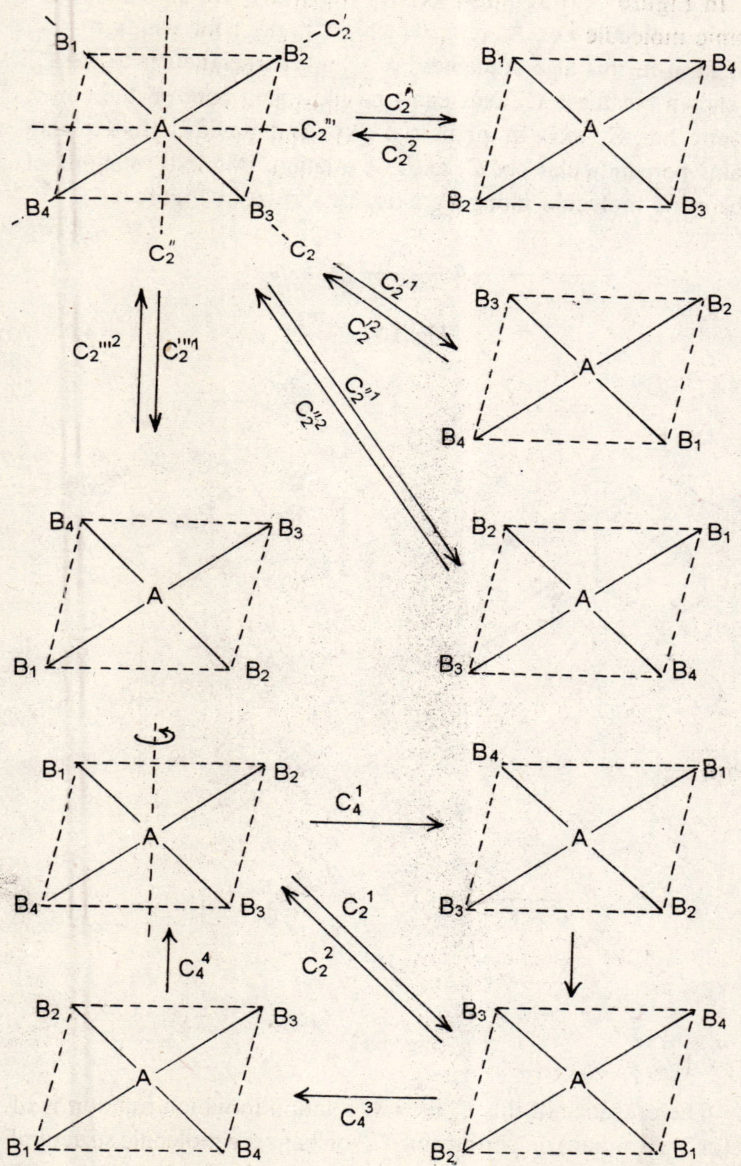
B1
B2
B3
B4
A
C2'
C2'''
C2''
C2
C2'1
C2^2
C2'''2
C2'''1
C2'1
C2'2
C2''1
C2''2
C4^1
C2^1
C2^2
C4^4
C4^3

Fig. 1.10

In Figure 1.11 rotation axis is illustrated for homo-nuclear diatomic molecule i.e., A_2 (e.g., H_2, N_2, Cl_2 etc.,) for which n = ∞. Apart from it, this also containes ∞ C_2 are perpendicular to the C_∞ axis shown but these C_2 axes can't be illustrated here on the paper. Benzene has C_6 axis as principal axis and besides that it also contains perpendicular six C_2 axes of rotation. Possible rotations of the benzene molecule along C_6 axis are shown in Figure 1.12.

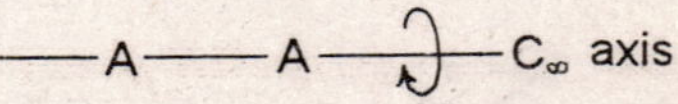

—A——A——C_∞ axis

Fig. 1.11

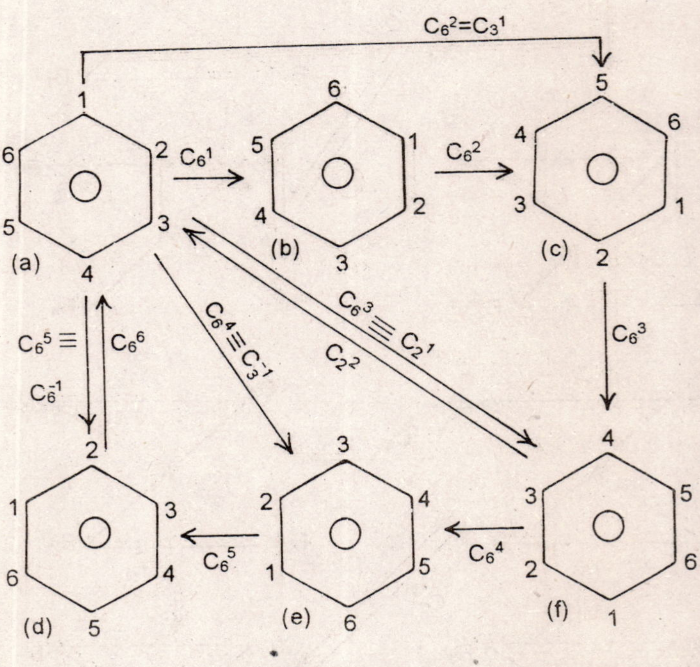

Fig. 1.12

The presence of this C_6 axis of rotation in which rotation is of 60° (or 2π/6 radian) on performing C_6^1 on benzene molecule structural representation equivalent to b is obtained. On C_6^2 operation representation c is obtained. This representation thus obtained from a structural representation in two successive C_6 rotations i.e., 60° followed by 60° rotations or in total 120° rotation (which is C_3^1 rotation axis). So, this can be written as

$$C_6^1 \cdot C_6^1 = C_6^2 = C_3^1$$

Structural representation d results after the three successive rotations on C_6 axis of rotation i.e., another 60° rotation from C_6^2 which can be directly obtained by a two fold rotation about the same axis i.e., $C_6^3 = C_2^1$ similarly following general algebric relations can be obtained i.e.,

$$C_6^4 = C_3^{-1}; C_6^5 = C_6^{-1} \text{ and } C_6^6 = C_1 = E$$

In allene molecule (Figure 1.13) there are three C_2 axis which are perpendicular to each other, one passing through the three carbon atoms and other two are perpendicular to this axis. In this case all the three axes are of two fold symmetry, so, the axis which is passing through the three carbon atoms (i.e., the molecular axis) is considered as principal axis. In trans and cis isomers of $C_2H_2Cl_2$,C_2 axis is there but their existence is different (Figure 1.14). In trans isomer this axis is perpendicular to the molecular plane and passes through C = C and in cis isomer it is in the molecular plane.

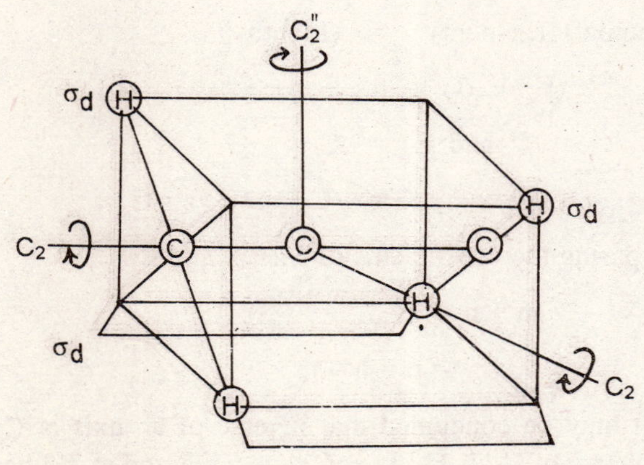

Fig. 1.13

(perpendicular to mol. plane)
C_2
Cl H
C=C
H Cl
Trans – $C_2H_2Cl_2$
(i)

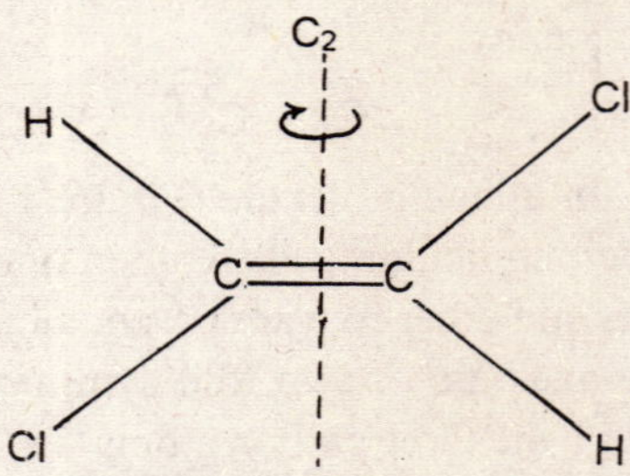

Fig. 1.14

Inverse of a proper axis of rotation: Consider that C_n^m is proper axis of rotation. Let C_n^p be its inverse. According to the algebric rule

$$(\text{Element})\,(\text{Element})^{-1} = (\text{Identity})$$

$$\text{i.e. } (C_n^m)\,.\,(C_n^p) = E$$

$$\text{and } E = C_n^n$$

$$= (C_n^m)\,.\,(C_n^p) = C_n^n$$

Comparing the coefficients of L.H.S. and R.H.S.

$$m + p = n$$

$$= p = n - m$$

So, it may be concluded that inverse of an axis is C_n^{n-m} for example inverse of C_3^2 is C_3^{3-2} i.e. C_3^1 (as n = 3 and m = 2 here). So, inverse of rotation operation related to C_3^2 is the structural representation related to C_3^1.

So, it is clear for proper axis or rotation or axis of rotation that:

(i). Leaving the atom/s at C_n axis all other atoms in the molecule must be in integral multiple of n.

(ii) Identity in any axis of rotation is obtained after completion of all rotations i.e., $C_n^n = E$.

(iii) The product of two or more symmetry operations is defined by the successive performance of individual operation.

(iv) A n-fold rotation axis, C_n implies the existence of all powers of rotation operations i.e., C_n, C_n^2, C_n^3 ---------- C_n^{n-1} and $C_n^n = E$.

3. Inversion centre (i): It is an imaginary point which lies inside the molecular body either on an atom or in space, through which if inversion is carried out would result in the similar or equivalent structure. It is designated by the symbol 'i'. Operation related to this symmetry element is inversion through a point. Obviously, equivalent structure can be obtained after performing the inversion operation if and only if this point lies at the centre of the molecule. So, it is termed as inversion centre. This point can't be considered anywhere in the molecule other than the centre. This is also known as centre of symmetry.

Let us try to find out this symmetry element in various molecules. Consider a water molecule. In this there is no possibility of locating any such point, so, centre of symmetry is absent in it. In the case of CO_2 molecule, 'i' lies on the central carbon atom. In homo-nuclear diatomic A_2 molecules (e.g., H_2, N_2, Cl_2 etc.,) this point lies in between the A-A bond i.e., in space. C_2H_4 possesses this symmetry element and the point lies in the mid of C = C i.e., in space. Same is the case with trans-$C_2H_2Cl_2$ molecule. In regular, AB_4 type square planar molecules (e.g., $[Ni(CN)_4]^{2-}$, XeF_4, $[PtCl_4]^{2-}$ etc.) it lies at the central A-atom, but in the regular AB_4 tetrahedron, it is absent i.e. there is no such symmetry element present in the case of regular AB_4 tetrahedral molecules (such as CH_4, CCl_4 etc.). In regular AB_6 octahedral molecules (such as $CO(NH_3)_6$, $(COCl_6)^{3-}$ etc.), this symmetry element lies on the central atom. It is absent in linear molecules like C = O (AB type) but is present in linear symmetrical AB_2 type molecules like CO_2 and again it is absent in unsymmetrical AB_2 type molecules like N-N-O. This symmetry element is present in benzene on the mid-point of the molecule i.e. in space, but on mono-substitution of benzene this symmetry element is lost i.e. in mono-substituted benzenes it is absent. In case of di-substituted benzenes only p-di-substituted benzenes have this symmetry element but it is not present in case of ortho or meta di-substituted benzenes. Figure 1.15 shows some examples related to this symmetry element along with related symmetry operations.

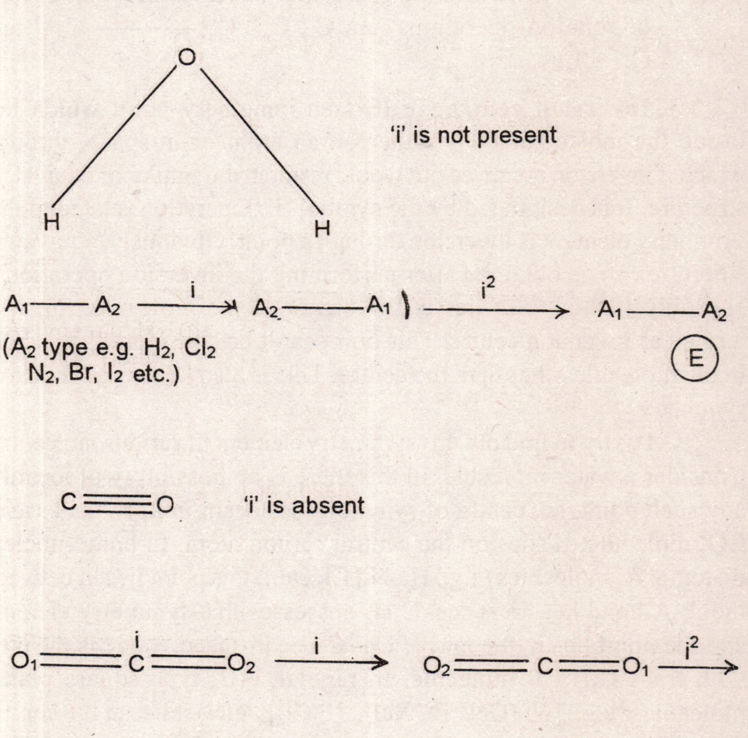

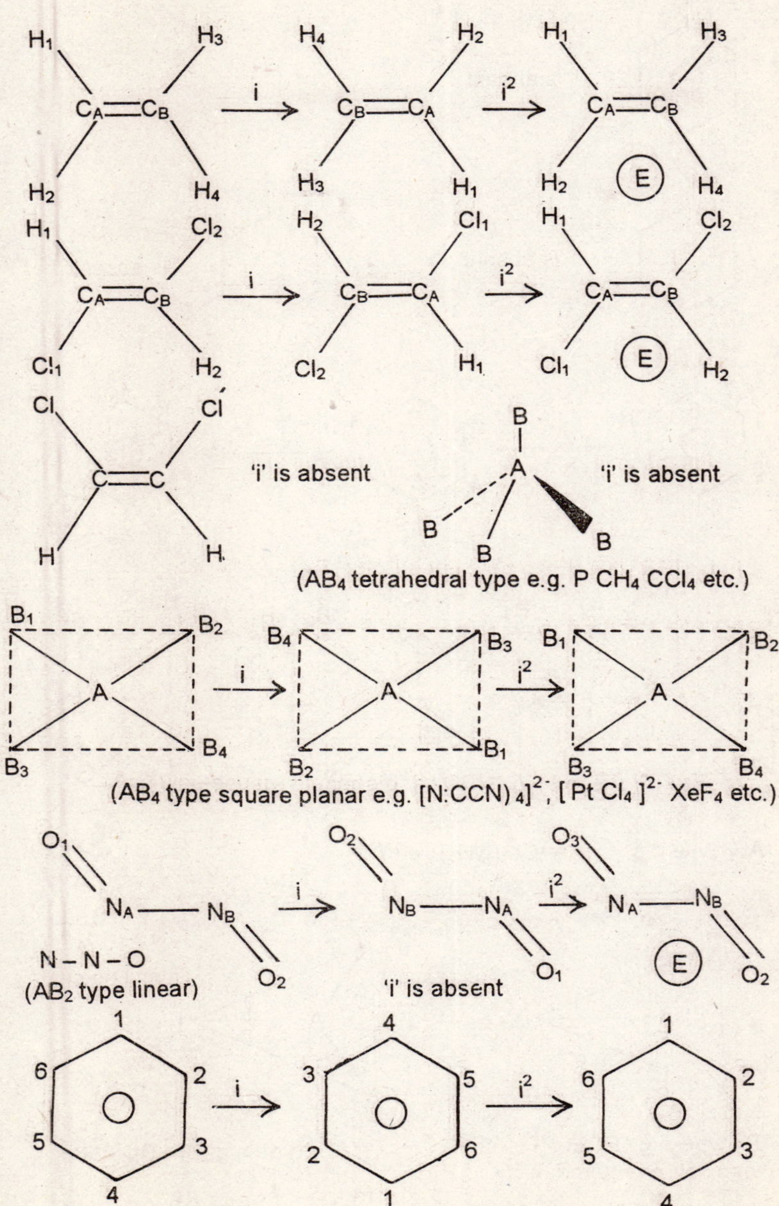
H1 H3 CA=CB H2 H4
i
H4 H2 CB=CA H3 H1
i2
H1 H3 CA=CB H2 E H4
H1 Cl2 CA=CB Cl1 H2
i
H2 Cl1 CB=CA Cl2 H1
i2
H1 Cl2 CA=CB Cl1 E H2
Cl Cl C=C H H
'i' is absent
B A B B B
'i' is absent
(AB4 tetrahedral type e.g. P CH4 CCl4 etc.)
B1 B2 A B3 B4
i
B4 B3 A B2 B1
i2
B1 B2 A B3 B4
(AB4 type square planar e.g. [N:CCN) 4]2-, [Pt Cl4]2- XeF4 etc.)
O1 NA NB O2
N – N – O
(AB2 type linear)
i
O2 NB NA O1
'i' is absent
i2
O3 NA NB E O2
1 2 3 4 5 6
i
4 5 6 1 2 3
i2
1 2 3 4 5 6

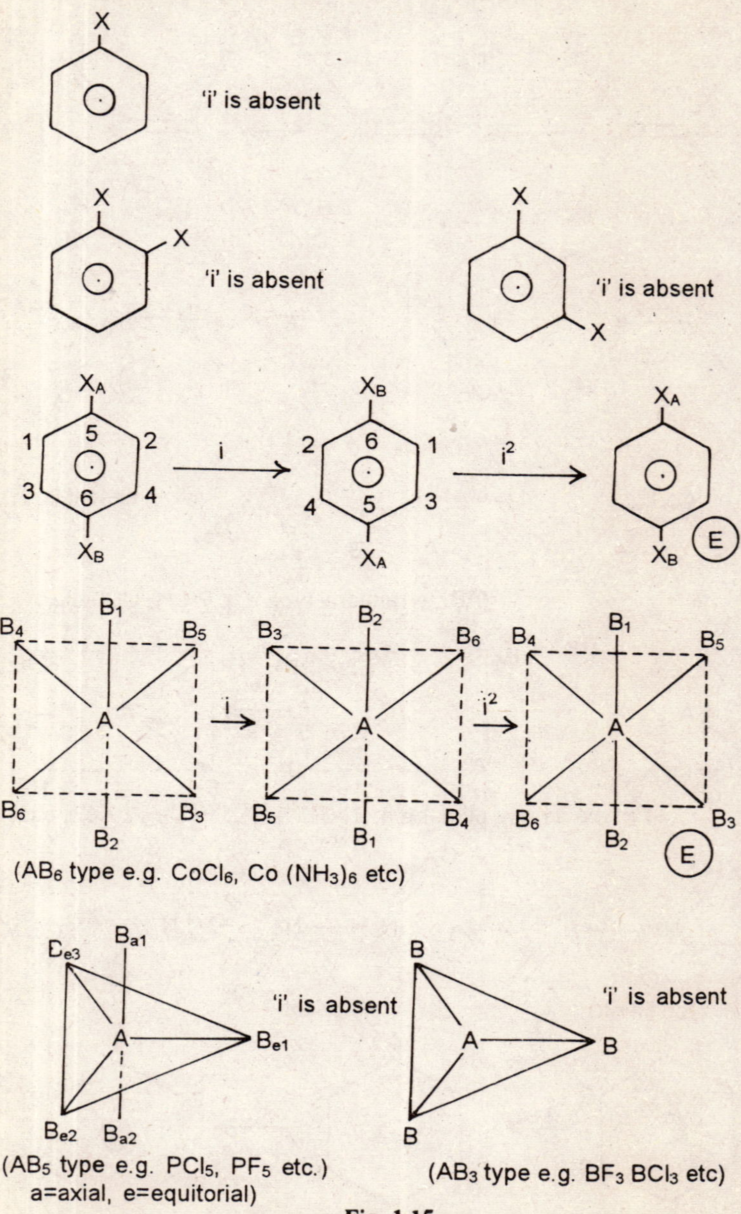

Fig. 1.15

It is clear from the operations shown in Figure 1.15 for this symmetry element that:

(i) All other atoms, except the atom which is the inversion centre itself, get inverted during inversion operation.

(ii) In the inversion operation-in first operation 'i' equivalent structure is obtained but in second operation 'i^2'identity is obtained. So, mathematically we can write

i^n (n = odd) = i (i.e. structure equivalent to inversion operation)

i^n (n = even) = E (i.e. structure equivalent to identity)

(iii) Inverse of inversion operation (or inversion centre) is inversion centre (Figure 1.16). Mathematically,

i^{-1} = i (i.e. the structure equivalent to the inversion centre operation)

(iv) Leaving the atom at centre, presence of all other atoms in pairs, is not necessary for the presence of centre of symmetry (e.g., AB_4 type tetrahedral molecules, cis-$C_2H_2Cl_2$)

$$O_1 = \overset{i}{C} = O_2 \xrightarrow{i} O_2 = C = O_1 \xrightarrow{i^{-1}} O_1 = C = O_2$$

Fig. 1.16

These are the properties of this symmetry element i.e. the properties of inversion centre.

4. Plane of Symmetry (σ): It is defined as an imaginary plane which passes through molecular body which may contain some atoms or which may be in space, through which if reflection is carried out, would result in the similar or equivalent structure. It is designated by 'σ' (sigma). Operation related to this symmetry element is reflection through this plane.

Let us try to find out this symmetry element in various molecules. Consider water molecules, this molecule has to planes as shown in Figure 1.17. Both of these planes are vertical planes as they contain the principal axis (C_n axis). One plane passes through the oxygen atom and other plane is the plane of the molecule or here it is the plane of the paper.

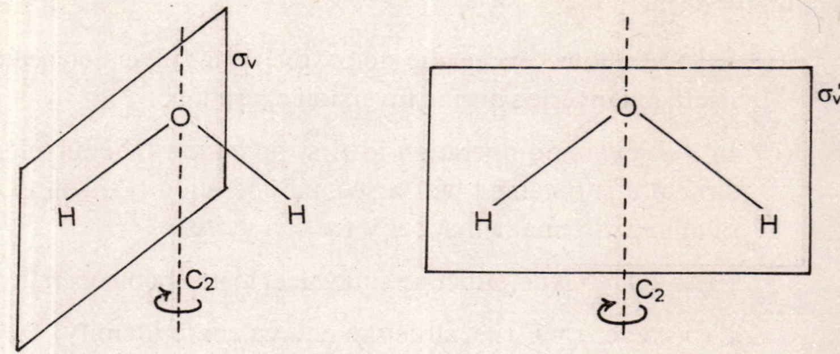

Fig. 1.17

A vertical plane is designated by σ_v and is defined as plane which contains the principal axis or which is parallel to principal axis. Operation related to σ_v of water molecule are given in Figure. 1.18.

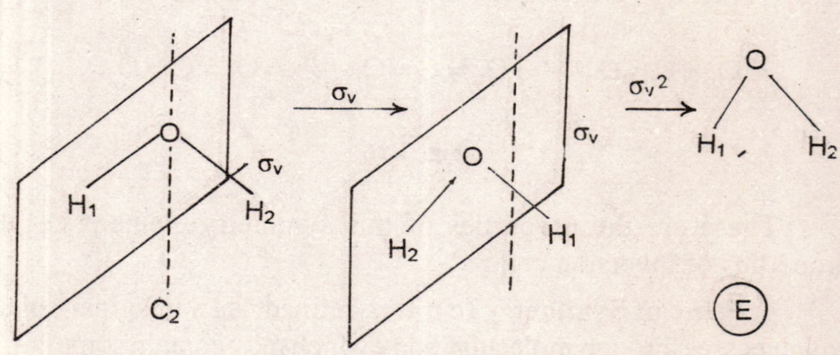

Fig. 1.18

There is another type of plane i.e. horizontal plane, which is designated by σ_h and which is defined as the plane which is perpendicular to the principal axis of rotation. Figure 1.19 shows the horizontal plane (σ_h) in case of PCl_5.

It is clear from the Figure 1.19 that in PCl_5 molecule the PCl_3 triangle is itself the horizontal plane which is perpendicular to the principal axis. If reflection is carried out through this σ_h plane (i.e.

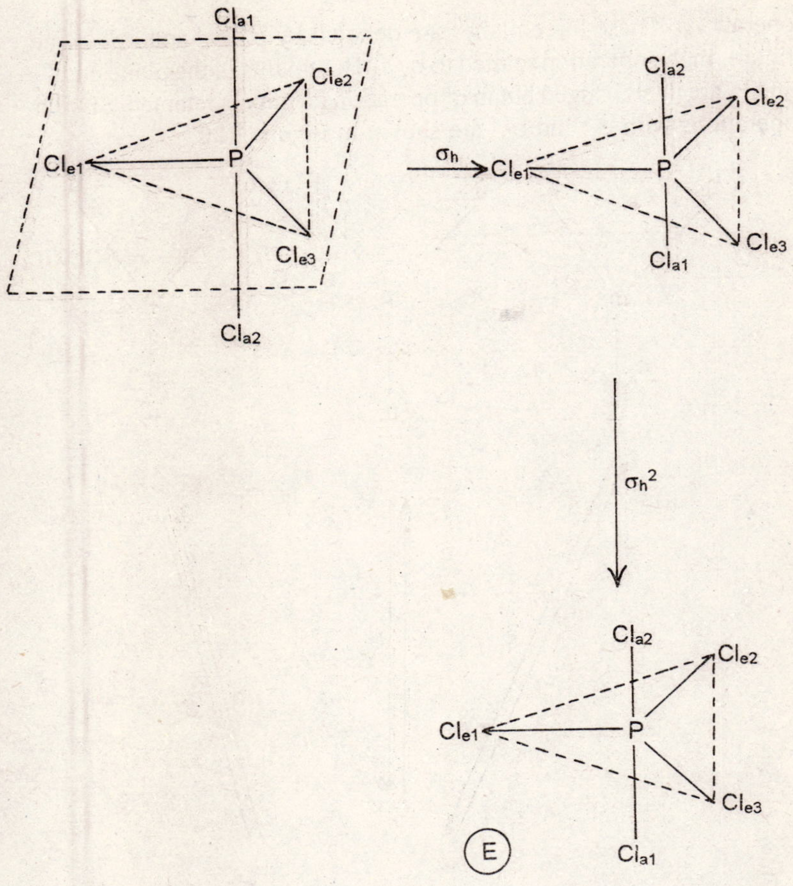

Fig. 1.19 : (a-axial, e-equitorial).

operation of reflection of axial chlorines) only these axial chlorines are exchanged in this reflection operation. Apart from it, PCl_5 also contains three σ_v i.e. planes passing through PCl_{a1}, Cl_{a2}, Cl_{e1}; PCl_{a1}, Cl_{a2}, Cl_{e2} and PCl_{a1}, Cl_{a2}, Cl_{e3}. In BF_3 also σ_h is present which is the molecular plane itself i.e. the BF_3 triangular plane as principal axis of rotation is perpendicular to this palne. Apart from this BF_3 also contains vertical planes of BF_3 are shown in Figure 1.20 with their related operations. These three planes are denoted by σ_v, σ_v' and σ_v'' in the figure. In the operation related to σ_v F_1-B, remains in the plane and F_2 and F_3 are interchanged but in σ^2 operation identity is retained. Similar operations with σ_v' and σ_v'' are shown in Figure 1.20.

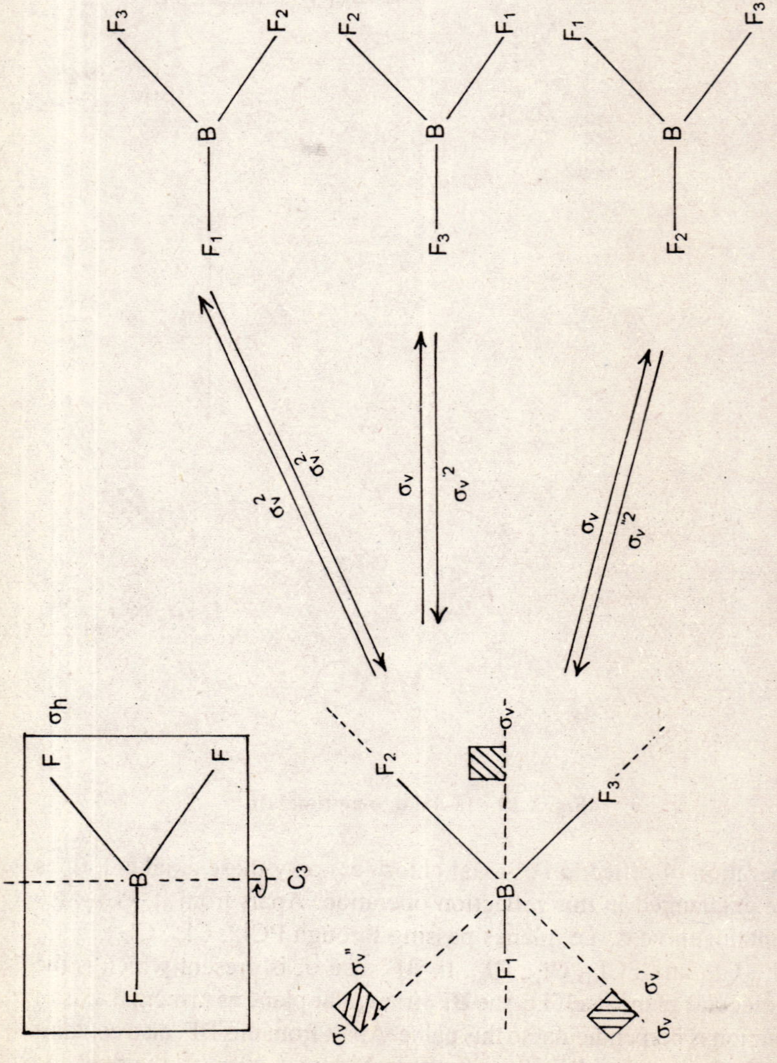

Fig. 1.20

In the case of NH_3 molecule there are three σ_v's (i.e. three vertical planes). Each one of these contains N and one of the H-atom. Figure 1.21. illustrates the operations related to three vertical planes of NH_3.

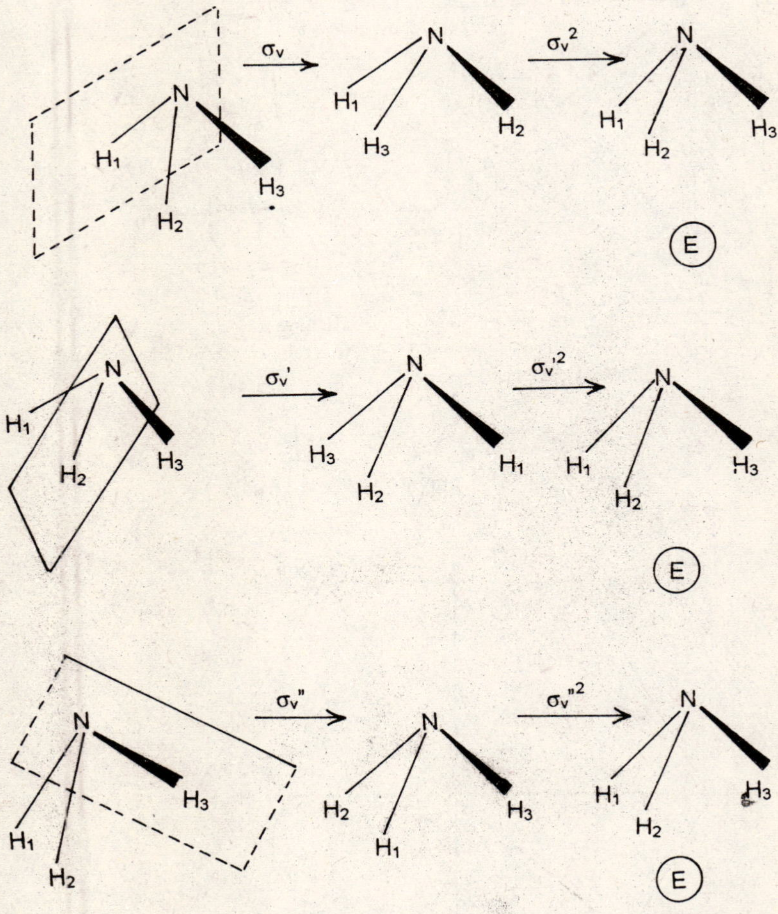

Fig. 1.21

A third type of plane is dihedral plane which is designated by σ_d. Dihedral plane is a plane which bisects the angles between the σ_v planes. This plane also contains the principal axis of rotation.

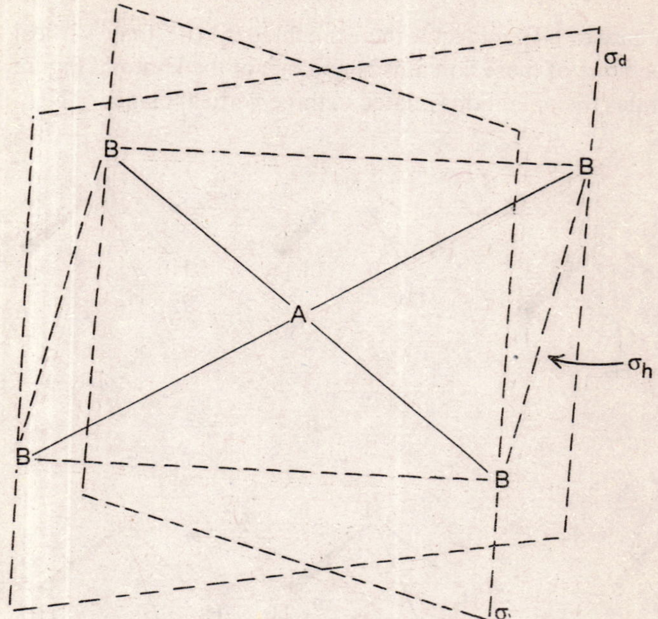

Fig. 1.22

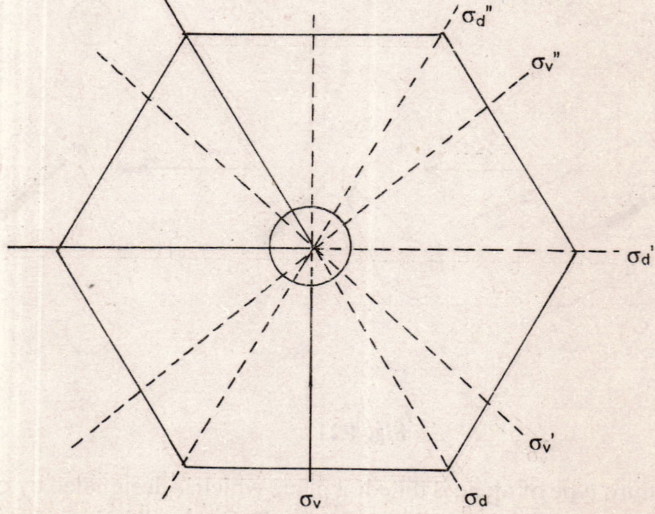

Fig. 1.23

This plane can be easily illustrated in the case of AB_4 square planar molecule such as XeF_4 (Figure 1.22). In benzene also all the three types of planes i.e. σ_v, σ_h and σ_d are present. Figure 1.23 shows σ_v and σ_d of benzene. σ_h is the plane of the molecule itself. In trans-$C_2H_2Cl_2$ only plane i.e. σ_h or horizontal plane is present which is the plane of molecule itself, but in case of cis-$C_2H_2Cl_2$, two vertical planes are present that contain the principal axis of rotation. Out of these one is the molecular plane and other is shown in the Figure 1.24.

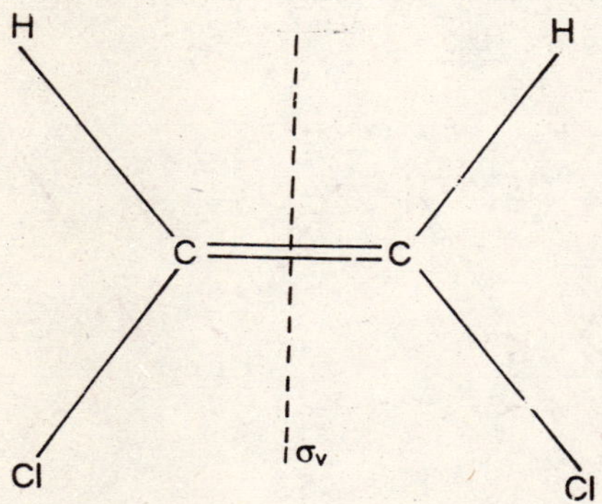

Fig. 1.24

So, it is clear that the plane of symmetry has the following properties:

(i) All other atoms, leaving those atoms which are in the plane, must be in even number and they get reflected during the related operation.

(ii) During first reflection operation through any plane equivalent structure is obtained while in the second operation identity is retained or in other words in odd operations equivalent structure is obtained while in even operations related to reflection, identity is retained; mathematically

σ^n (n = odd) = σ (i.e. structure related to reflection operation is obtained and

σ^n (n = even) = E (i.e. structure related to identity is obtained)

(iii) With reference to principal axis of rotation, three types of planes are there viz. σ_v – vertical plane, plane passing through the principal axis σ_h –horizontal plane which is perpendicular to the principal axis and σ_d – dihedral plane, plane passing through the principal axis and bisect the angles between the σ_v.

(vi) Inverse of reflection operation through a plane is structure equivalent to reflection i.e. inverse of a plane is a plane or mathematically $\sigma^{-1} = \sigma$ (Figure 1.25)

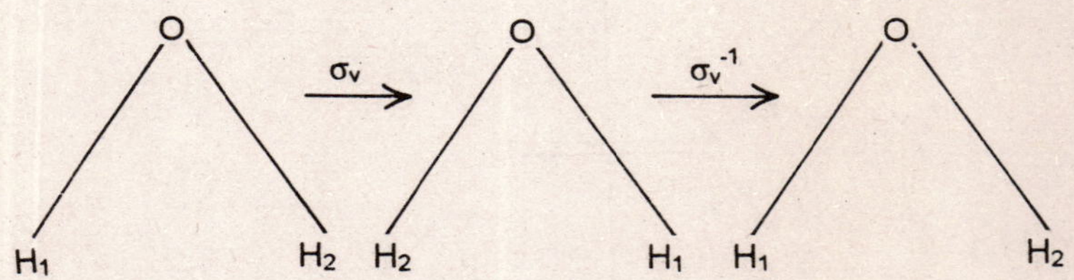

Fig. 1.25

5. Improper axis of rotation (S_n): It is defined as an imaginary line which passes through the molecular body on or along which the rotation by an angle θ (θ = 360°/n or 2π/n) followed by the reflection through a plane which is perpendicular to the axis, would result in the equivalent structure. It is designated by the symbol 'S_n' and it is the combination of two operations that are rotation followed by reflection. This operation can be written as the combination of two i.e.

$$S_n = \sigma . C_n \text{ or}$$

$$S_n = C_n \text{ followed by } \sigma$$

The plane which is perpendicular to proper axis of rotation is the horizontal plane. So, it may be written that $S_n = C_n$ followed by σ_h. Figure 1.26 shows the S_6 operations in staggered form of ethane. In this representation ethane molecule is represented with the help of two triangles with hydrogen atoms are labelled as 1 to 6.

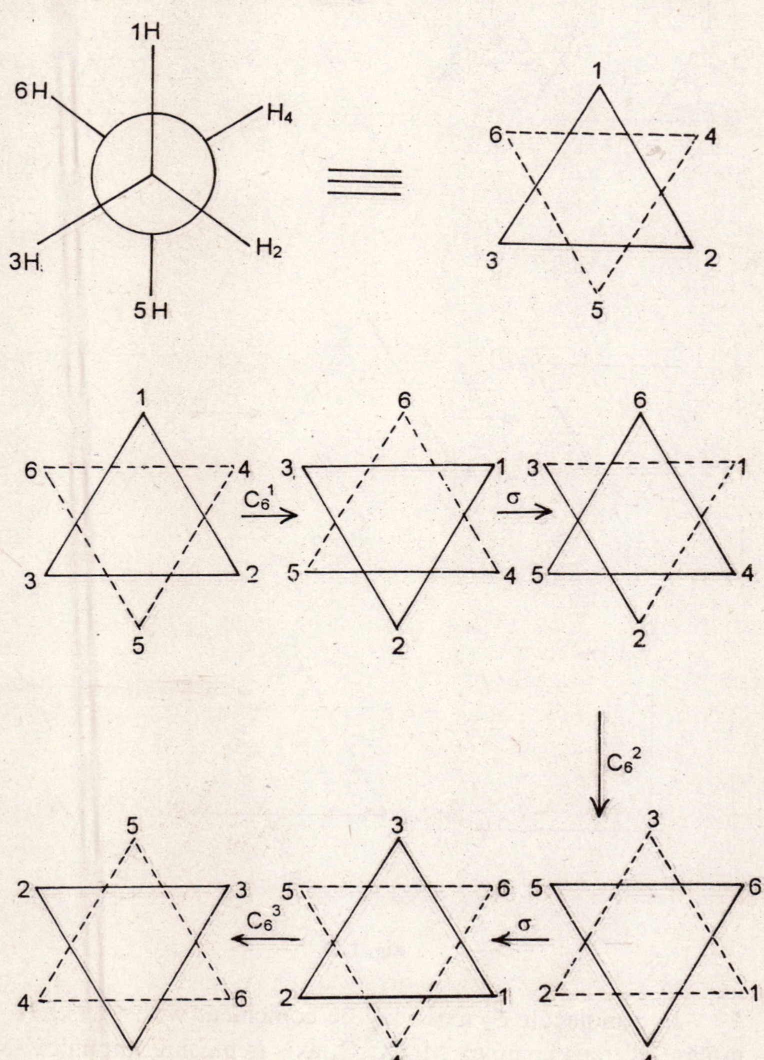

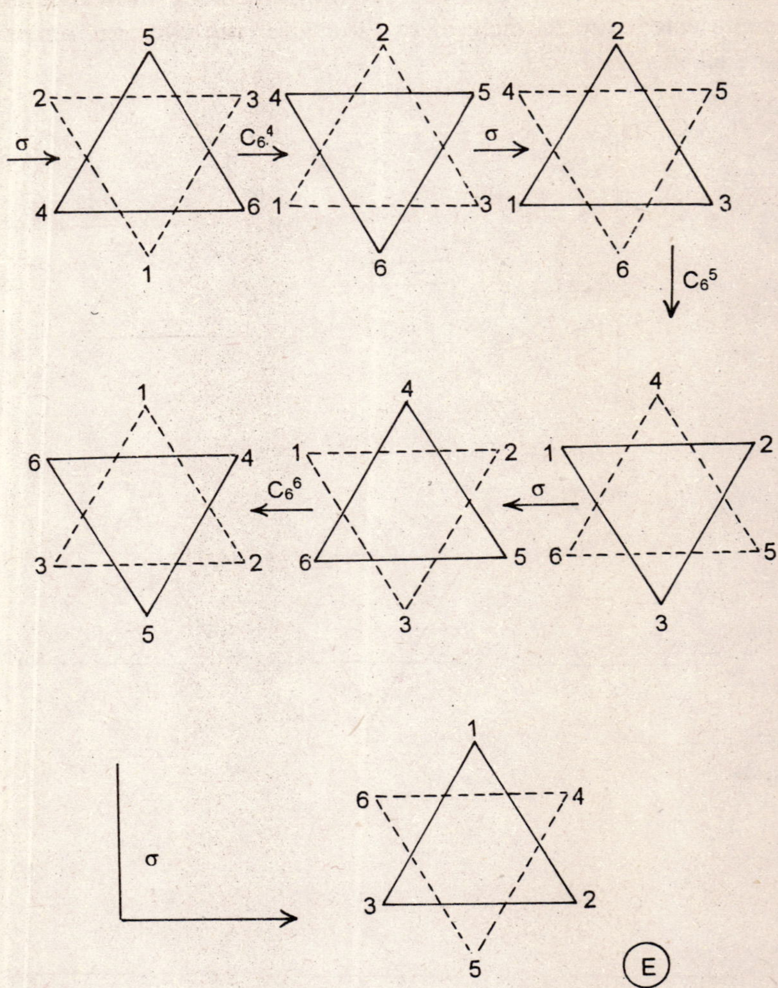

Fig. 1.26

In a molecule S_n axis may be coincident with C_n axis, e.g. in case of trans complex ML_4X_2, C_4 axis is passing through X-M-X which is coincident with S_4 axis (Figure 1.27). In this molecule σ or σ_h is ML_4 square plane. It is clear from Figure 1.27 that identity is obtained after S_4 operation. So, it can be concluded that

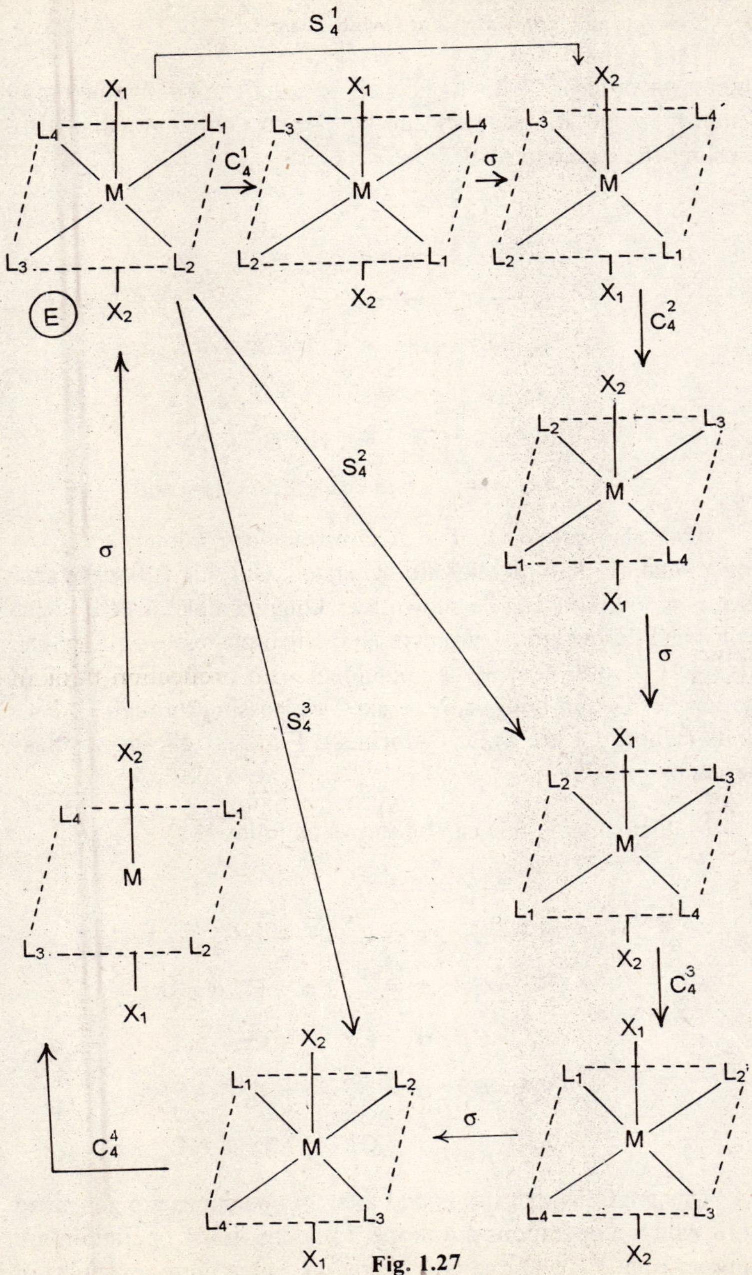

S 1 4
X1
L4
L1
M
L3
L2
X2
E
C 1 4
L3
L4
L2
L1
σ
X2
L3
L4
M
L2
L1
X1
C 2 4
X2
L2
L3
M
L1
L4
X1
S 2 4
σ
σ
S 3 4
X1
L2
L3
M
L1
L4
X2
C 3 4
X1
L1
L2
M
L4
L3
X2
σ
X2
L1
L2
M
L4
L3
X1
C 4 4
X2
L4
L1
M
L3
L2
X1

Fig. 1.27

when n (or order) of the axis is even, identity can be obtained after n improper operations. This can be shown mathematically also. Consider the example of S_6.

$$S_6^1 = C_6^1 . \sigma = S_6^1$$

$$S_6^2 = C_6^2 . \sigma^2 = C_6^2 . E = C_6^2$$

$$S_6^3 = C_6^3 . \sigma^3 = C_6^3 . \sigma = S_6^3$$

$$S_6^4 = C_6^4 . \sigma^4 = C_6^4 . E = C_6^4$$

$$S_6^5 = C_6^5 . \sigma^5 = C_6^5 . \sigma = S_6^5$$

$$S_6^6 = C_6^6 . \sigma^6 = E . E = E$$

(as σ^n = (n = even) = E; σ^n (n = odd) = σ)

This also proves that in n improper operations, (n/2) – 1 proper rotations and identity are generated. But this is not the case when n = odd. This can be shown by taking example of PCl_5 which has trigonal-bi-pyramidal structure and which possesses S_3 improper axis. Here also S_n (i.e. S_3) is coincident with principal C_n axis of rotation i.e. C_3 and both of these axes are passing through Cl-P-Cl bonds (where Cl are axial Chlorines). Figure 1.28 shows these operations in PCl_5.

Mathematically this can be shown as follows:

$$S_3^1 = C_3^1 . \sigma = S_3^1$$

$$S_3^2 = C_3^2 . \sigma^2 = C_3^2 . E = C_3^2$$

$$S_3^3 = C_3^3 . \sigma^3 = C_3^3 . \sigma = E . \sigma = \sigma$$

$$S_3^4 = C_3^4 . \sigma^4 = C_3^1 . E = C_3^1$$

$$S_3^5 = C_3^5 . \sigma^5 = C_3^2 . \sigma = S_3^2$$

$$S_3^6 = C_3^6 . \sigma^6 = C_3^3 . E = E . E = E$$

This confirms that if n is odd total 2n operations are generated out of which n operations are proper rotations, n – 1 operations are improper rotations and one is σ operation. Some more examples of molecules which contain S_n axis are given in Figure 1.29.

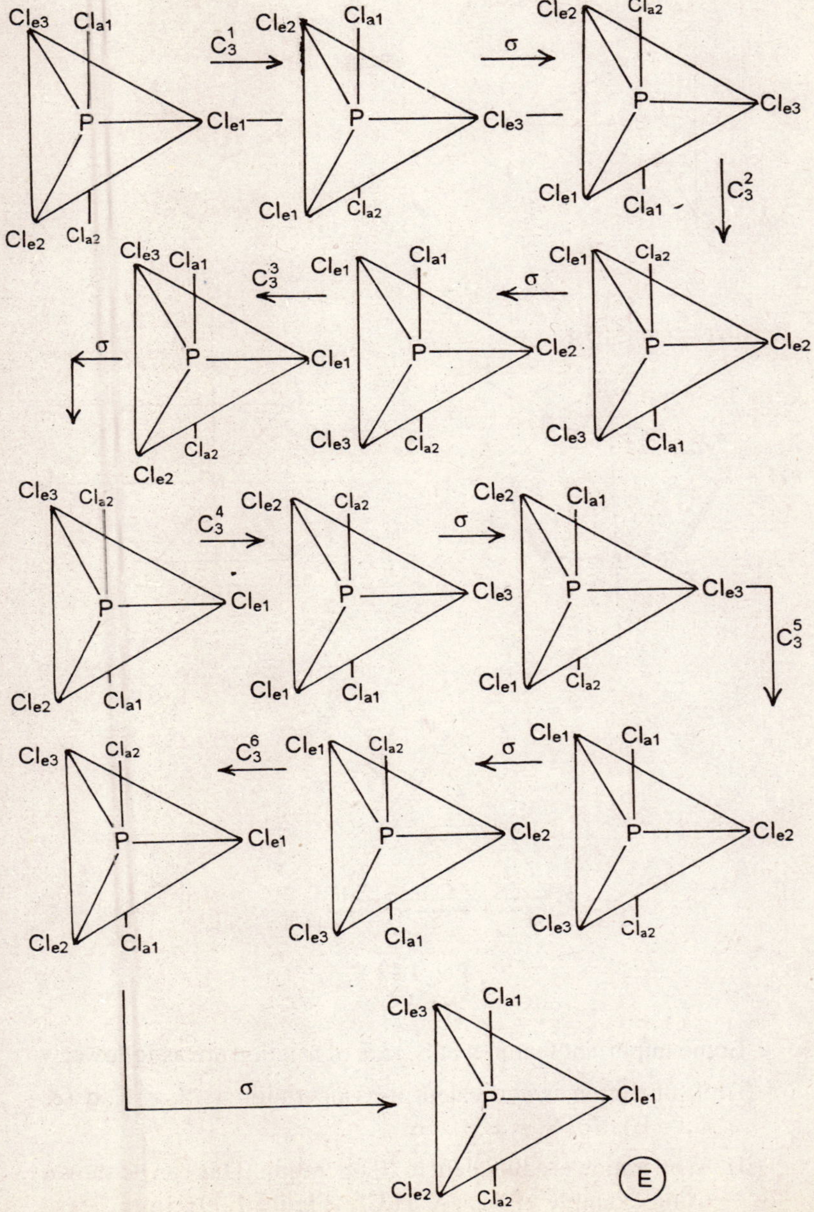

Cle3
Cla1
P
Cle1
Cle2
Cla2
C31
Cle2
Cla1
P
Cle3
Cle1
Cla2
σ
Cle2
Cla2
P
Cle3
Cle1
Cla1
C32
Cle1
Cla2
P
Cle2
Cle3
Cla1
σ
Cle1
Cla1
P
Cle2
Cle3
Cla2
C33
Cle3
Cla1
P
Cle1
Cla2
Cle2
σ
Cle3
Cla2
P
Cle1
Cle2
Cla1
C34
Cle2
Cla2
P
Cle3
Cle1
Cla1
σ
Cle2
Cla1
P
Cle3
Cle1
Cla2
C35
Cle1
Cla1
P
Cle2
Cle3
Cla2
σ
Cle1
Cla2
P
Cle2
Cle3
Cla1
C36
Cle3
Cla2
P
Cle1
Cle2
Cla1
σ
Cle3
Cla1
P
Cle1
Cle2
Cla2
E

Fig. 1.28

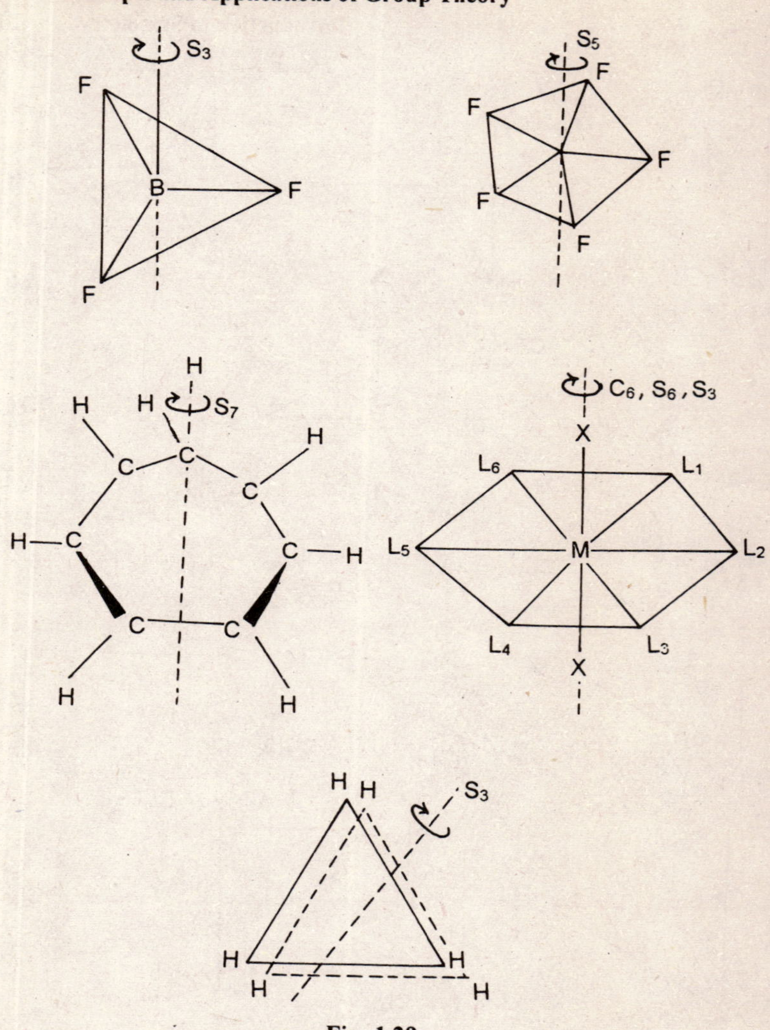

Fig. 1.29

Some important features of S_n axis of rotation are as follows:

(i) S_1 operation is equivalent to σ operation as $S_1 = C_1.\sigma$ (& $C_1 = E$). So, $S_1 = E.\sigma = \sigma$.

(ii) S_2 operation is equivalent to 'i' operation. This can be shown using example of trans-$C_2H_2Cl_2$ (Figure 1.30). In this case

C_n is perpendicular to the molecular plane and passes through $C = C$ and σ_h is the molecular plane itself.

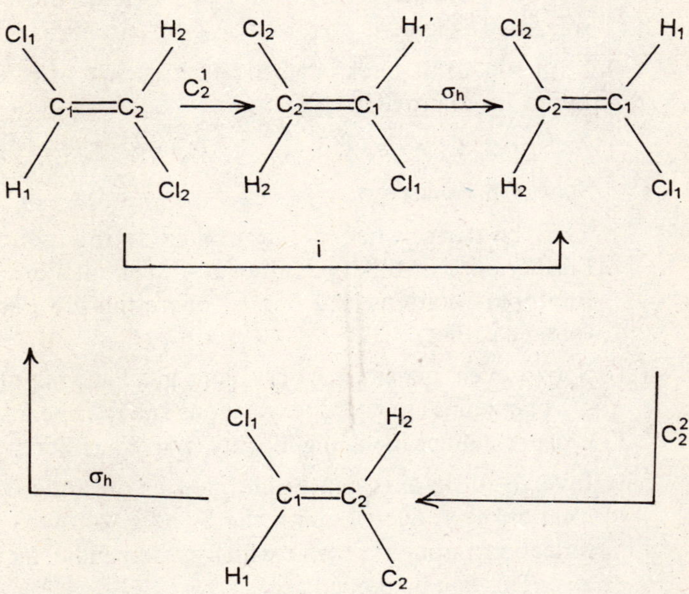

Fig. 1.30

(iii) If C_n and σ_h are present in a molecule S_n should also be present. This can be explained using example of eclipsed ethane in which C_n and S_n are coincident and both of these axes pass through C-C (Figure 1.31).

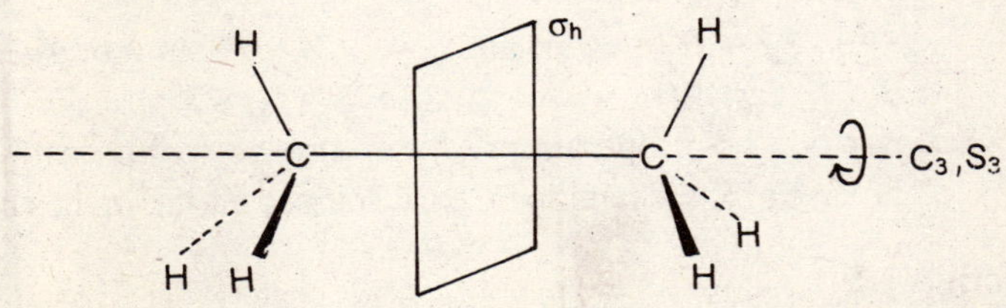

Fig. 1.31

(iv) If C_n and σ_h are not present in the molecule S_n may or may not be there. This can be seen in the case of staggered ethane.

(v) S_n (n = odd) requires independent existence of σ_h and C_n. Let us try to prove it.

As $S_n = \sigma_h^1.C_n^1 \rightarrow \sigma_h^2.C_n^2..........\sigma_h^n.C_n^n = \sigma_h^n.E = \sigma_h^n$ (n = odd)

So, σ_h^n (n = odd) = σ_h

This confirms that σ_h should exist independently. Furthermore in this operation, n proper rotations, n – 1 improper rotations and one σ_h operation are generated (proved earlier).

(vi) S_n (n = even) generates n operations in all and out of which n/2 operations are improper rotations and n/2 operations are proper rotations including identity (proved earlier).

(vii) Inverse of S_n axis can be defined as an anticlockwise rotation by (360°/n) about the S_n axis followed by the reflection through a plane which is perpendicular to this axis. This can be proved as

$$S_n^{-1} . S_n^1 = (C_n^{-1} \sigma_h)(C_n^1 \sigma_h)$$

Since $C_n^1 \sigma_h$ car be written as $\sigma_h C_n^1$

$$= (C_n^{-1} \sigma_h)(\sigma_h C_n^1) = C_n^{-1} C_n^1 = E$$

As it is always possible to express the inverse operation in terms of clockwise rotations about the same axis and it is proved earlier that S_n^n (n = even) = E and S_n^{2n} (n = odd and m = 2n) = E. So,

$S_n^{n-1} . S_n^1 = E$ (n even) and $S_n^{2n-1} . S_n^1 = E$ (n = odd) and so, $S_n^{-1} . S_n^{n-1}$ (n = even) and $S_n^{-1} . S_n^{m-1}$ (m = 2n, n = odd). So, operation S_3^5 can also be written as S_3^{-1} the inverse of S_3^1 (example PCl_5), discussed earlier in the chapter.

Section 1.3

Symmetry and Optical Activity

Optical activity is an important property that is discussed usually for organic molecules. Symmetry is related to this property of optical activity of the molecules. It is a well known fact that if the mirror image of a molecule can't be superimposed on the original configuration, then the molecule is optically active. If the mirror image is superimposed on the original configuration then the molecule is optically inactive. Consideration of symmetry helps in deciding the existence of optical activity in a molecule.

Molecules whose mirror images are not super-imposable on them are known as dissymmetric molecules i.e. these are the molecules have some symmetry. These molecules differ from those which lack in any symmetry i.e. assymmetric ones. A molecule which has no improper rotation axis will be dissymmetric one and will show optical activity. As improper axis of rotation includes S_1 which is equivalent to σ (proved earlier) and S_2 which is equivalent to 'i' (proved earlier). Therefore, it may be concluded that optical activity exists in a molecule if and only if it lacks in the plane and the centre of symmetry. Furthermore, if a molecule possesses only C_n it may be dissymmetric and may be optically active. In brief $S_1 = \sigma$ and S_2 = 'i', if they are present in a molecule, molecule is optically inactive.

In case of trans-1, 2-dichloro cyclopropane there is no $S_1 = \sigma$ and S_2 = 'i' and its mirror image is not super-imposable on the original molecule, so, it is optically active. (Figure 1.32).

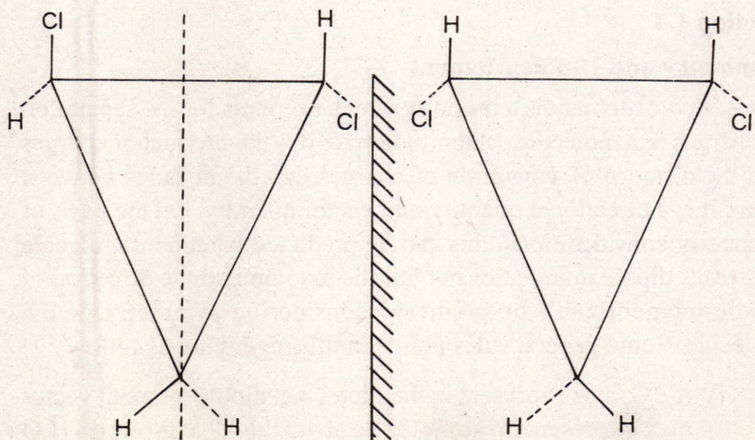

Fig. 1.32

In Figure 1.33 lactic acid is shown with its mirror image. This compound is also optically active as it does not contain $S_1 = \sigma$ and $S_2 = $ 'i'.

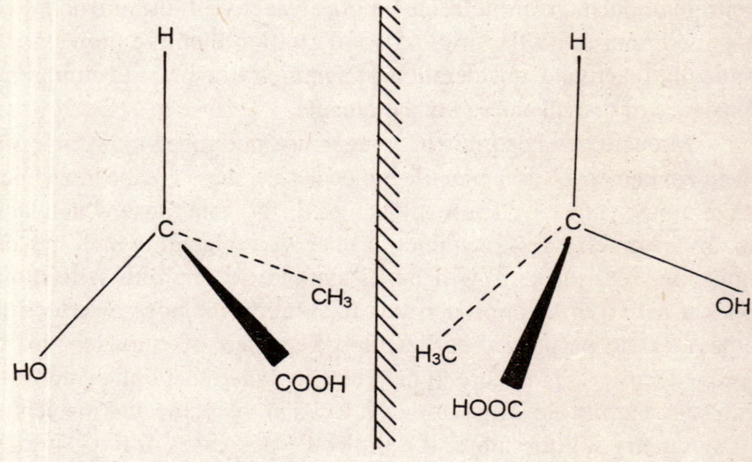

Fig. 1.33

So, a general rule can be framed that a molecule which possesses S_n axis, is optically inactive and a molecule without S_n axis is optically active.

Section 1.4

Symmetry and Dipole Moment

Dipole moment is a property which accounts for the separation of charges in a molecule. Magnitude wise it is the product of charges and the distance of separation of charges i.e., the distance between them. It is a directional quantity or a vector quantity. On the basis of symmetry considerations, this can be predicted whether a molecule possesses dipole moment or not but the amount and the direction of dipole moment can't be predicted considering symmetry of the molecule. Some general rules/points in this regard are as follows:

(i) if a C_n axis is present in the molecule, dipole moment vector may be present in the molecule along this C_n axis (e.g. in H_2O dipole moment vector is along C_2 axis which is principal axis).

(ii) If there are more rotation axes present coinciding with C_n (principal axis), then also dipole moment vector lies along the C_n axis.

(iii) If besides C_n axis (the principal axis) more rotation axes are present in the molecule and these subsidiary axes are perpendicular to the principal C_n axis, then there is no chance of existence of dipole moment in the molecule as a vector can't lie along all the axes (e.g., in BF_3, PCl_5 etc. where there is $C_3 \perp 3C_2$ axis, dipole moment is absent).

(iv) If molecule possesses a plane of symmetry, dipole moment vector lies along this plane and if there are more planes in the molecule which are intersecting with each other, the dipole moment vector lies along the line of intersection of planes (e.g., in case of H_2O molecule dipole moment vector lies along the line of intersection of two σ_v's).

(v) In centre of symmetry or 'i' is present in the molecule, dipole moment should be absent as a vector can't lies at a point (e.g., in all centro-symmetry molecules such as CO_2, AB_4 type square planar molecules dipole moment is absent).

Some examples are given in Figure 1.34 showing existence and absence of dipole moment. In each molecule dipole moment vector is shown by 'μ' vector.

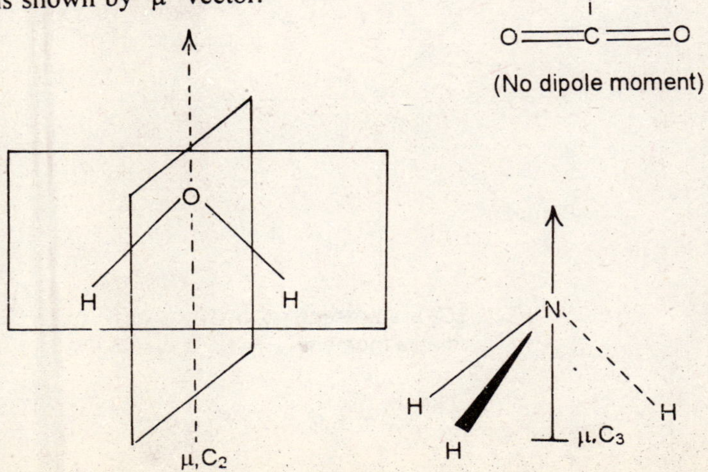

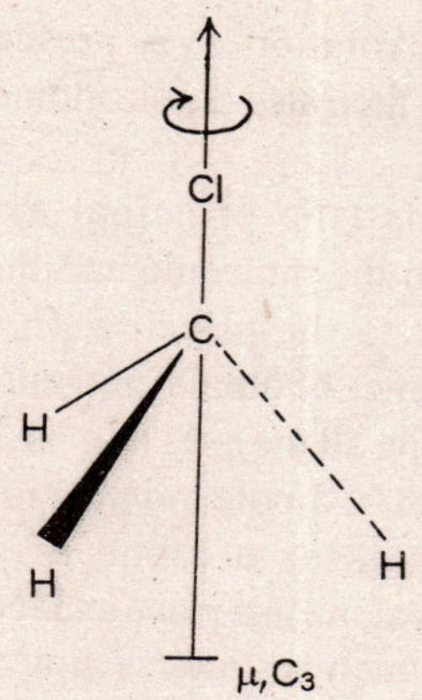
Cl
C
H
H
H
μ, C3

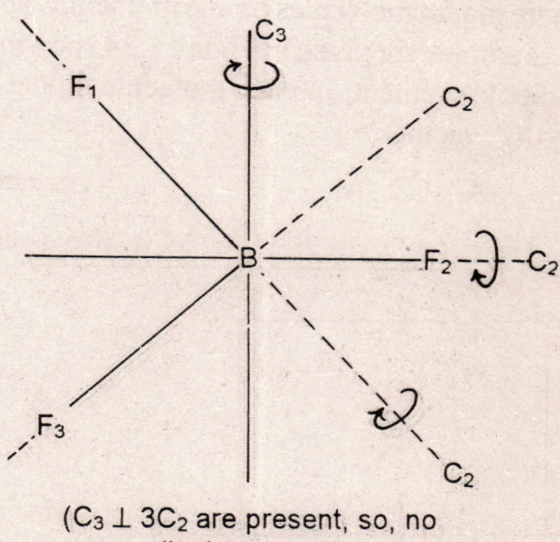
C3
F1
C2
B
F2
C2
F3
C2
(C3 ⊥ 3C2 are present, so, no
dipole moment)

Fig. 1.34

Problems

Q.1: List all the symmetry operations related to following:

(i) C_6 (ii) S_4 (iii) S_5 (iv) C_3

Ans: (i) C_6^1, $C_6^2 = C_3^1$, $C_6^3 = C_2^1$, $C_6^4 = C_3^2$, C_6^5 and $C_6^6 = E$

(ii) S_4^1, $S_4^2 = C_4^2$, S_4^3, $C_4^4 = E$ as shown below

$S_4^1 = C_4^1 . \sigma = S_4^1$

$S_4^2 = C_4^2 . \sigma^2 = C_4^2$,

$S_4^3 = C_4^3 . \sigma^3 = S_4^3 . \sigma = S_4^3$

$S_4^4 = C_4^4 . \sigma^4 = E. E = E$

Total four operations, out of which two are improper rotations and two are proper rotations including identity, are there.

(iii) $S_5^1 = C_5^1 . \sigma = S_5^1$

$S_5^2 = C_5^2 . \sigma^2 = C_5^2$

$S_5^3 = C_5^3 . \sigma^3 = C_5^3 . \sigma = S_5^3$

$S_5^4 = C_5^4 . \sigma^4 = C_5^4 . E = C_5^4$

$S_5^5 = C_5^5 . \sigma^5 = E . \sigma = \sigma$

$S_5^6 = C_5^6 . \sigma^6 = C_5^6 . E = C_5^1$

$S_5^7 = C_5^7 . \sigma^7 = C_5^2 . \sigma = S_5^2$

$S_5^8 = C_5^8 . \sigma^8 = C_5^3 . E = C_5^3$

$S_5^9 = C_5^9 . \sigma^9 = C_5^4 . \sigma = S_5^4$

$S_5^{10} = C_5^{10} . \sigma^{10} = C_5^5 . E = E . E = E$

Total ten operations are generated, out of which five are improper rotations including σ and five are proper rotations including identity.

(iv) C_3^1, C_3^2 and $C_3^3 = E$

Q 2: Explain the terms symmetry elements and symmetry operations.

Ans : Please refer to text section 1.1.

Q. 3 : Which of the following molecules possess inversion centre.

(i) Trans-H_2O_2 (ii) SO_2 (iii) HCl (iv) CHClBr-CHClBr (staggered) (v) CCl_4

Ans : (i)

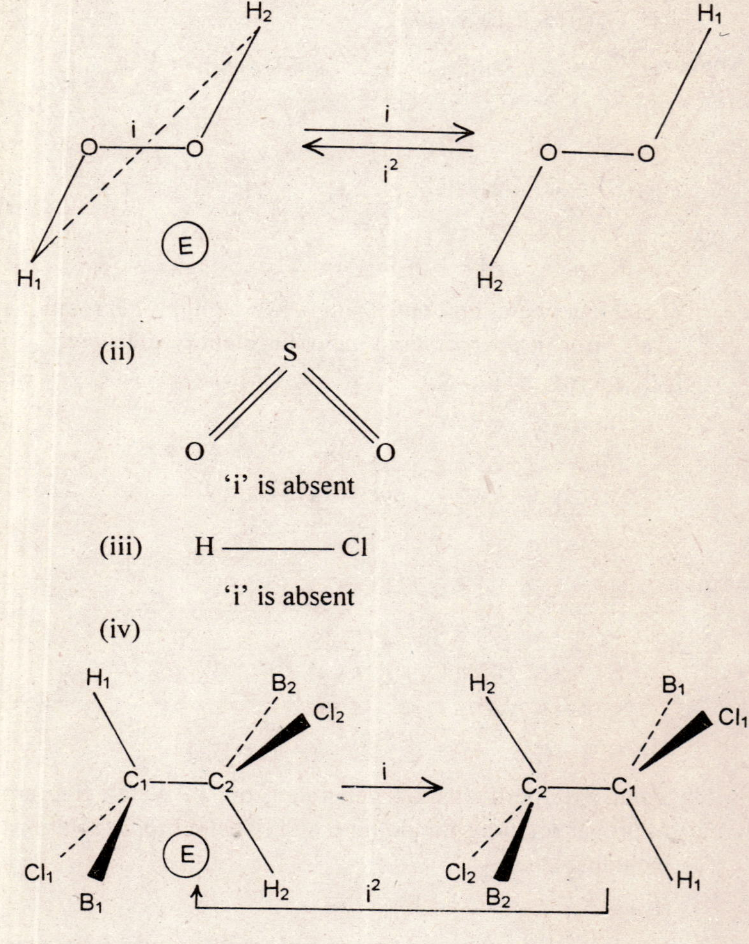

(v)

'i' is absent

Q. 4 : Identify and give the list of symmetry elements belonging to the following molecules:

(i) $PtCl_4^{2-}$ (ii) CO_3^{2-} ion (iii) $Ni(CO)_4$ (iv) $POBrCl_2$ (v) $SOCl_2$ and (vi) CH_3COOH

Ans : (i) List of symmetry elements in $PtCl_4^{2-}$ is as follows:

a. E

b. C_4 passing through Pt and perpendicular to molecular plane,

c. Four C_2 axes as shown in figure,

d. σ_h i.e., $PtCl_4$ square,

e. Two σ_v along C_2' and C_2''' and two σ_d along C_2 and C_2'',

f. 'i' at Pt-atom

g. S_4 coincident with C_4

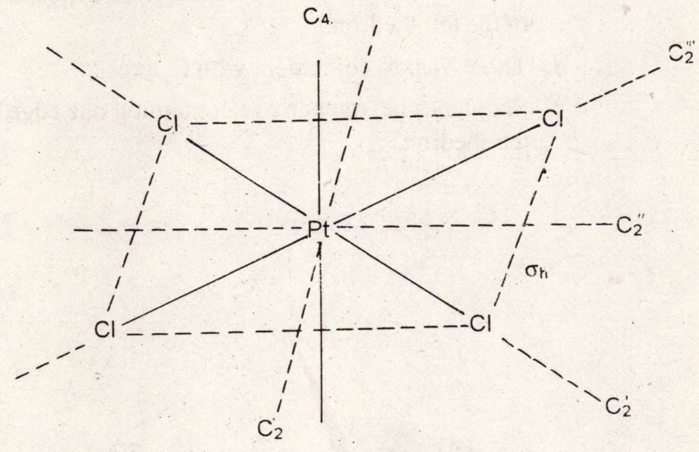

(ii) List of symmetry elements in CO_3^{2-} ion is as follows:

a. E

b. C_3 axis perpendicular to molecular plane and passing through central carbon atom,

c. Three C_2 axes perpendicular to the principal C_3 axis and passing through each C-O bond,

d. One σ_h i.e., the molecular plane itself,

e. Three σ_v's that are coincident with three C_2 axis

f. One S_3 which is coincident with C_3 principal axis.

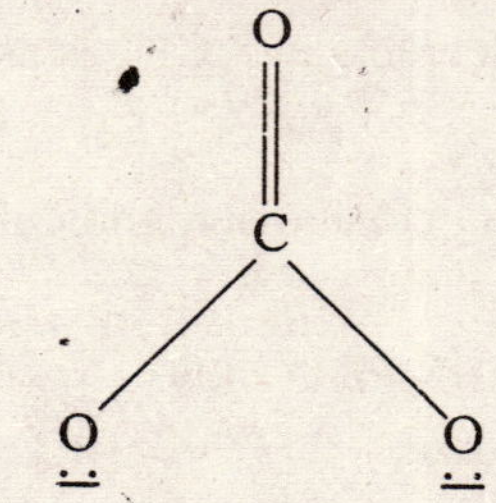

(iii) List of symmetry elements in $Ni(CO)_4$ is as follows:

a. E

b. Four C_3 axes one along each Ni-CO bond,

c. Three C_2 axes each one bisecting two opposite edges of the tetrahedron,

d. Three S_4 axes coincident with C_2 axes,

e. Six planes i.e. σ_d each one containing one edge of the tetrahedron,

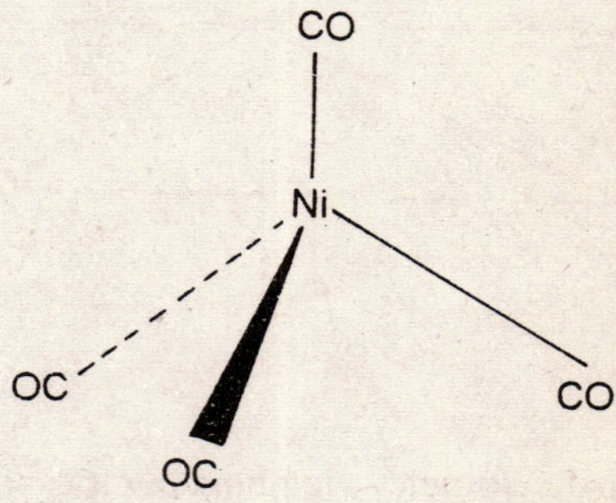

(iv) List of symmetry elements in $POBrCl_2$ is as follows:

a. E

b. One plane i.e. plane containing P,O,Br

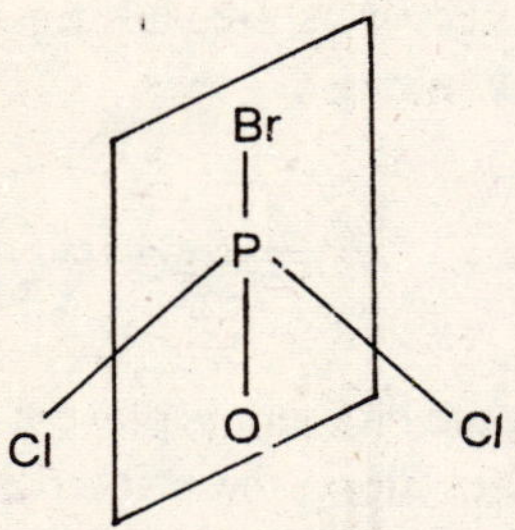

(v) List of symmetry elements in $SOCl_2$ is as follows:

a. E

b. C_2 axis passing through S & O,

e. One σ_v passing through S & O and one σ_v i.e. the molecular plane

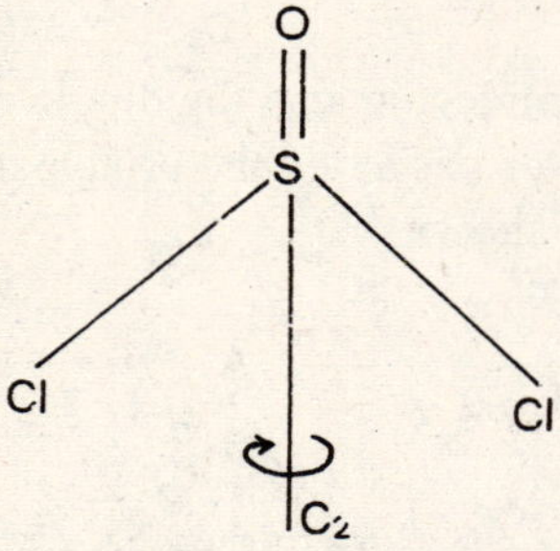

(vi) This molecule has only $C_1 = E$ (identity)

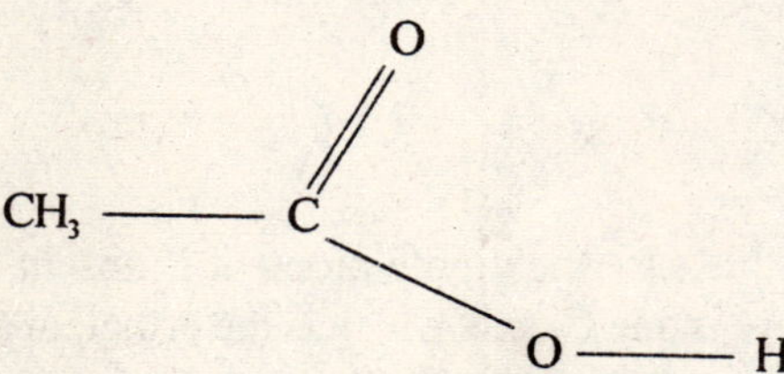

Q. 5 : Which of the following molecule contains the dipole moment?

(i) CO (ii) H_2O (iii) SO_2 (iv) PCl_5 (v) XeF_4

Ans : (i) In CO dipole moment is present and the dipole moment vector is lies along the principal axis of rotation i.e. along C_x axis.

$C \equiv O$ — C_{∞}, μ

(ii) In H_2O dipole moment is present and the dipole moment vector is lies along the principal axis i.e., along the C_2 axis as shown in figure. This axis can also be considered as the line of intersection of two σ_v's.

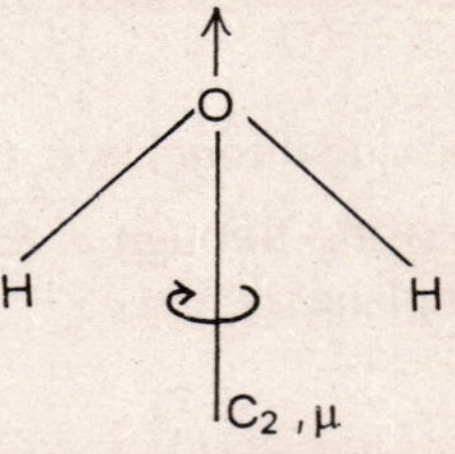

(iii) In this molecule also the dipole moment is present and the vector lies along the principal axis i.e., along the C_2 axis as shown.

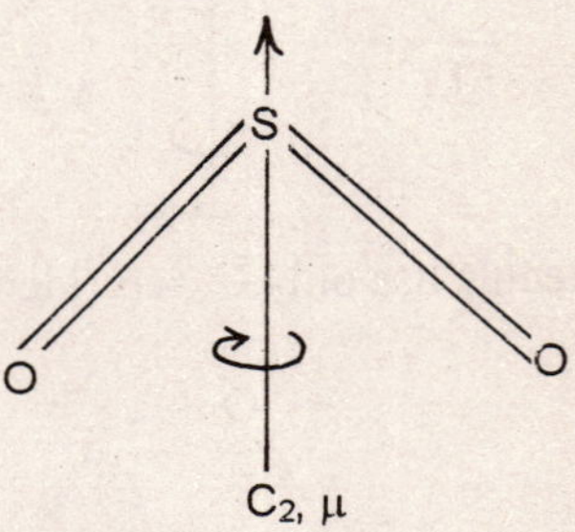

(iv) In this case the dipole moment is absent as in this case there is one C_3 axis which is the principal axis and which passes through Cl-P-Cl bonds and this molecule also contains three C_2 axes which are perpendicular to the principal, C_3 axis. So, dipole moment vector can't lie along all the axes.

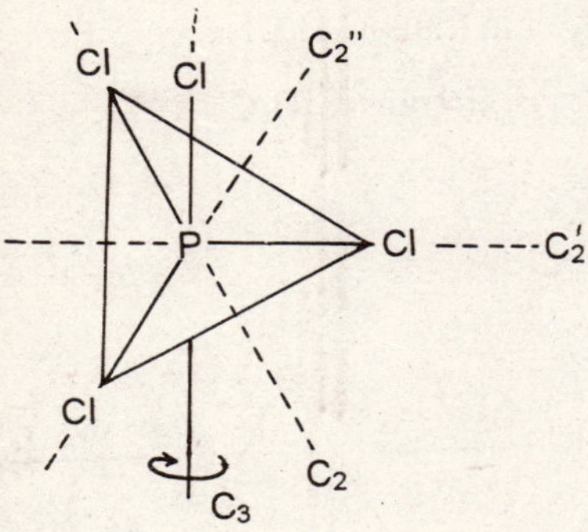

(v) In this molecule also the dipole moment is absent as in this case C_4 axis is the principal axis and four subsidiary C_2 axes are also present which are perpendicular to this principal axis. So, dipole moment vector can't lie along all the axes. Moreover this molecule also possesses the inversion centre 'i' at Xe atom and dipole moment vector can't lie at a point.

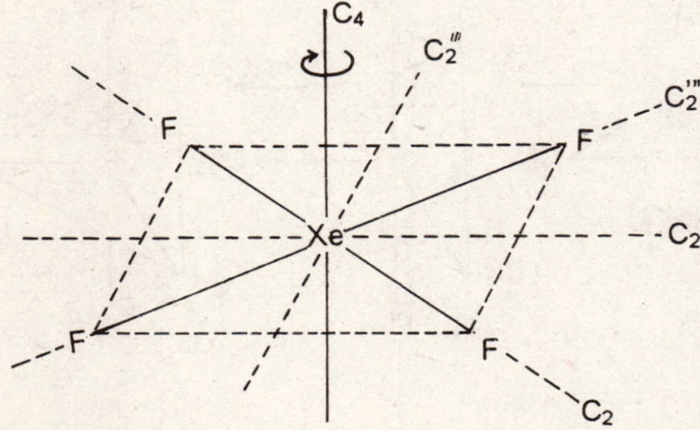

Q. 6 : Give the equivalent operation to the following combination of symmetry operations.

(i) C_2^1 followed by σ_v in case of H_2O molecule.

(ii) $C_3^1\,\sigma_h$ in PCl_5

(iii) $C_2^1\,.i$ in trans-$C_2H_2Cl_2$

(iv) $C_2\,.\,\sigma_h$ in trans-$C_2H_2Cl_2$

(i)

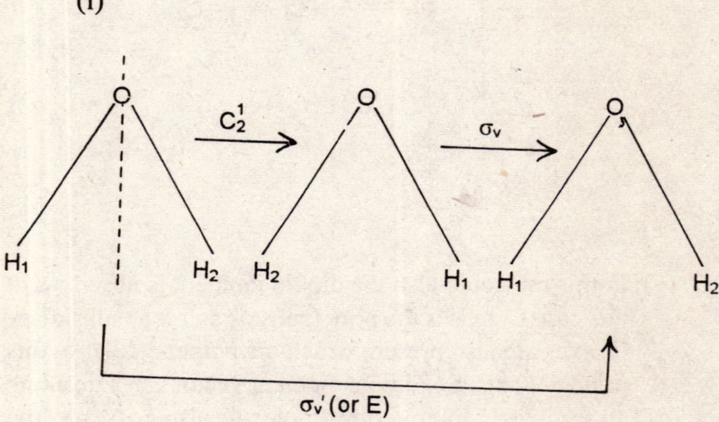

(ii)

Cle3
Cla1
Cle1
Cle2
Cla2.
C3 1
Cle2
Cla1
Cle3
Cle1
Cla2
σv
Cle2
Cla2.
Cle3
Cle1
Cla1
P
S3 1

(iii)

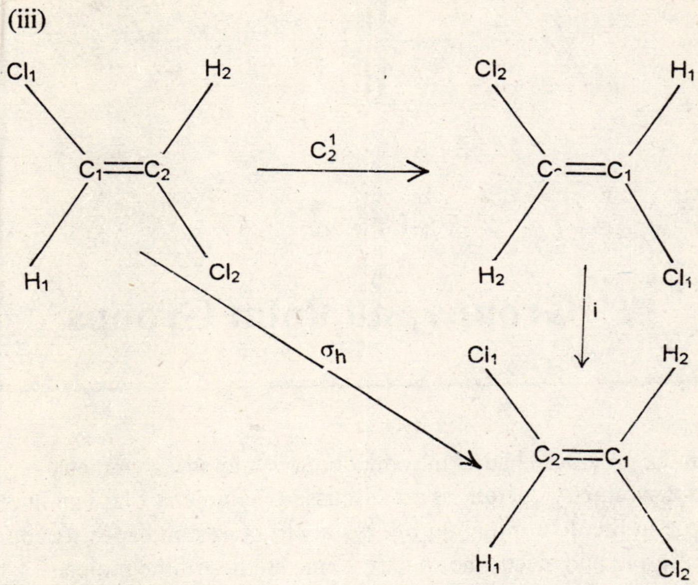
Cl1
H2
C1=C2
H1
Cl2
C2^1
Cl2
H1
C=C1
H2
Cl1
i
σh
Cl1
H2
C2=C1
H1
Cl2

(iv)

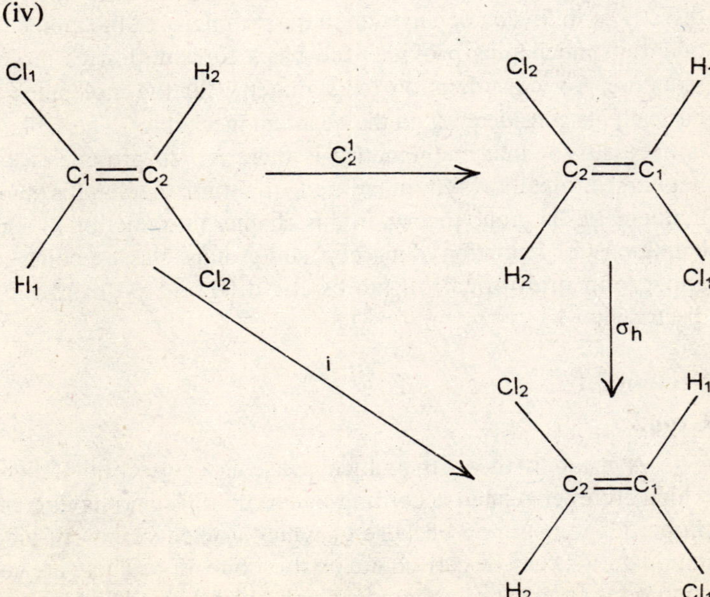
Cl1
H2
C1=C2
H1
Cl2
C2^1
Cl2
H1
C2=C1
H2
Cl1
σh
i
Cl2
H1
C2=C1
H2
Cl1

2

Groups and Point Groups

In the previous chapter introduction to symmetry, symmetry elements and symmetry operations are discussed. Symmetry plays an important role in the determination of crystal structures and spectroscopic i.e. infra-red and electronic spectroscopic studies of the molecules. Other physical properties like dipole-moment and optical activities etc., also, directly or indirectly depends upon the symmetry of the molecule. As quantum mechanics provides the basis for solution of molecular structure. So, consideration for symmetry for the molecules is the basically its consideration in the quantum mechanics. Therefore, there is necessity of that mathematics is there which provides a bridge between molecular symmetry and quantum mechanics and that mathematics is group theory. In this chapter introduction to a group, conditions for formation of a group, sub-groups, classes, point-groups and group multiplication tables for different point-groups are discussed.

Section 2.1

Group

A group, in the mathematical sense, is a collection of elements which are inter-related according to certain rules and having certain properties in common because of which a wide variety of algebric manipulations can be carried out on this collection. There are certain conditions for a set of elements which must be satisfied by them to form a group. The conditions are as follows:

(i) The product of any two elements in a group and the square of element must be the element of the same group.

Let us consider A, B, C be the three elements of a group. So, according to this condition of formation of a group AB = C or BC = A or A^2 = B or C.

AB may be equal to BA. In that case these elements said to show the law of commutation with each other. If in a group, elements show the law of commutation i.e. if elements of a group commute with each other then the group is known as abelian group.

(ii) A group should have an element that must commute with all other elements and leave them unchanged

i.e., EX = XE = X or this can be said that E (identity) is such an element.

(iii) In a group for the group elements associative law should hold good i.e. in a group of A, B and C three elements A(BC) = (AB)C

(iv) Every element should have a reciprocal [i.e. $(\text{element})^{-1}$ or inverse of the element] which should also be the member of the same group. Let in a group R is the reciprocal of the element A. So, RA = AR = E (identity). If A is the element of the group, R should also be another element of the group.

Apart from these necessary conditions for formation of group there is one another property of the group i.e., the reciprocal of the product of two or more elements in a group must be equal to the product of reciprocals taken in reverse order Let A, B and C are three elements of a group. So, according to this property $(ABC)^{-1} = C^{-1} B^{-1} A^{-1}$

This can be proved as follows:

Let ABC be the three elements of a group and their product is P (say).

So, (ABC) = P

If we right multiply both the sides with the inverse of each element i.e., with C^{-1} B^{-1} and A^{-1}

$$ABCC^{-1}B^{-1}A^{-1} = PC^{-1}B^{-1}A^{-1}$$

$\because\ CC^{-1} = E$

$$= ABEB^{-1}A^{-1} = PC^{-1}B^{-1}A^{-1}$$

$$= ABB^{-1}A^{-1} = PC^{-1}B^{-1}A^{-1}$$

[since $BB^{-1} = E$]

$$= AEA^{-1} = PC^{-1}B^{-1}A^{-1}$$

[since $AA^{-1} = E$]

$$= E = PC^{-1}B^{-1}A^{-1}$$

$$= P^{-1}E = C^{-1}B^{-1}A^{-1}$$

or $\quad P^{-1} = C^{-1}B^{-1}A^{-1}$

and since

$$P = ABC$$

So,

$(ABC)^{-1} = C^{-1}B^{-1}A^{-1}$ ....Proved

Order of the Group

Total number of elements in a group is called as its order. It is designated by 'h'. e.g., if a group has 3 elements viz. A, B and C, its order is three. In case of molecules if symmetry is taken into consideration, number of possible symmetry operations is the order of the molecular point group.

Subgroup

It is the smaller group that may be formed within the larger group. A subgroup must follows the necessary conditions for formation of a group. The order of the subgroup, taken as 'k' must be the divisor of the order of the larger group i.e., 'h' or in other words h/k must be an integer. Let us consider E(identity), A, B, C, D and F be the elements of a group. Subgroups of the order two i.e. EA, EB, EC, ED and EF can be formed from this larger group and similarly of the order three i.e. EAB, EAC, EAD, and ..... so on can be formed as a subgroup with a condition that all these subgroups must follows the requirements of formation of a group.

Classes

This is another way of expressing a smaller group from a larger group. It is necessary to consider an operation i.e., similarity transformation order to understand a class.

Let us consider a group with members X, A, B and C. For these elements the following relationship may hold good:

$$B = X^{-1} AX.$$

If it is so, then B is termed as similarity transform of A using X and this can be said that A and B are conjugate to each other. Properties of conjugate elements are given below:

(i) Each and every element must conjugate to itself, i.e. $A = X^{-1} AX$. This may hold good if X is identity operation.

(ii) If A is conjugate to B then B is conjugate to A, i.e., if $A = X^{-1} BX$ then their must be some element Y present in the group so that, $B = Y^{-1} AY$.

(iii) If A is conjugate with B and C then B and C must be conjugate with each other.

Now, class can be defined. A complete set of those elements in a larger group which are conjugate to each other is known as a class.

Symmetry elements as group

A molecule can have various symmetry elements and each one of them may be associated with one or more symmetry operations. This can be easily proved that a collection of symmetry elements (in terms of their related operations) constitute a group. For this it is necessary to show that the collection of a particular set of symmetry operations follow the necessary conditions for formation of a group. Let us try to prove this with the simplest case i.e., for the case of water molecule. Here there are four symmetry elements viz. E, C_2, σ_v and σ_v' and they constitute a group (say X).

(i) First condition of formation of a group is that any two of these four symmetry elements should combine to give another element of the same group. This can be proved by taking combination of any of the two elements. Let us take C_2 and σ_v (Figure 2.1).

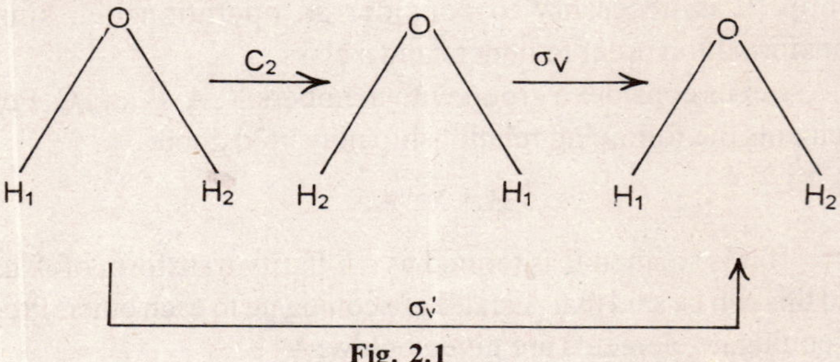

Fig. 2.1

So, C_2 and σ_v give rise to new symmetry operation i.e. σ_v' which is also the member of the same group (X). Let us try another combination $C_2.C_2$ (Figure 2.2 which shows that these elements combine to give another symmetry element (i.e. E or identity) which is also a member of the same group. Similar combinations can be shown for other elements also.

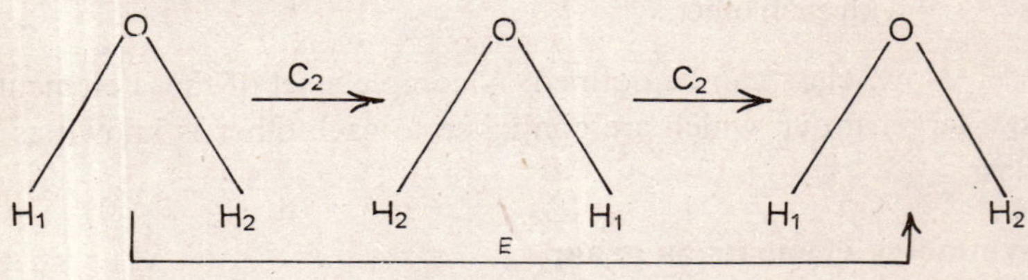

Fig. 2.2

(ii) Commutation of any two elements can be checked in the same manner. Let us try the following commutation $C_2 . \sigma_v$ and $\sigma_v . C_2$ (Figure 2.3).

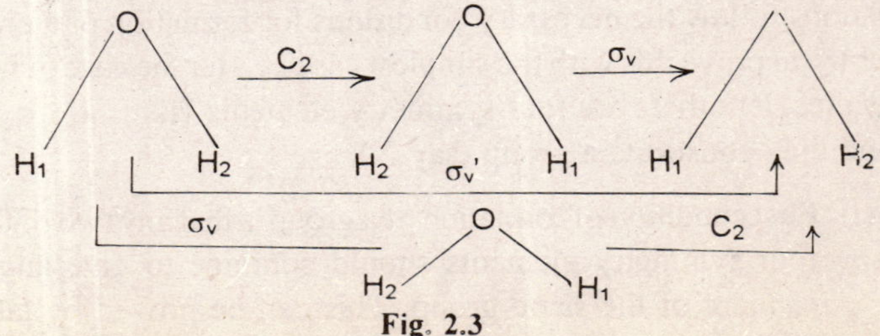

Fig. 2.3

So, $C_2 . \sigma_v$ and $\sigma_v . C_2$ both are equal to σ_v'. Furthermore, it is necessary condition that in a group there should be an element which can commute with all other elements of the group. Identity is that element which can commute with all other elements of the group X and leaves them unchanged.

As
$$E . C_2 = C_2$$
$$E . \sigma_v = \sigma_v$$
$$E . \sigma_v' = \sigma_v'$$

(iii) Third condition is to check whether the associative law holds good for these elements or not. Let us try $(C_2.\sigma_v) \sigma_v' = C_2(\sigma_v . \sigma_v')$ (Figure 2.4)

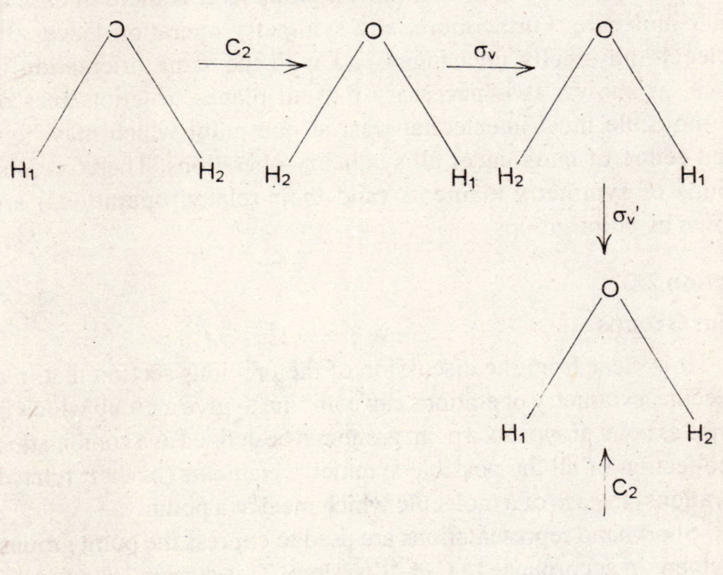

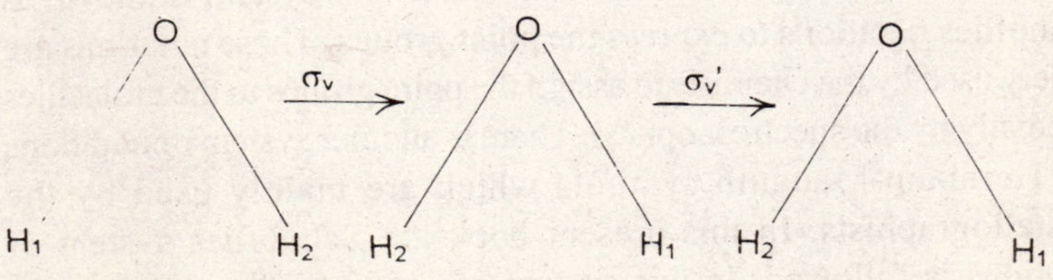

Fig. 2.4

Likewise this can be proved for other elements of the group also.

(iv) Fourth condition is the presence of inverse of every element. The inverse operations clearly correspond to the inverse elements required by any collection of elements which is to be in the same group. This can be proved for any of the four elements as

$C_2^{-1} = C_2$ and
$\sigma_v^{-1} = \sigma_v$ etc.

Therefore, this can be concluded that symmetry operations related to symmetry elements form a group as it is there in case of water molecule. Furthermore, any symmetry operation leaves the molecule physically unchanged and with the same orientation in space, as shown. It is necessary that all planes, rotation axes of the molecule must intersect at-least at one point which may be a fixed centre of mass under all symmetry operations. Therefore, the groups of symmetry elements (and their related operations) are known as point groups.

Section 2.2

Point Groups

It is clear from the discussion of the previous section that in a molecule, symmetry operations can combine to give a group which is termed as point group. So, a point group can be defined as a combination or collection of all the possible symmetry elements (or their related operations) in case of a molecule which meet at a point.

Short hand representations are used to express the point groups which are in accordance to C-σ-'i' system. This system is known as Schnoflies notations to express the point groups. These notations are largely used by the chemists to assign the point groups to the molecules (specially by the spectroscopists). There is another system of notations i.e. Hermann-Mauguin symbols which are mainly used by the crystallographists. In this present book the Schnoflies system of notations is followed. In this system of notations there are certain rules on the basis of which point group symbol can be assigned to a molecule. These rules are given below one by one in sequence.

(i) In the first step, check for the special point groups i.e. $C_{\infty v}$, $D_{\infty h}$, T_d, O_h and I_h. These are the groups of the molecules belonging to high symmetry, e.g., in case of linear molecules axis of rotation is C_∞ axis which means that any small rotation about this axis will bring a small change in the molecule into coincidence with itself. Molecules like HCl, CO etc. belong to $C_{\infty v}$ point group (Figure 2.5)

Fig. 2.5

These molecules also contain σ_v's which are the plane of the molecule itself and contains the C_∞ axis. IInd type of linear molecules are those which possess the centre of symmetry i.e., 'i' along with the C_∞ axis and perpendicular to the principal C_∞ axis there are $\infty\, C_2$'s axes coincident with the $\infty\, \sigma_v$'s. These molecules belong to $D_{\infty h}$ group e.g. CO_2, H-H, Cl-Cl etc. T_d, O_h and I_h are the special point groups assigned to the molecules of high symmetry i.e., to the molecules belonging to regular tetrahedron, octahedron and icosahedron geometries. T_d group (e.g., methane, Figure 2.6) has four C_3 axes or rotations and six σ_d's that bisect the six different H-C-H angles. In total there are 24 elements in this group.

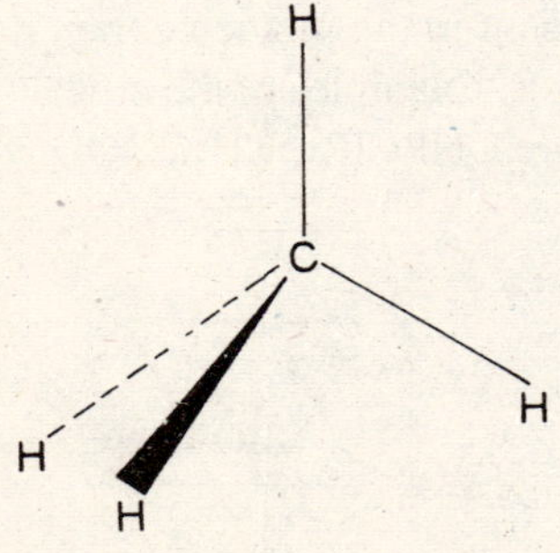

Fig. 2.6

O_h group (e.g., complexes with regular octahedron geometry i.e., $Co(NH_3)_6$, figure 2.7) has 48 elements as group members. List is E, $6C_4$, $3C_2$ (= C_4^2), $6S_4$, $8C_3$, $8S_6$, $6C_2$, $6\sigma_v$ $3\sigma_n$ and i.

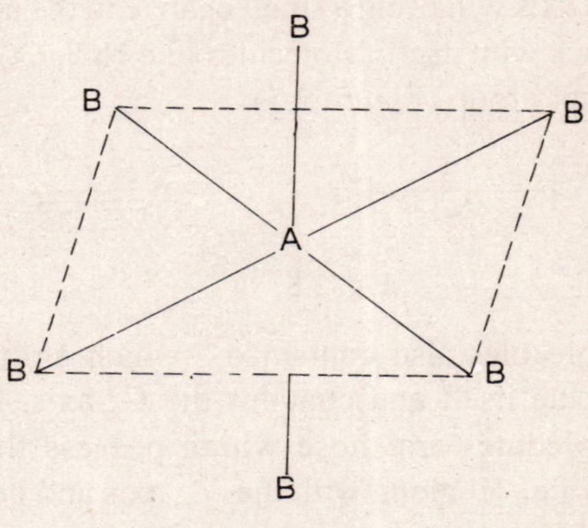

Fig. 2.7

In I_h (Figure 2.8) point group there are total 120 number of operations as group members viz. E, $24C_5$, $24S_{10}$, $20C_3$, $20S_3$, $15C_2$, 15σ and i.

(ii) Suppose it is not so, then look for the proper axis of rotation. If it is present in the molecule then look for next point otherwise the point group may be C_1, C_i or C_s. Molecules belonging to point groups C_1, C_i and C_s are the point groups of low symmetry. If a molecule possess only identity i.e., if only E is present in the molecule then it belongs to C_1 point group as C_1 = E. Examples of the molecules belonging to this point group are CHFClBr, CH_3COOH etc.

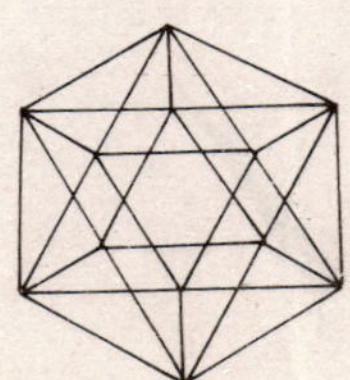

Fig. 2.8

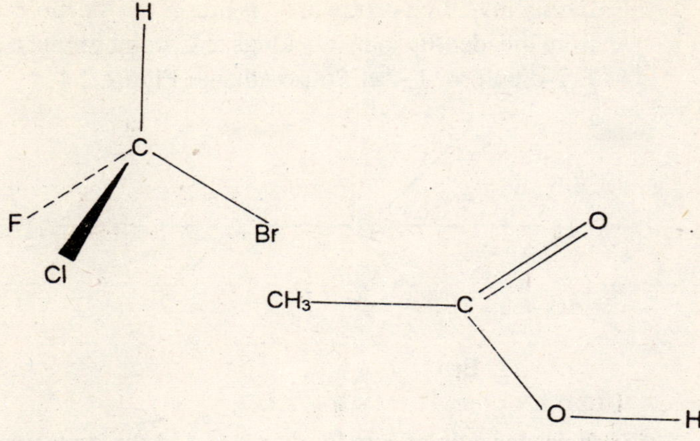

Fig. 2.9

If only σ is present in the molecule, apart from identity then it belongs to C_s point group. Some examples of this point group are SO_2BrF, NOCl, HOCl and non-symmetrical aromatic compounds like quinoline, mono-substituted naphathalene etc. (Figure 2.10)

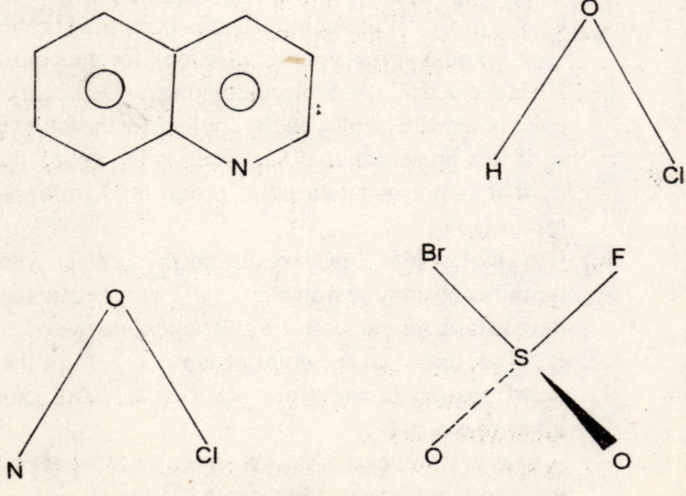

Fig. 2.10

If only inversion centre or 'i' is present in the molecule apart from the identity then it belongs to C_i point group, e.g., trans-1, 2-dichloro, 1, 2-diflouro ethane (Figure 2.11)

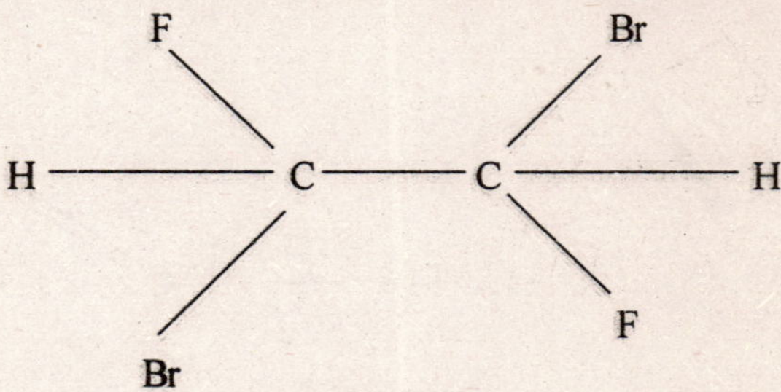

Fig. 2.11

(iii) If proper axis of rotation is present in the molecule and the molecule does not belong to the special point groups, referred above in the point (i) and (ii) then look for the highest order axis which serves as the principal axis and then look for S_{2n} coincident with C_n. If it is there then the point group is S_{2n}, otherwise look for the next point (iv).

(iv) Next step is to look for the nC_2 which are perpendicular to the principal C_n axis. If these axes are there then point group belongs to the dihedral or D category i.e., the point group may be D_n, D_{nh} or D_{nd}. If it is not so then the point group may be C_n, C_{nv} or C_{nh}. If the point group is from the D category then look for next point (v) otherwise look for the point (vi).

(v) If it is decided that the molecule belongs to D category i.e., molecule possesses $C_n \perp nC_2$, one by one look for the following planes:
σ_h if it is present then point group is D_{nh} if not then
σ_d if it is present then point group is D_{nd} otherwise point group is D_n.

(vi) If in a molecule $C_n \perp nC_2$ axes are not present then point group of the molecule may be from the C_n, C_{nv} or C_n. For this decision we have to look for planes in the molecules one by one:
σ_h, if present then the point group is C_{nh} if not then
σ_v, if present in the molecule then to point group is C_{nv} otherwise it is C_n.
A simple flow chart is drawn for assignment of point groups to various molecules (Figures 2.12).

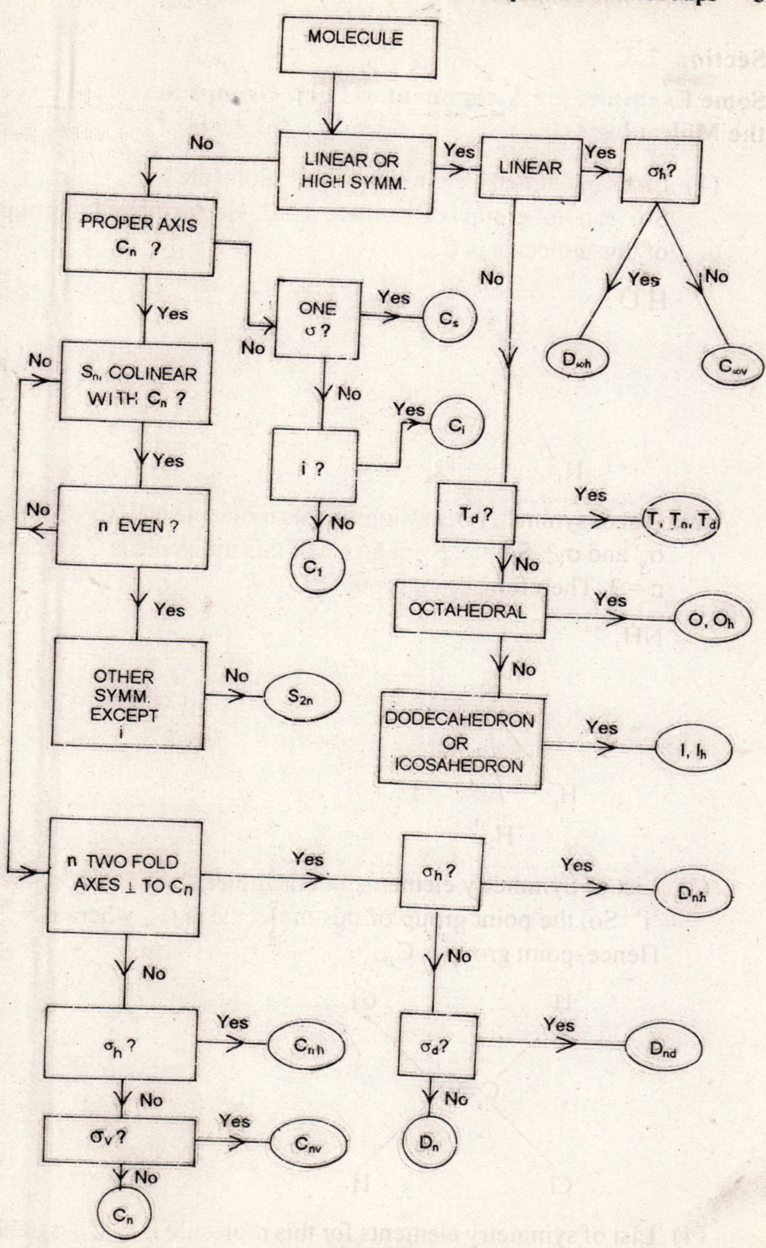
MOLECULE
LINEAR OR HIGH SYMM.
No
Yes
LINEAR
Yes
σh?
Yes
No
D∞h
C∞v
No
PROPER AXIS Cn ?
ONE σ ?
Yes
Cs
No
Yes
No
Sn, COLINEAR WITH Cn ?
No
i ?
Yes
Ci
No
C1
Yes
No
n EVEN ?
Td ?
Yes
T, Th, Td
No
OCTAHEDRAL
Yes
O, Oh
Yes
No
OTHER SYMM. EXCEPT i
No
S2n
DODECAHEDRON OR ICOSAHEDRON
Yes
I, Ih
n TWO FOLD AXES ⊥ TO Cn
Yes
σh?
Yes
Dnh
No
No
σh?
Yes
Cnh
σd?
Yes
Dnd
No
No
σv?
Yes
Cnv
Dn
No
Cn

Fig. 2.12

Section 2.3

Some Examples for Assignment of Point Groups to the Molecules:

(1) List of symmetry elements in this molecule is E, C_2, σ_v,σ_v'. So, its point group is C_{nv} where n is 2. Hence the point group of this molecule is C_{2v}.

H_2O

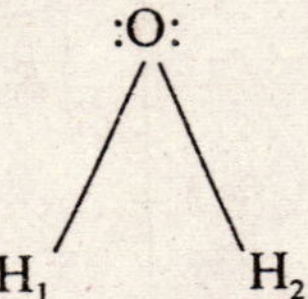

(2) List of symmetry operations in this molecule is E, C_3^1,C_3^2, σ_v, σ_v' and σ_v''. So, the point group of this molecule is C_{nv} where n = 3. Therefore, point group is C_{3v}.

NH_3

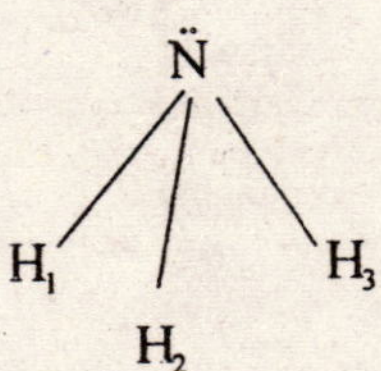

(3) List of Symmetry elements in this molecule is E, C_2, σ_h and 'i'. So, the point group of this molecule is C_{nh} where n = 2. Hence, point group is C_{2h}.

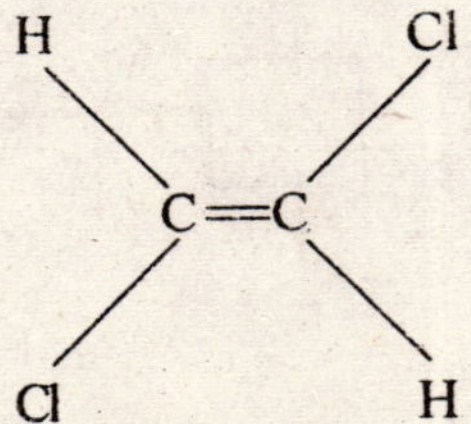

(4) List of symmetry elements for this molecule is E, C_2, σ_v, and σ_v'. So, the point group is C_{nv} where n = 2. Hence, the point

group is C_{2v}.

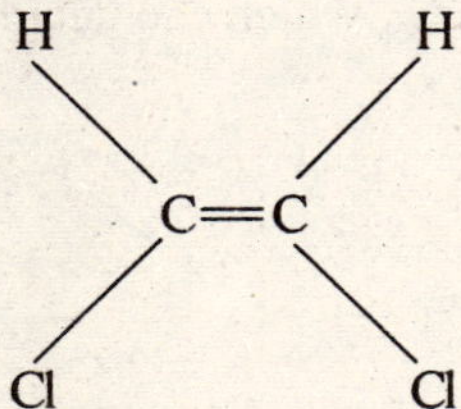

(5) List of symmetry elements for this molecule is E, $2C_3 \perp 3C_2$, $3\sigma_v$, σ_h and $2S_3$. So, the point group of this molecule is D_{nh} where n = 3. Hence, the point group is D_{3h}.

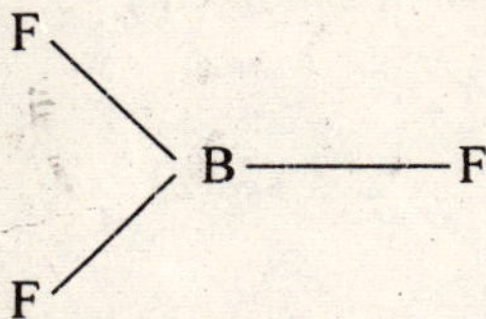

(6) List of symmetry elements for this molecule is E, $2C_3 \perp 3C_2$, $3\sigma_v$, σ_h and $2S_3$. So, the point group of this molecule is D_{nh} where n = 3. Hence, the point group is D_{3h}.

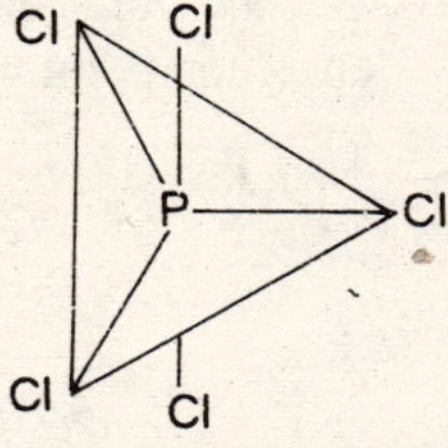

(7) List of symmetry elements in this molecule is E, C_2 σ_h and 'i'. So, the point group of this molecule is C_{nh} where n = 2. Hence, point group is C_{2h}.

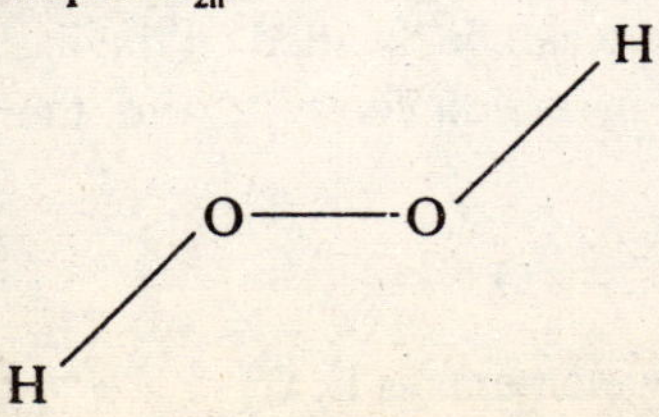

(8) List of symmetry elements in this molecule is E, C_3^1, C_3^2, σ_h S_3. So, the point group is C_{3h} which can decided on the basis of presence of C_3 and σ_h.

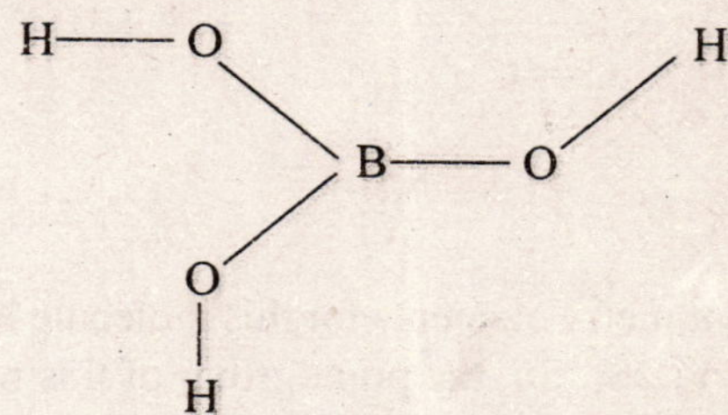

(9) Only identity is present in this molecule i.e. E (= C_1). So, the point group is C_1.

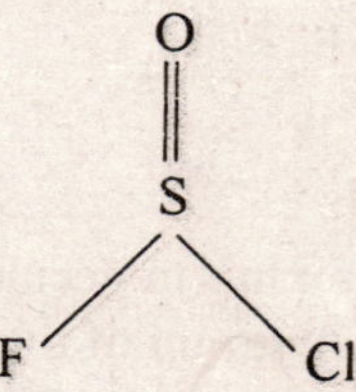

(10) List of symmetry elements for this molecule is E, $C_4 \perp 4C_2$, $4\sigma_v$, σ_h, S_4 and 'i'. So, the point group of this molecule is D_{nh} where n = 4. Hence, the pcint group is D_{4h}.

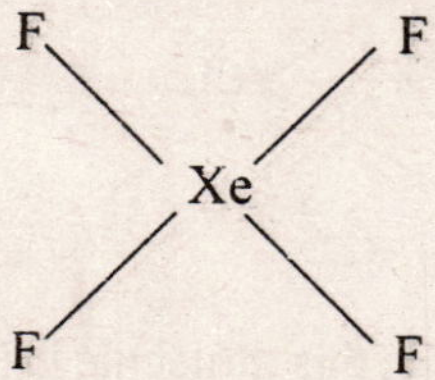

(11) List of symmetry element is E, C_∞ ... ... $\infty\,\sigma_v$, 'i', S_∞ ... ... $\perp\infty C_2$. Point group can be decided on the basis of $C_\infty \perp \infty C_2$ and presence of inversion centre 'i'. So, the point group is $D_{\infty h}$.

O=C=O

(12) List of symmetry elements is E, C_∞ ... ... $\infty\,\sigma_v$. So, the point

group is $C_{\infty v}$.

$$C\equiv O$$

(13) List of symmetry elements for this molecule is E, $C_2 \perp 2C_2$, $2\sigma_d$, $2S_4$. So, the point group of this molecule is D_{nd} where n = 2. Hence, the point group is D_{2d}.

Allene

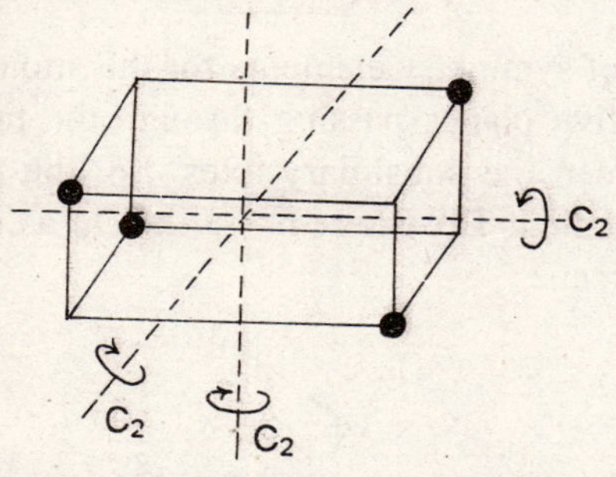

(14) List of symmetry elements for this molecule is E, $2C_6 \perp 6C_2$, $2C_3$, $3\sigma_v$, σ_h, $2S_6$, $2S_3$ 'i'. So, the point group of this molecule is D_{nh} where n = 6 i.e. D_{6h}.

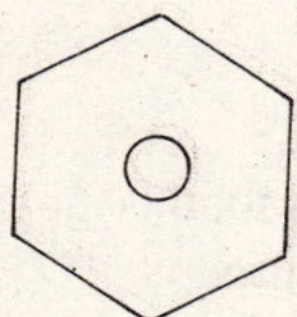

(15) List of symmetry elements for this molecule is E, $C_4 \perp 4C_2$, $4\sigma_v$, σ_h, S_4, and 'i'. So, the point group of this molecule is D_{nh} where n = 4. Hence, the point group is D_{4h}.

$PtCl_4^{-2}$

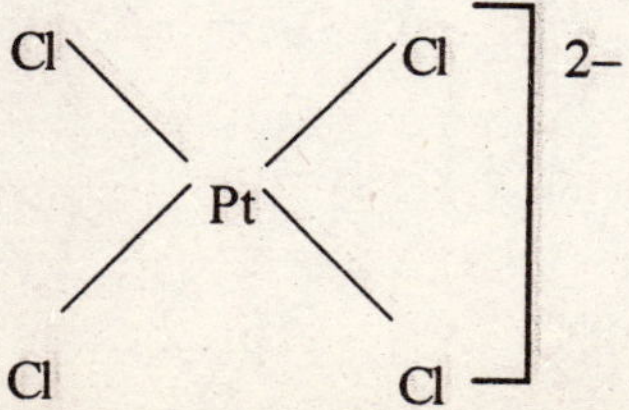

(16) List of symmetry elements for this molecule is E, $C_5 \perp 5C_2$, $5\sigma_v$, σ_h, S_5. So, the point group of this molecule is D_{nh} where

n = 5. Hence, the point group is D_{5h}.

$C_5H_5^-$ ion

(17) List of symmetry elements for this molecule is E, $C_5 \perp 5C_2$ and five planes passing through the principal axis but in between the subsidiary axes. So, the point group of this molecule is D_{nd} where n = 5 i.e. D_{5d} as it also contains σ_d.

Ferrocene

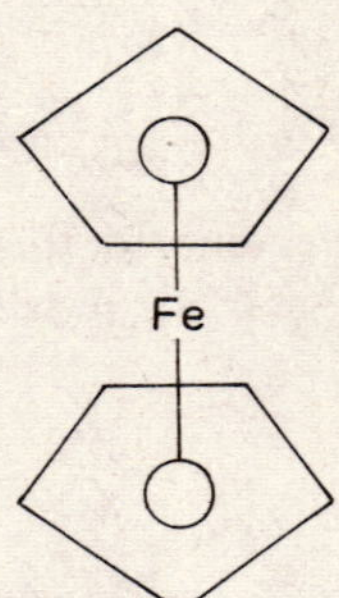

(18) List of symmetry elements for this molecule is E, C_3^1, C_3^2, σ_v, σ_v' and σ_v''. So, the point groups is C_{nv} where n = 3. Hence, the point group is C_{3v}.

$POCl_3$

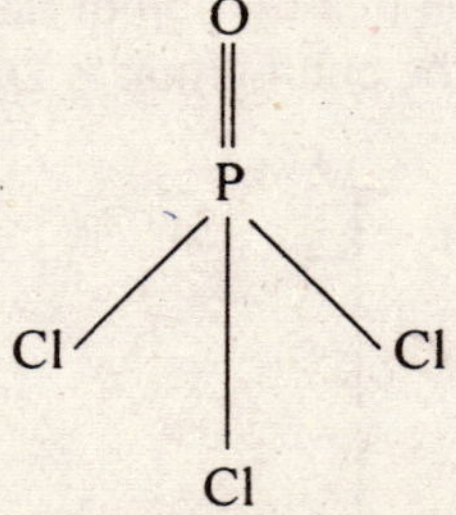

(19) List of symmetry elements for this molecule is E, C_2, σ_v, and σ_v'. So, the point groups is C_{nv} where n = 2. Hence, the

point group is C_{2v}.

Cl_2O

O

Cl Cl

(20) In this molecule, the principal axis is C_4 and no S_8 or perpendicular $4C_2$ axes are present but $4\sigma_v$'s are there in this molecule, Therefore, the point group is C_{4v}.

SF_5Cl

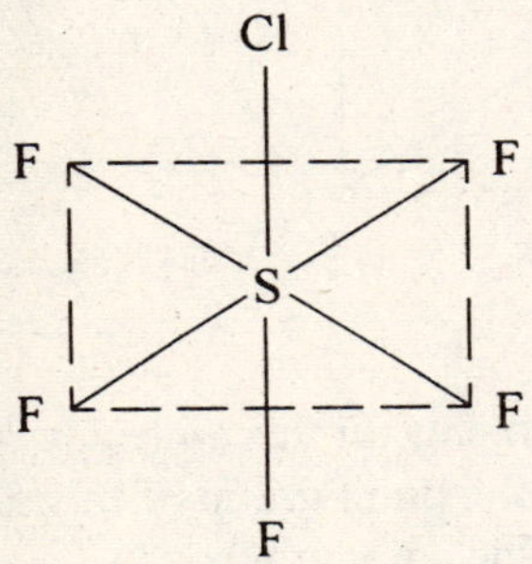

(21) It is a non linear molecular with principal axis C_2 and perpendicular $2C_2$ axes with a horizontal plane. So, this molecule belongs to D_{2h} point group.

B_2H_6

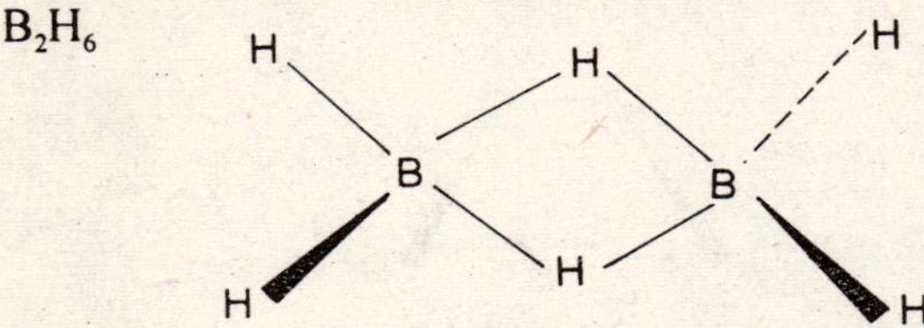

(22) It is highly symmetrical molecule with regular tetrahedral geometry and list of symmetry elements is E, $8C_3$, $3C_2$, $6S_4$, 6σ. Therefore, its point group is T_d.

CH_4

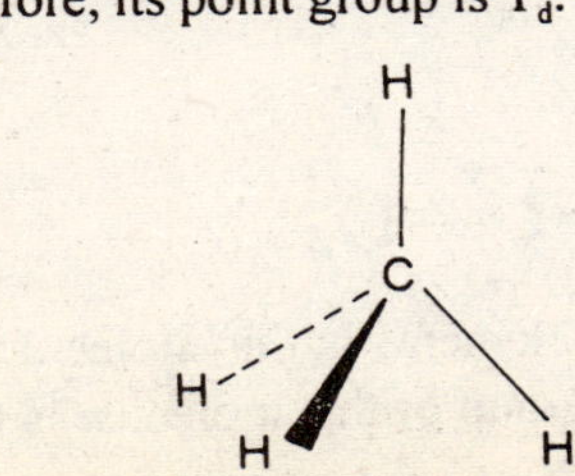

(23) It is also highly symmetrical molecule with regular octahedral geometry. The list of symmetry elements for this molecule is E, $6C_4$, $3C_2$ (C_4^2), $6S_4$, $8C_3$, $8S_6$, $6C_2$, 'i', $3\sigma_h$ and $6\sigma_v$. The point group is O_h.

SF_6

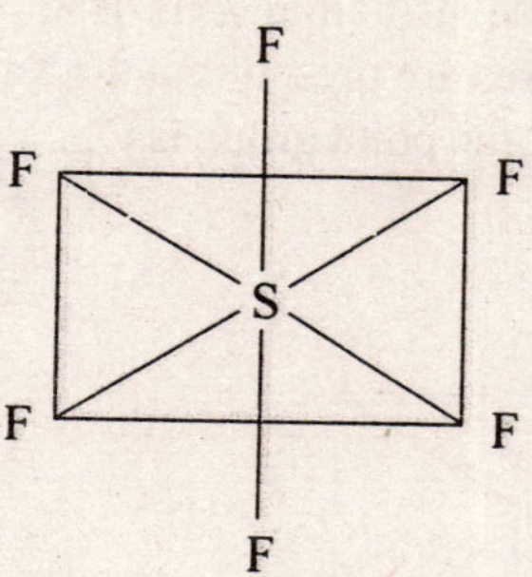

(24) List of symmetry elements for this molecule is E, $2C_3 \perp 3C_2$, $3\sigma_v$, σ_h and $2S_3$. So, the point group of this molecule is D_{nh} where n = 3. Hence, the point group is D_{3h}.

Eclipsed Ethane

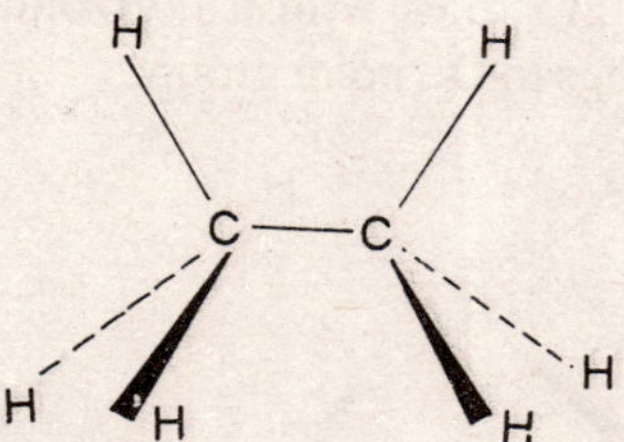

(25) List of symmetry element is E, C_∞ ... ... $\infty\ \sigma_v$, 'i', S_∞ ... ... ∞C_2. Point group can be decided on the basis of $C_\infty \perp \infty C_2$ and presence of inversion centre 'i'. So, the point group is $D_{\infty h}$.

Acetylene

H—C≡C—H

(26) List of symmetry elements in this molecule is E, C_2, σ_h and 'i'. So, the point group of this molecule is C_{nh} where n = 2. Hence, point group is C_{2h}.

Trans N_2F_2

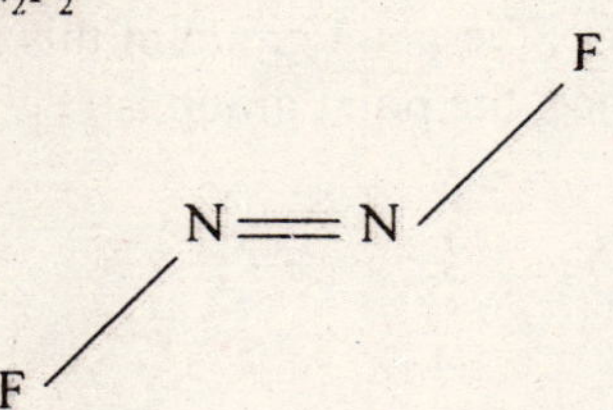

(27) This molecule is similar to NH_3 molecule so, far as symmetry is concerned. Therefore, its point group is C_{3v} with the list of elements is E, $3C_3$ and $3\sigma_v$.

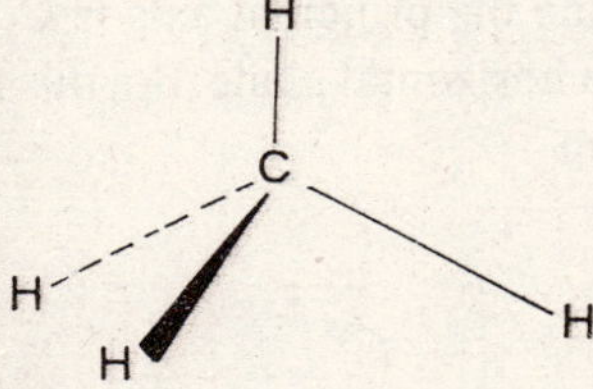

(28) This is a linear molecule with absence of 'i'. So, its point group is $C_{\infty v}$.

H——X

(X = Cl, Br etc.)

(29) List of symmetry elements for this molecule is E, $C_4 \perp 5C_2$, $5\sigma_v$, σ_h, S_5. So, the point group of this molecule is D_{nh} where n = 5. Hence, the point group is D_{5h}.
IF_7

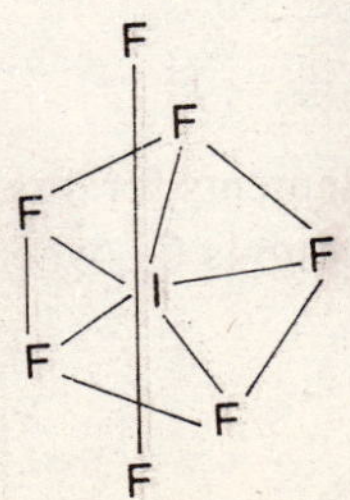

(30) List of symmetry elements for this molecule is E, $2C_3 \perp 3C_2$, $3\sigma_v$, σ_h and $2S_3$. So, the point group of this molecule is D_{nh} where n = 3. Hence, the point group is D_{3h}.
NO_3^- ion

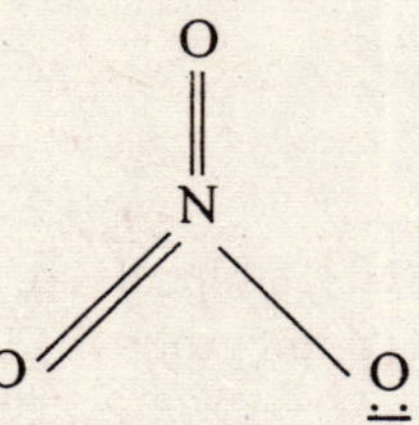

(31) In this molecule the principal axis is C_2 with perpendicular $2C_2$ axes and a horizontal plane. So, this molecule belongs to D_{2h} point group.
Naphathalene

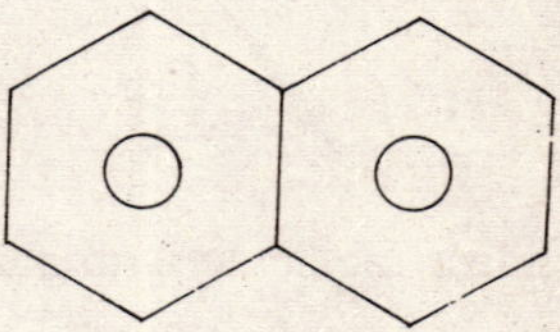

(32) In this molecule also the principal axis is C_2 with perpendicular $2C_2$ axes and a horizontal plane. So, this molecule belongs to D_{2h} point group.

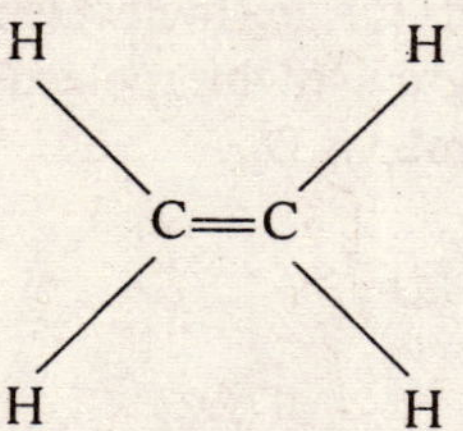

(33) List of symmetry elements for this molecule is E, C_2, σ_v, and σ_v'. So, the point group is C_{nv} where n = 2. Hence, the point group is C_{2v}.
NO_2

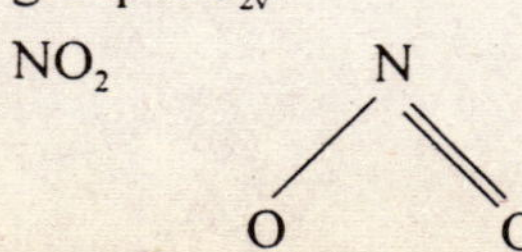

(34) This is a linear molecule with absence of 'i'. So, its point group is $C_{\infty v}$.

N_2O $\qquad N \equiv \overset{+}{N} - \overset{-}{O}$

(35) Principal axis in this molecule is C_3 with perpendicular $3C_2$ axes with σ_d. So, the point group of this molecule is D_{3d}.
Staggered ethane

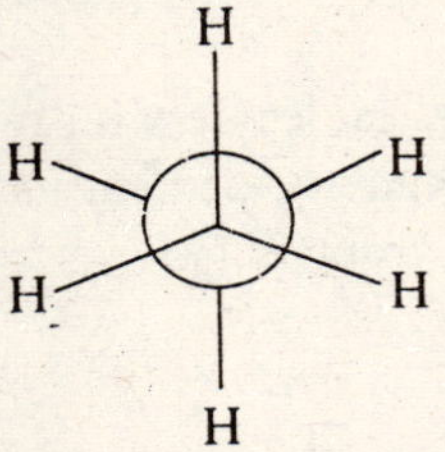

H
H H
H H
H

(36) In this case one hydrogen is replaced by deuterium. So, there is no C_2 axis here as it is in case of H_2O molecule, no σ_v is there. Only one plane is there in this molecule i.e., the molecular plane. So, its point group is C_s.

HDO

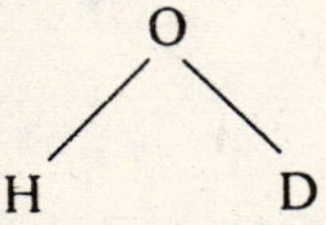

O
H D

(37) List of symmetry elements for this molecule is E, C_2, σ_v, and σ_v'. So, the point group is C_{nv} where n = 2. Hence, the point group is C_{2v}.

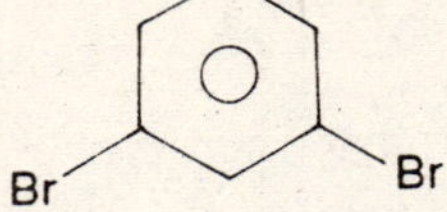

Cl
Br Br

(38) This molecule is different from acetylene. On substitution of both the hydrogen atoms by Cl & Br respectively, its centre of symmetry is lost. So, its point group changes from $D_{\infty h}$ to $C_{\infty v}$.

$Cl - C \equiv C - Br$

(39) List of symmetry elements for this molecule is E, $2C_3 \perp 3C_2$, $3\sigma_v$, σ_h and $2S_3$. So, the point group of this molecule is D_{nh} where n = 3. Hence, the point group is D_{3h}.

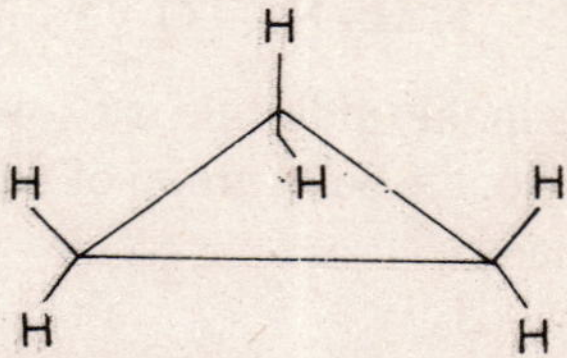

(40) It is highly symmetrical molecule with regular tetrahedral geometry and list of symmetry elements is E, $8C_3$, $3C_2$, $6S_4$, 6σ. Therefore, its point group is T_d.

$SnCl_4$

(41) It is also highly symmetrical molecule with regular octahedral geometry. The list of symmetry elements for this molecule is E, $6C_4$, $3C_2$ (C_4^2), $6S_4$, $8C_3$, $8S_6$, $6C_2$, 'i', $3\sigma_h$ and $6\sigma_v$. The point is O_h.

$Cr(CO)_6$

(42) In this molecule also the principal axis is C_2 with perpendicular $2C_2$ axes and a horizontal plane. So, this molecule belongs to D_{2h} point group.

Pt $(NH_3)_2$ Cl_2 trans (ignore H)

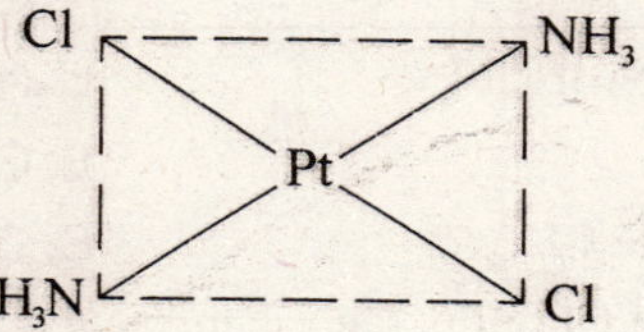

(43) In this molecule, the principal axis is C_4 and no S_8 or perpendicular $4C_2$ axes are present but $4\sigma_v$'s are there in this molecule, Therefore, the point group is C_{4v}.

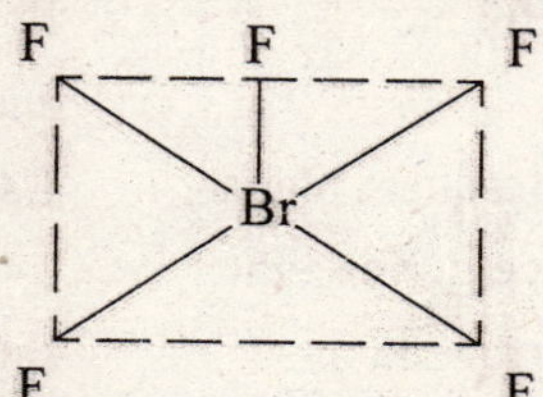

(44) 1,1,1-triflouro ethane in Gauche form has E and C_3 operations possible. Hence, its point group is C_3.

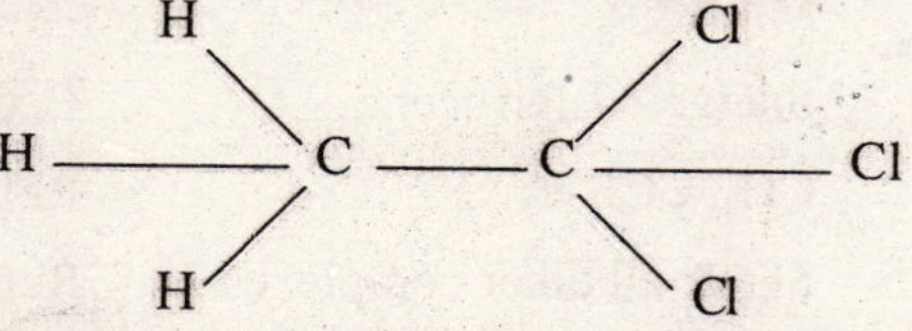

(45) This molecule is similar to H_2O molecule, so, far as symmetry is concerned. So, its point groupo is C_{2v}.

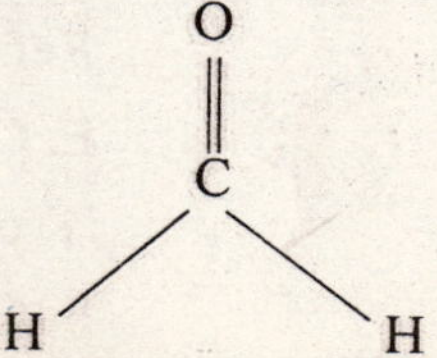

Table 2.1: Shows some common point groups with the suitable examples.

Point Group	*Examples*	*Symmetry Elements*
1	2	3
C_1	SOFCl, SiBrClFI	C_1 (= E)
C_2	H_2O_2	E, C_2
C_{2h}	trans-$C_2H_2Cl_2$	E, C_2, σ_h and 'i'
C_{2v}	H_2O, SO_2Cl_2, SO_2, ClO_2^{-1} etc.	E, C_2 & $2\sigma_v$
C_{3v}	NH_3, CH_3Cl, $POCl_3$ etc.	E, C_3 and $3\sigma_v$
D_{2h}	N_2O_4 (planar), $C_2O_4^{2-}$	E, $C_2 \perp 2C_2$, $2\sigma_v$, σ_h and i
D_{3h}	BF_3, PCl_5, PF_5 etc.	E, $C_3 \perp 3C_2$, $3\sigma_v$, and σ_h
D_{4h}	Regular AB_4 square planar Compounds e.g. XeF_4, $PtCl_4^{2-}$	E, $C_4 \perp C_2$, C_4^3, S_4-S_4^3, $\perp 4C_2$, 'i', σ_h, $2\sigma_v$, σ_d
D_{5h}	Eclipsed ferrocene	E, C_5, C_5^2...C_5^3...C_5^4 S_5, $5\sigma_v$, σ_h, $\perp 5C_2$
D_{2d}	Allene	E, $C_2 \perp 2C_2$, $2\sigma_d$, $2S_4$
D_{4d}	$Mn_2(CO)_{10}$	----------
D_{5d}	Staggered ferrocene	E, $C_5 \perp 5C_2$, σ_d etc...
T_d	CH_4, CCl_4 etc.	Ref. Previous text
O_h	SF_6 & all other complexes in regular octahedral geometry.	Ref. Previous text
I_h	Dodecaborane	Ref. Section 2.2
$D_{\infty h}$	CO_2, H_2, N_2 etc.	E, C_∞, ∞C_2, $\infty\sigma_v$, σ_h, 'i' etc.
$C_{\infty v}$	HCl, CO etc.	E, C_∞, σ_v, etc.

Section 2.4

In the Section 2.1 earlier, it is shown that symmetry elements constitute a group which is termed as point group. This was shown by taking example of H_2O. Now, it is clear that E, C_2, σ_v and σ_v' together constitute a point group which is C_{2v}. This can be shown for other point groups also. Let us try to prove it for C_{2h} point group i.e. for the symmetry elements E, C_2, σ_h and 'i'.

For this let us consider a suitable example i.e. trans-$C_2H_2Cl_2$ (Figure 2.13).

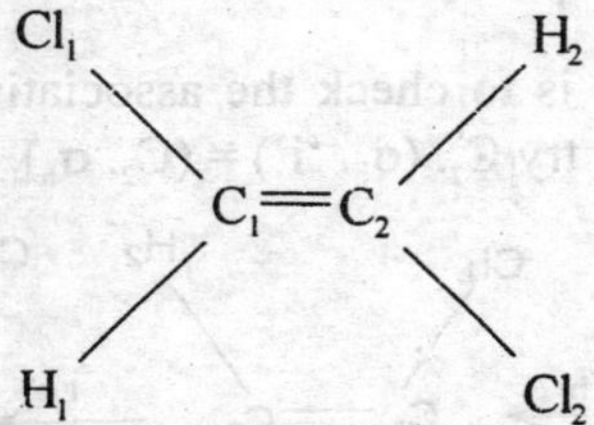

Fig. 2.13

In order to prove that the symmetry operations i.e. E, C_2, σ_h and 'i' which are present in this molecule, constitute a group, we have to prove all the four properties of formation of a group must be satisfied. Let us try to prove them one by one.

(i) First condition or the property is that any two of these elements should combine to give another element which should be the member of the same group. Let us take the combination of C_2. σ_h (Figure 2.14).

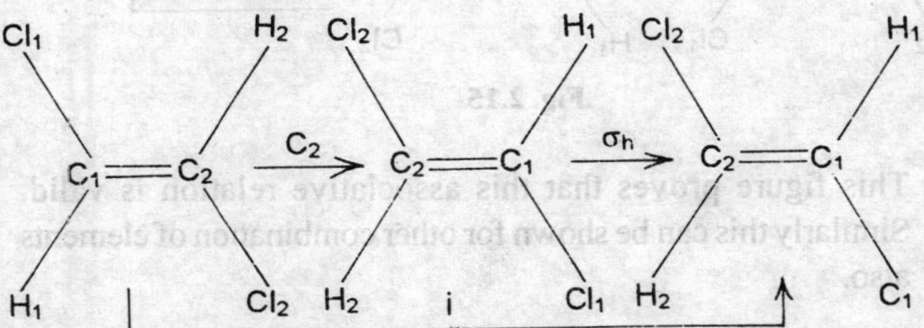

Fig. 2.14

So, it is clear that this combination i.e. $C_2 . \sigma_h$ = 'i' which is the element of the same group. Similarly this can be shown for other combinations (i.e. $\sigma_h . C_2$, C_2, 'i', 'i'. C_2 etc.)

(ii) Second condition is the existence of an element which should combine with all other elements and that element is identity i.e. E.

i.e. $E . C_2 = C_2$

$E . \sigma_h = \sigma_h$

E. 'i' = 'i'

(iii) Third condition is to check the associative law for these elements. Let us try $C_2 . (\sigma_h . \text{'i'}) = (C_2 . \sigma_h) . i$

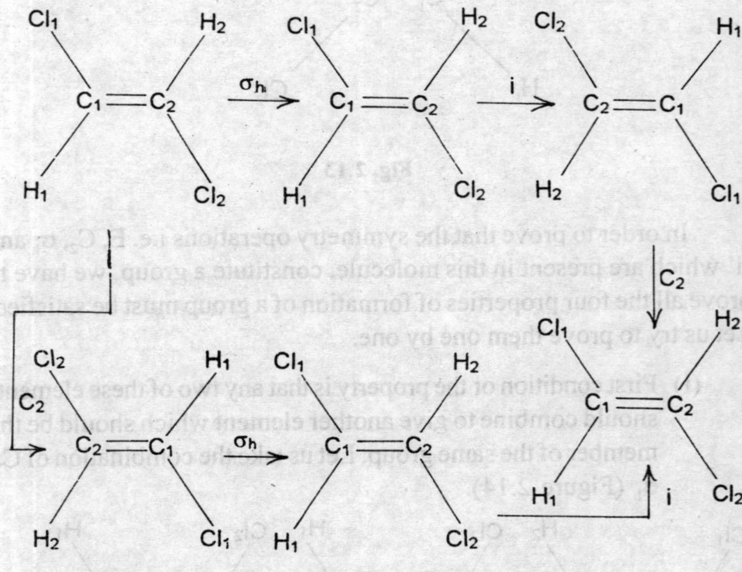

Fig. 2.15

This figure proves that this associative relation is valid. Similarly this can be shown for other combination of elements also.

(vi) Next condition is to check presence of inverse of every element which should be member of the same group i.e., the inverse operation should clearly correspond to inverse elements required by any collection of elements which is to be a member of the same group. This can be proved easily for any of these four elements, as

$$E^{-1} = E$$
$$C_2^{-1} = C_2$$
$$\sigma_h^{-1} = \sigma_h \text{ and}$$
$$i^{-1} = i$$

Therefore, it is proved that these elements together constitute a point group and the notation according to Schnoflies symbols for this point group is C_{2h}.

Section 2.5

So, it is clear from the discussion given in the previous sections that symmetry elements together constitute a group which is termed as point group. In this section let us try to define subgroups and classes in a larger point group, taking suitable examples.

In C_{2v} point group there are four symmetry elements viz. E, C_2, σ_v and σ_v' and in all four operations are there in this group. Total two subgroups are possible for C_{2v} point group viz. C_2 (= E. C_2) i.e. C_2 which is the combination of identity and C_2 and C_s = (= E.σ_v or E.σ_v') i.e., combination of identity with plane of symmetry. In both of these groups total number of operations are two. So, the order of the subgroups is 2 and that of the main group is 4. According to the condition of formation of a subgroup h/k should be an integer. Here it is verified i.e. 4/2 is an integer.

Let us take another example of C_{3v} point group. In this point group the order of the group is 6, and the possible subgroups is C_3 (= E, C_3^1, C_3^2) with order 3 and h/k is 6/3 i.e., an integer. Similarly, in D_{6h} point group (order 24), subgroups of order 6 (i.e. E, 5C_5) is possible.

In order to locate classes in a larger group we have to check that the elements of that combination should conjugate to each other. In the case of C_{2v} point group this can be proved easily that all operations are conjugate to each other. So, all the elements belong to same class.

i.e. $(\sigma_v^{-1})\,(C_2)\,(\sigma_v) = C_2$ and

$$(C_2^{-1})\,(\sigma_v)\,(C_2) = \sigma_v$$

So, C_2 and σ_v are conjugate to each other. Similarly this can be shown for other elements also. Therefore, they belong to same class and the order of this class is 4. So, is the order of the C_{2v} point group.

In the case of NH_3 molecule (i.e. C_{3v} point group), C_3^1 and C_3^2 belong to a similar class while $3\sigma_v$'s constitute another class. This can be shown as

$$(\sigma_v^{-1})\,(C_3^1)\,(\sigma_v) = C_3^2$$

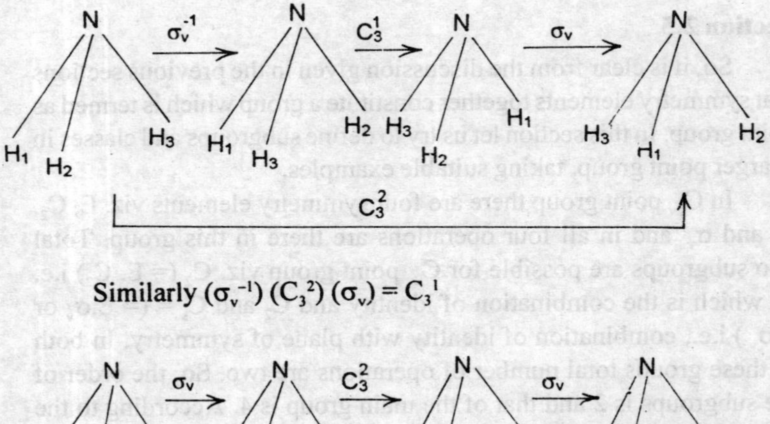

Similarly $(\sigma_v^{-1})\,(C_3^2)\,(\sigma_v) = C_3^1$

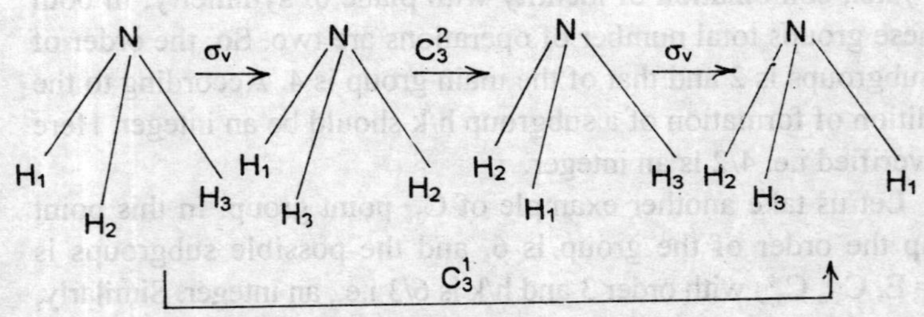

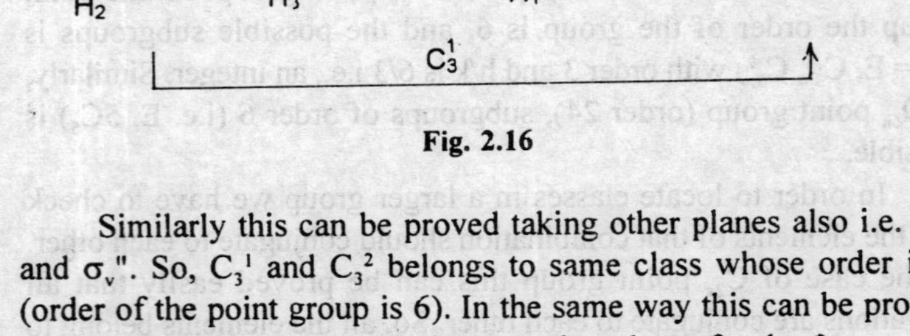

Fig. 2.16

Similarly this can be proved taking other planes also i.e. σ_v', and σ_v''. So, C_3^1 and C_3^2 belongs to same class whose order is 2 (order of the point group is 6). In the same way this can be proved that all the three planes constitute another class of order 3.

Section 2.6
Group Multiplication Tables

In this section, group multiplication tables are given. If for group a complete list of the elements is known (equal to the order of the group i.e. h). Their possible products (i.e. h^2 number of elements) can be arranged in the form of a table known as group multiplication table. This table for a point group consists of 'h' columns and 'h' rows. Each row and column is labeled with the group elements. As multiplication may not be commutative always, there is a general rule to construct a group multiplication table that we shall take the factors in the order (column element) × (row element). For example let us take a group of order 3 with group elements as E (identity), P, Q. Group multiplication table can be constructed as follows.

	E	P	Q
E	E	P	Q
P	P	Q	E
Q	Q	E	P

If, let us say, PP = Q, QP = E, PQ = E, QQ = P, EP = P, EQ = Q, PE = P and QE = Q. Let us try construct group multiplication tables for some of the representative point groups viz. C_{2v} and C_{2h} etc.

C_{2v}	E	C_2	σ_v	σ_v'
E	E	C_2	σ_v	σ_v'
C_{2v}	C_{2v}			
σ_v	σ_v			
σ_v'	σ_v'			

List of symmetry elements of this point group is E, C_2, σ_v and σ_v' and since EE ≡ E, $EC_2 \equiv C_2$, $E\sigma_v \equiv \sigma_v$, $E\sigma_v' \equiv \sigma_v'$, $C_2.E=C_2$, $\sigma_v E \equiv \sigma_v$ and $\sigma_v' E \equiv \sigma_v'$ (as product of any symmetry element with identity means that symmetry element). In this way, a row and a column in the multiplication table can be easily filled and for rest of the elements of the group multiplication table, let us try to do some combination operations on the water molecule (i.e., on the exampleof C_{2v} point group).

$$C_2 . C_2 \equiv E$$

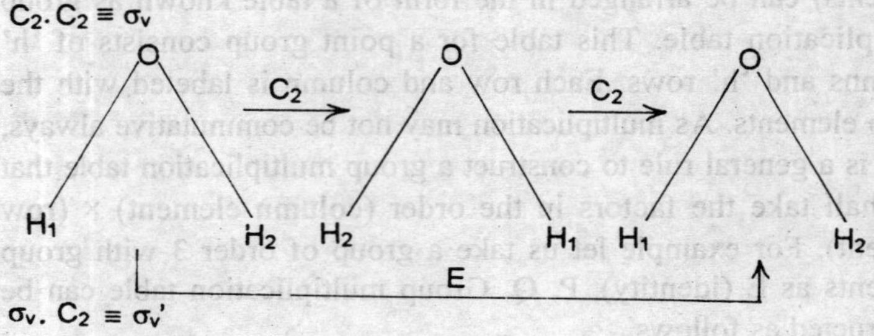

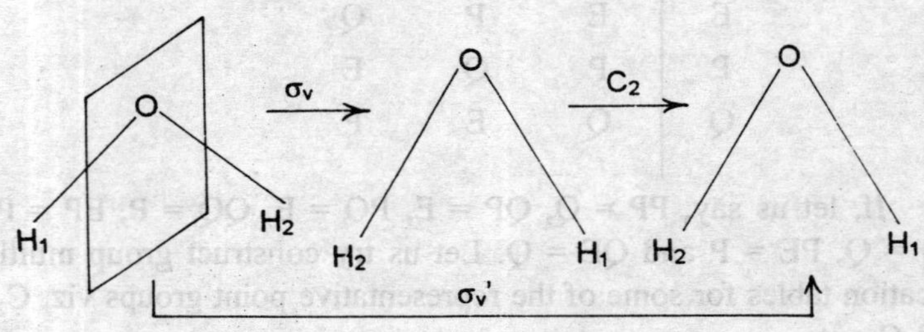

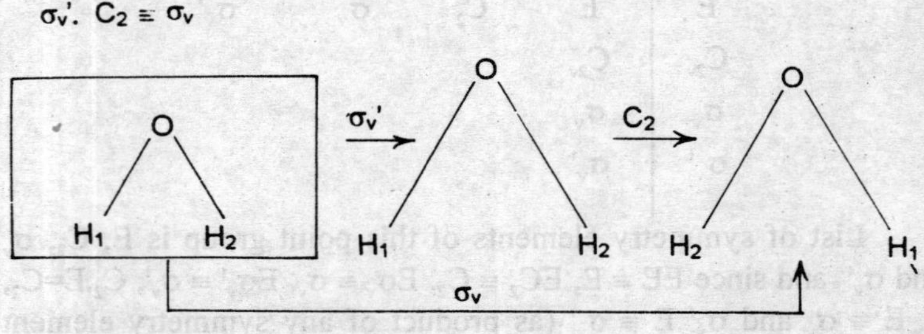

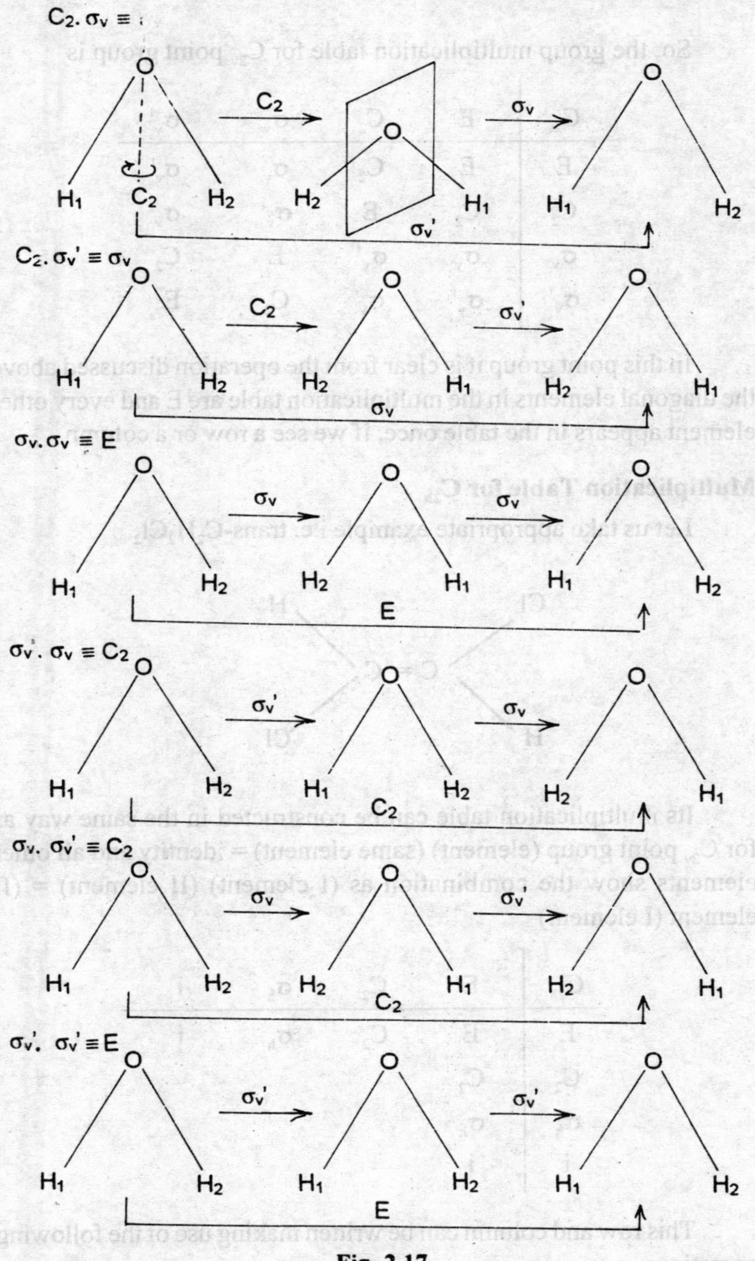

C2. σv ≡ :
O
H1
C2
H2
C2
H2
O
H1
σv
O
H1
H2
σv'
C2. σv' ≡ σv
C2
σv'
σv
σv. σv ≡ E
σv
σv
E
σv'. σv ≡ C2
σv'
σv
C2
σv. σv' ≡ C2
σv
σv'
C2
σv'. σv' ≡ E
σv'
σv'
E

Fig. 2.17

So, the group multiplication table for C_{2v} point group is

C_{2v}	E	C_2	σ_v	σ_v'
E	E	C_2	σ_v	σ_v'
C_2	C_2	E	σ_v'	σ_v
σ_v	σ_v	σ_v'	E	C_2
σ_v'	σ_v'	σ_v	C_2	E

In this point group it is clear from the operation discussed above the diagonal elements in the multiplication table are E and every other element appears in the table once, if we see a row or a column.

Multiplication Table for C_{2h}

Let us take appropriate example i.e. trans-$C_2H_2Cl_2$

```
Cl              H
   \           /
     C  =  C
   /           \
H               Cl
```

Its multiplication table can be constructed in the same way as for C_{2v} point group (element) (same element) = identity and all other elements show the combination as (I element) (II element) = (II element (I element)

C_{2h}	E	C_2	σ_h	i
E	E	C_2	σ_h	i
C_2	C_2			
σ_h	σ_h			
i	i			

This row and column can be written making use of the following operations

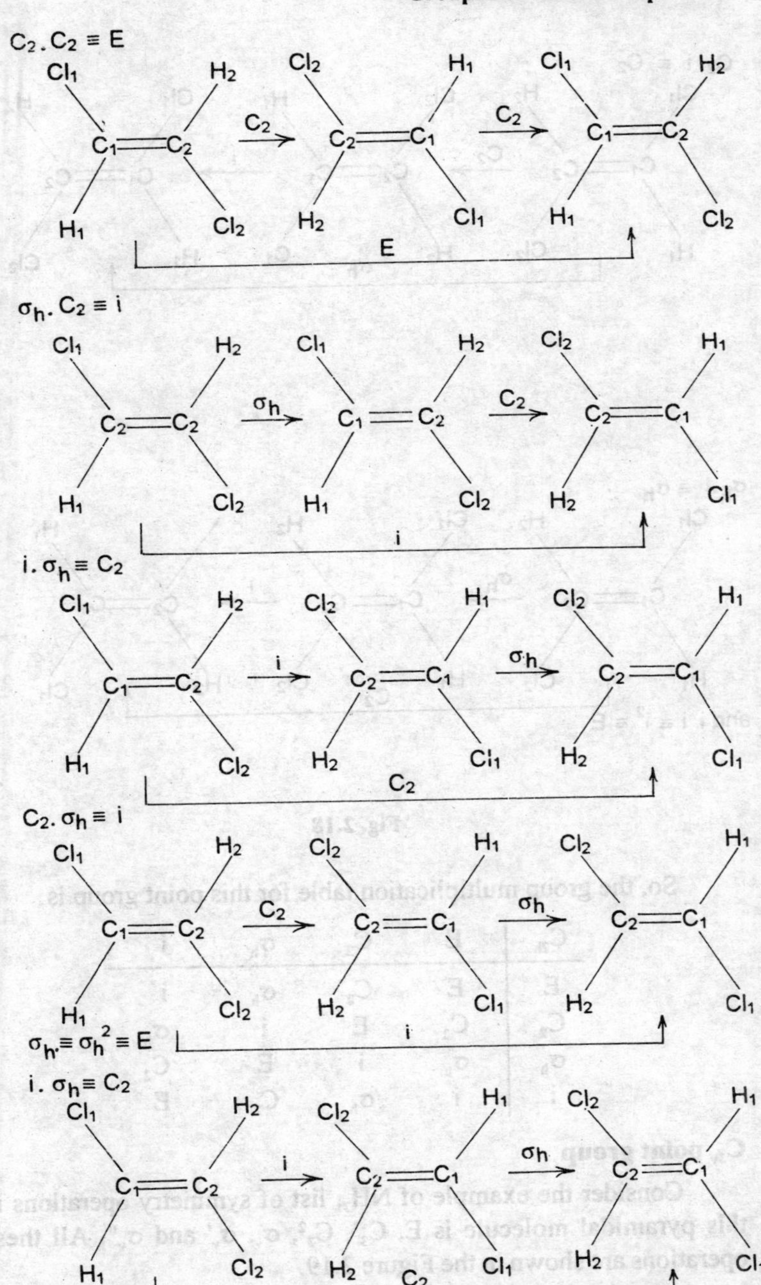
$C_2 . C_2 \equiv E$
$\sigma_h . C_2 \equiv i$
$i . \sigma_h \equiv C_2$
$C_2 . \sigma_h \equiv i$
$\sigma_h . \equiv \sigma_h^2 \equiv E$
$i . \sigma_h \equiv C_2$

$C_2 . i \equiv C_2$

$\sigma_h . i \equiv \sigma_h$

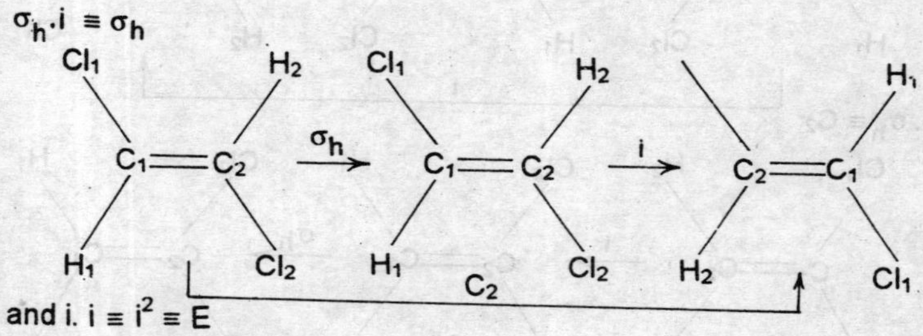

and $i . i \equiv i^2 \equiv E$

Fig. 2.18

So, the group multiplication table for this point group is

C_{2h}	E	C_2	σ_h	i
E	E	C_2	σ_h	i
C_2	C_2	E	i	σ_h
σ_h	σ_h	i	E	C_2
i	i	σ_h	C_2	E

C_{3v} point group

Consider the example of NH_3, list of symmetry operations in this pyramidal molecule is E, C_3^1, C_3^2, σ_v, σ_v' and σ_v''. All these operations are shown in the Figure 2.19.

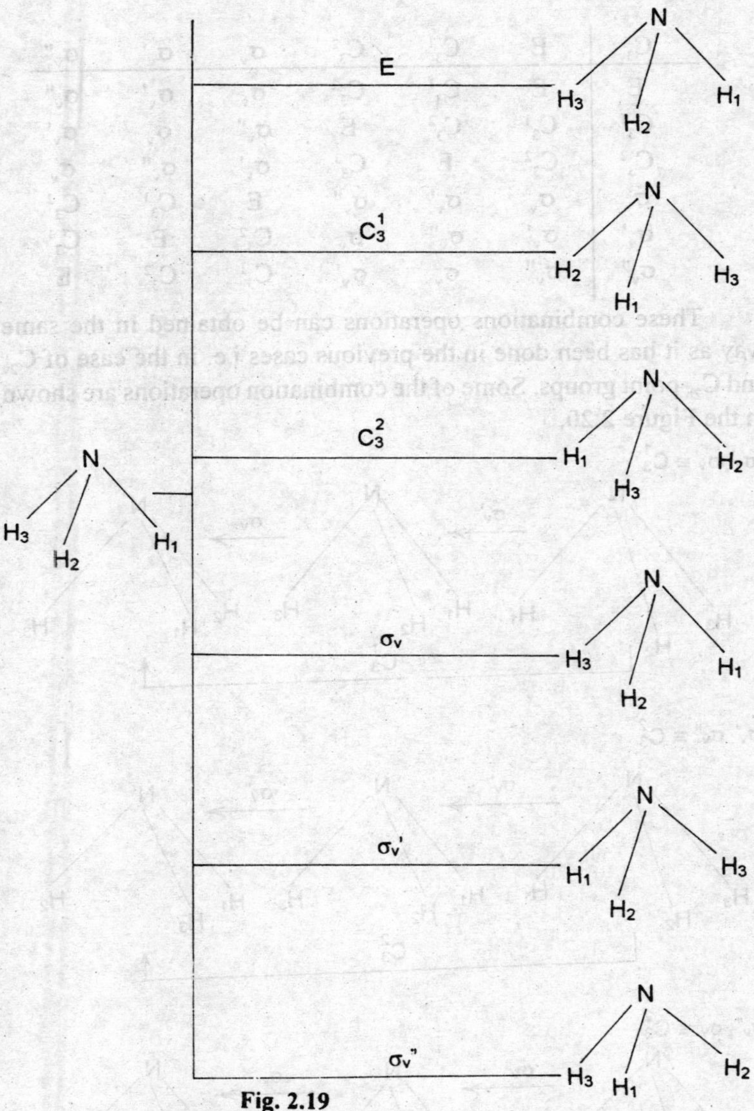

Fig. 2.19

Following the same rule that if these operations are written in a row and in a column and considering the combination operations, following table can be obtained as the group multiplication table for this case.

C_{3v}	E	C_3^1	C_3^2	σ_v	σ_v'	σ_v''
E	E	C_3^1	C_3^2	σ_v	σ_v'	σ_v''
C_3^1	C_3^1	C_3^2	E	σ_v''	σ_v	σ_v'
C_3^2	C_3^2	E	C_3^1	σ_v'	σ_v''	σ_v
σ_v	σ_v	σ_v'	σ_v''	E	C_3^1	C_3^2
σ_v'	σ_v'	σ_v''	σ_v	C_3^2	E	C_3^1
σ_v''	σ_v''	σ_v	σ_v'	C_3^1	C_3^2	E

These combinations operations can be obtained in the same way as it has been done in the previous cases i.e. in the case of C_{2v} and C_{2h} point groups. Some of the combination operations are shown in the Figure 2.20.

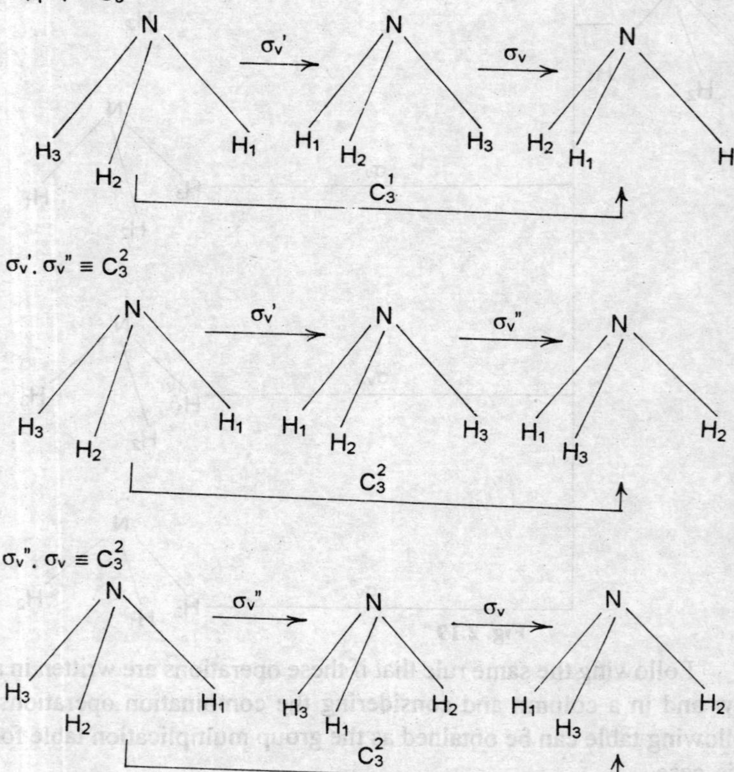

$\sigma_v''.\ \sigma_v' \equiv C_3^1$

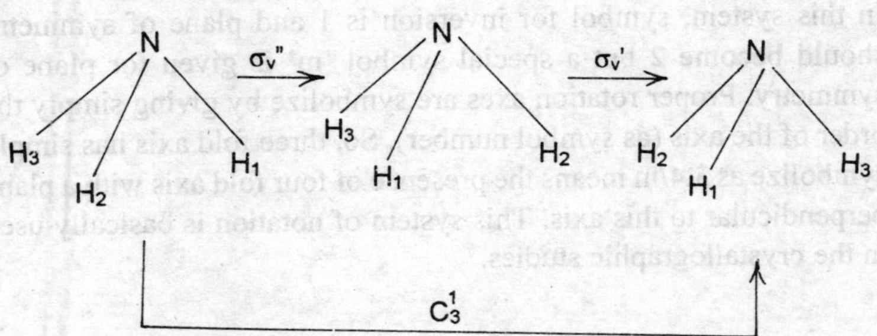

$C_3^1.\ C_3^1 \equiv C_3^2$

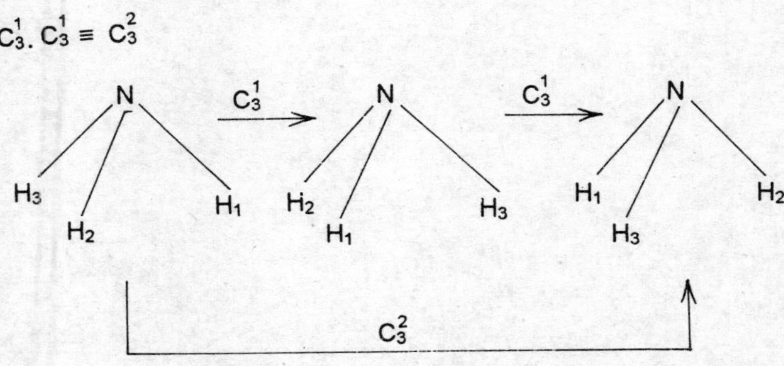

Fig. 2.20

Similarly all other combination operations can be shown for this case. In this way with the help of combination operations, group multiplication table for a point group can be constructed.

Section 2.7

Hermann-Mauguin Symbols

So, far we are discussing the point group symbols and the notations for various symmetry elements in C-σ-'i' system which is known as Schnoflies system of notation. There is another system of notation of symbolize the various symmetry elements and the point groups which is known as Hermann-Mauguin system of notations. In this system, symbol for inversion is 1 and plane of symmetry should become 2 but a special symbol 'm' is given for plane of symmetry. Proper rotation axes are symbolize by giving simply the order of the axis (as symbol number). So, three fold axis has simply symbolize as 3.4/m means the presence of four fold axis with a plane perpendicular to this axis. This system of notation is basically used in the crystallographic studies.

Problem

Q. 1: List the symmetry elements and name the point group for each of the following:

(a) 1, 2-Dichlorobenzene

(b) Eclipsed ferrocene

(c) Anthracene

(d) $[Pt(en)_2]^{2+}$ ion

(e) $Fe_2(CO)_9$

Ans: (a) List of symmetry elements in this molecule is E, C_2, σ_v and σ_v' as in the case of water molecule. So, the point group of this molecule is C_{2v}.

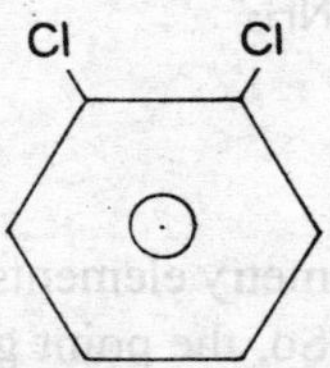

(b) List of symmetry elements in this molecule is E, $C_5 \perp 5C_2$, σ_h, S_5, $5\sigma_v$. So, the point group of this molecule is D_{5h}.

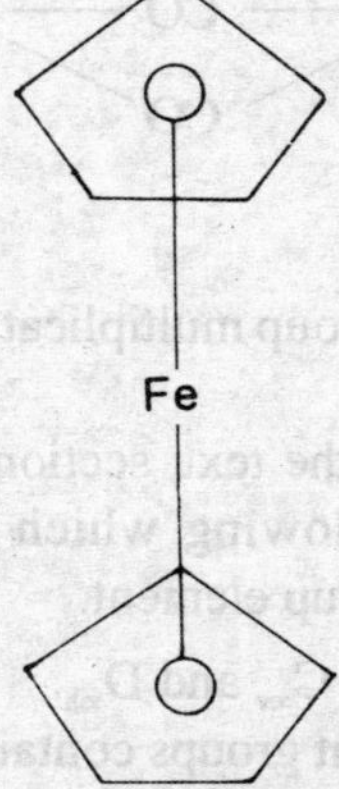

(c) List of symmetry elements in this molecule is E, $C_2 \perp$

$2C_2$, σ_h, 'i' and σ_v. So, the point group of this molecule is D_{2h}.

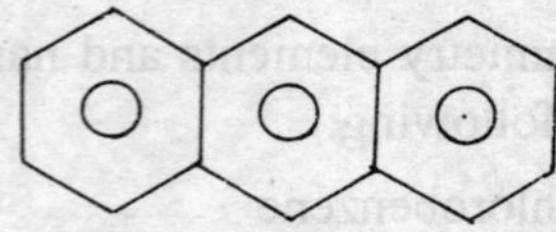

(d) List of symmetry elements in this molecule is E, $C_2 \perp 2C_2$. So, the point group of this molecule is D_2.

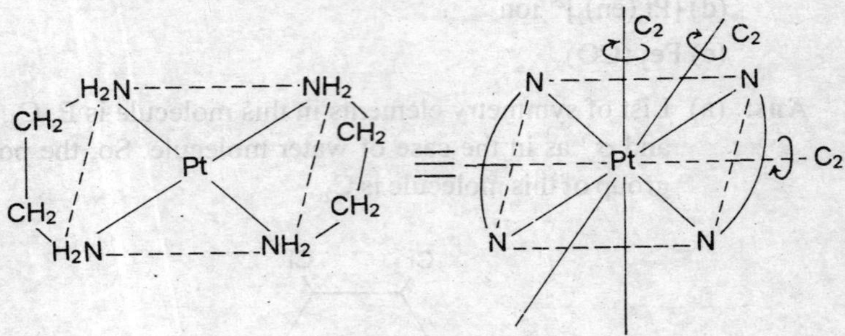

(e) List of symmetry elements in this molecule is E, C_3, C_2^2, σ_h, S_3, S_3^5. So, the point group is C_{3h}.

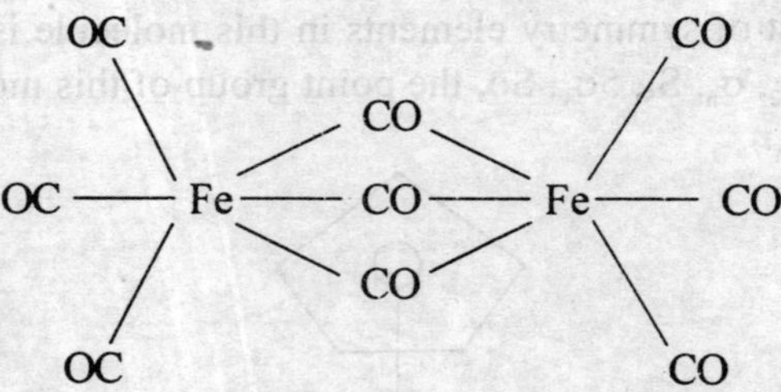

Q. 2: Construct the group multiplication table for C_{2v} and C_{2h} point groups.

Ans: Please refer to the text, section 2.6.

Q. 3: Out of the following which point group has centre of inversion as group element.

C_{2v}, C_{2h}, D_2, C_2, $C_{\infty v}$ and $D_{\infty h}$.

Ans: C_{2h} and $D_{\infty h}$ point groups contain inversion centre e.g. trans-$C_2H_2Cl_2(C_{2h})$ and $CO_2(D_{\infty h})$.

Q. 4: Molecules belonging to which of the following point groups are molecules of high symmetry and which are of low symmetry.

C_1, O_h, C_i, C_s, T_d and C_{2v}. Give one suitable example in each case.

Ans: Point groups for the molecules of high symmetry are O_h, T_d and I_h e.g. O_h (all complexes in the regular octahedral geometry e.g. SF_6) T_d (molecules in the regular tetrahedral geometry e.g. CH_4, CCl_4 etc.) I_h (e.g. dodecaborane)

Point groups for the molecules with low symmetry are C_1, C_i and C_s e.g. C_1(CH_3COO^{-1} ion); C_i(trans staggered 1,2-dichloro 1, 2-dibromo ethane); C_s (α-naphthol).

Q. 5: Find out the point group for $[Pt\ (en)_2]^{2+}$ ion. Co.struct the group multiplication table for this point group.

Ans: Point group for this molecule is D_2 [please refer to Q1. (d)]. Group multiplication for this point group is

D_2	E	C_2	C_2'	C_2''
E	E	C_2	C_2'	C_2''
C_2	C_2	E	C_2''	C_2'
C_2'	C_2'	C_2''	E	C_2
C_2''	C_2''	C_2'	C_2	E

As the combinations of the various symmetry operations on the molecule are

$$E.C_2 \equiv C_2.E \equiv C_2$$

$$E.C_2' \equiv C_2'.E \equiv C_2'$$

$$E.C_2'' \equiv C_2''.E \equiv C_2''$$

$$C_2.C_2 \equiv C_2'.C_2' \equiv C_2''.C_2'' \equiv E$$

and

$$E.E. \equiv E$$

$$C_2.C_2' \equiv C_2.C_2' \equiv C_2''$$

These can be shown easily by performing various combinations of symmetry operations on the referred molecule.

3

Features of Point Groups and Character Tables

In the previous chapter, it has been shown that how symmetry elements constitute a group which is termed as point group, what is a group, sub-group and a class. This chapter discusses the details of features of point groups, such as representations of symmetry elements in a point group, character tables for point groups, great orthogonality theorem.

Section 3.1

Introduction to vectors and matrices

A vector is a physical quantity which has magnitude and a direction in three dimensional space. It is represented by A.

$$\vec{A} = A_x\hat{i} + A_y\hat{j} + A_z\hat{k}$$

Where $\hat{i}$, $\hat{j}$ & $\hat{k}$ are unit vectors in x, y and z directions respectively.

Some important properties of vectors

It is often require to have a product of two vectors. Product of vectors may be of two types (i) Scalar product and (ii) Vector product.

Scalar or dot product is that in which the result is a number (or scalar) not a vector. It is given by

$$\vec{A}.\vec{B} = A.B \cos\theta$$

$$\vec{A}.\vec{B} = A_x B_x + A_y. B_y + A_z. B_z$$

If two vectors are considered as

$$A = A_x\hat{i} + A_y\hat{j} + A_z\hat{k} \text{ and}$$

$$B = B_x\hat{i} + B_y\hat{j} + B_z\hat{k}$$

If $\theta = 0°$ i.e. $\cos\theta = 1$, so, $\vec{A}.\vec{B}$ = A.B (simply) but if $\theta = 90°$ (or $\cos\theta = 0$), so, A.B = 0, in this case vectors are orthogonal to each other.

Another type of vector product is vector or cross product in which result is a vector

i.e. $\vec{A}\times\vec{B}$ and it is represented by the following determinant

$$A \times B = \begin{vmatrix} \hat{i} & \hat{j} & \hat{k} \\ A_x & A_y & A_z \\ B_x & B_y & B_z \end{vmatrix}$$

Third important operation for vectors is to find out their projections, which can be find out with the help of trigonometry.

Let a vector is

$$\vec{A} = A_x\hat{i} + A_y\hat{j}$$

The components of this vector are

$$A_x = A\cos\theta \text{ and } A_y = A\sin\theta$$

Where A is the magnitude or the scalar length of a vector.

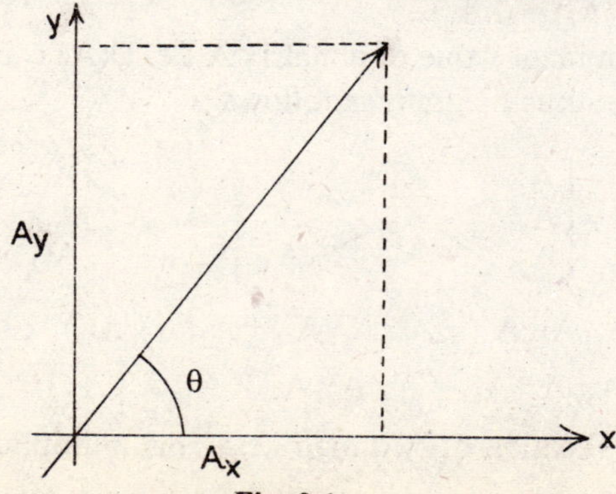

Fig. 3.1

Matrices

A matrix is a rectangular array of numbers or symbols which are called as matrix elements, it's general form is

$$\begin{bmatrix} A_{11} & A_{12} & A_{13} \\ A_{21} & A_{22} & A_{23} \\ A_{31} & A_{32} & A_{33} \end{bmatrix}$$

Here in this case all elements of the matrix may be defined by a common symbol A_{ij} where i = number of row and j = number of column. Above is the 3 × 3 matrix i.e. it has 3 rows and 3 columns. If in a matrix number of rows and number of columns are same as it is in there in this case the matrix is called as square matrix and in this matrix elements in which i = j i.e. A_{11}, A_{22}, A_{33} are diagonal elements. A matrix in which all diagonal elements are equal to 1 and all other elements are zero, is called as unit matrix.

$$\begin{bmatrix} 1 & 0 & 0 \\ 0 & 1 & 0 \\ 0 & 0 & 1 \end{bmatrix}$$

The sum of all diagonal elements of a matrix is its trace or character.

Properties of a Matrix

(i) Multiplication by a scalar: A matrix can be multiplied by a scalar (a number). In this multiplication all the elements of the matrix are to be multiplied with this scalar (number). i.e. $k(A_{ij}) = (kA_{ij})$

(ii) Determinant value of a matrix A i.e. D(A) can be obtained using Surus diagram as follows

$$D(A) = \begin{bmatrix} A_{11} & A_{12} & A_{13} \\ A_{21} & A_{22} & A_{23} \\ A_{31} & A_{32} & A_{33} \end{bmatrix} \begin{matrix} A_{11} & A_{12} \\ A_{21} & A_{22} \\ A_{31} & A_{32} \end{matrix}$$

i.e. $(A_{11} A_{22} A_{33} + A_{12} A_{23} A_{31} + A_{13} A_{21} A_{32})$

$- (A_{13} A_{22} A_{31} + A_{11} A_{23} A_{32} + A_{12} A_{21} A_{33})$

(iii) Multiplication of two matrices: This multiplication can be obtained by multiplying the ith row of the first matrix by j^{th} column of the second matrix i.e. row by column. So, the elements of the resultant matrix are given by the formula

$$C_{ik} = \sum_{j=1}^{n} A_{ij} B_{jk}$$

e.g.

$$\begin{bmatrix} A_{11} & A_{12} & A_{13} \\ A_{21} & A_{22} & A_{23} \\ A_{31} & A_{32} & A_{33} \end{bmatrix} \begin{bmatrix} B_{11} & B_{12} & B_{13} \\ B_{21} & B_{22} & B_{23} \\ B_{31} & B_{32} & B_{33} \end{bmatrix}$$

$$= \begin{bmatrix} C_{11} & C_{12} & C_{13} \\ C_{21} & C_{22} & C_{23} \\ C_{31} & C_{32} & C_{33} \end{bmatrix}$$

Where $C_{11} = A_{11} B_{11} + A_{12} B_{21} + A_{13} B_{31}$

$C_{12} = A_{11} B_{12} + A_{12} B_{22} + A_{13} B_{32}$

$C_{13} = A_{11} B_{13} + A_{12} B_{23} + A_{13} B_{33}$

$C_{21} = A_{21} B_{11} + A_{22} B_{21} + A_{23} B_{31}$

$C_{22} = A_{21} B_{12} + A_{22} B_{22} + A_{23} B_{32}$

$C_{23} = A_{21} B_{13} + A_{22} B_{23} + A_{23} B_{33}$

$C_{31} = A_{31} B_{11} + A_{32} B_{21} + A_{33} B_{31}$

$C_{32} = A_{31} B_{12} + A_{32} B_{22} + A_{33} B_{32}$

$C_{33} = A_{31} B_{13} + A_{32} B_{23} + A_{33} B_{33}$

(iv) In general matrices do not commute with each other i.e.

$$AB \neq BA$$

(v) Matrix multiplication is associative.

(vi) The product of a matrix and its inverse is commutative and is equal to unit matrix.

$$A.A^{-1} = A^{-1}.A = I \text{ (identity)}$$

(vii) Transpose of a matrix A is obtained by changing its rows into columns (or vice versa) and this is given the symbol A (or A').

$$A = \begin{bmatrix} A_{11} & A_{12} & A_{13} \\ A_{21} & A_{22} & A_{23} \\ A_{31} & A_{32} & A_{33} \end{bmatrix}$$

So,

$$A' = \begin{bmatrix} A_{11} & A_{21} & A_{31} \\ A_{12} & A_{22} & A_{32} \\ A_{13} & A_{23} & A_{33} \end{bmatrix}$$

If A = A' the matrix is called as symmetric.

(viii) A matrix is orthogonal if its transpose is equal to its inverse.

(xi) Inverse of a matrix

$$A = \begin{bmatrix} A_{11} & A_{12} & A_{13} \\ A_{21} & A_{22} & A_{23} \\ A_{31} & A_{32} & A_{33} \end{bmatrix}$$

is given by

$$A^{-1} = \begin{bmatrix} A_{11}/|A| & A_{21}/|A| & A_{31}/|A| \\ A_{12}/|A| & A_{22}/|A| & A_{32}/|A| \\ A_{13}/|A| & A_{23}/|A| & A_{33}/|A| \end{bmatrix}$$

where A_{ij} are co-factors of the matrix.

Section 3.2

Symmetry operations as matrix representations

Previous section gives the brief introduction to vectors and matrices. Let us try to express these symmetry operation as matrix representations.

Identity: Let us consider a point x, y, z in the reference frame of axes.On identity operation the point remains unchanged i.e. no effect can be seen on this point on identity operation. This can be represented by matrix equation as

$$\begin{pmatrix} x' \\ y' \\ z' \end{pmatrix} = \begin{pmatrix} 1 & 0 & 0 \\ 0 & 1 & 0 \\ 0 & 0 & 1 \end{pmatrix} \begin{pmatrix} x \\ y \\ z \end{pmatrix}$$

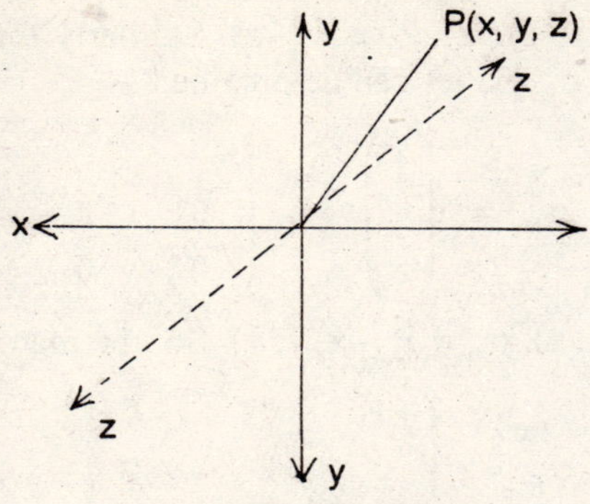

Fig. 3.2

or P(x, y, z). E = P (x', y', z'). So, the matrix representation equivalent to identity is the unit matrix i.e.

$$E = \begin{pmatrix} 1 & 0 & 0 \\ 0 & 1 & 0 \\ 0 & 0 & 1 \end{pmatrix}$$

or identity matrix.

Reflections: Let us consider a point P (x, y, z) in the reference frame of axep. If reflection is carried out in xy plane then only z co-ordinate changes to -z and x, y co-ordinates remain same after reflection. This reflection operation through σ_{xy} plane can be represented in the matrix as

$$\begin{pmatrix} x' \\ y' \\ z' \end{pmatrix} = \begin{pmatrix} 1 & 0 & 0 \\ 0 & 1 & 0 \\ 0 & 0 & -1 \end{pmatrix} \begin{pmatrix} x \\ y \\ z \end{pmatrix}$$

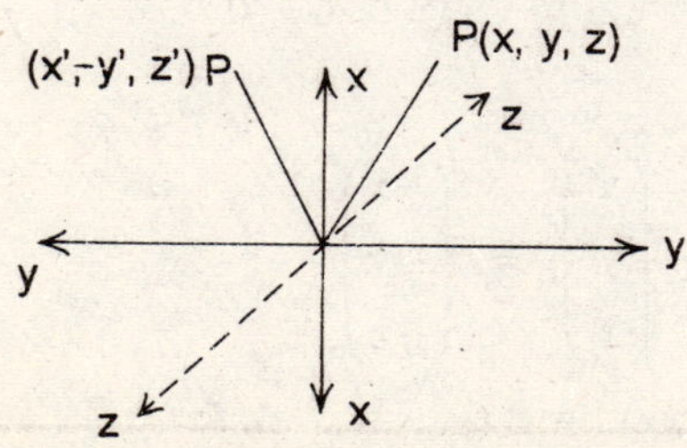

Fig. 3.3

i.e. P (x, y, z). σ_{xy} P (x, y, –z). Similarly for the reflections through σ_{yz} and σ_{xz} planes can be obtained as

$$\sigma_{yz} = \begin{pmatrix} x' \\ y' \\ z' \end{pmatrix} = \begin{pmatrix} -1 & 0 & 0 \\ 0 & 1 & 0 \\ 0 & 0 & 1 \end{pmatrix} \begin{pmatrix} x \\ y \\ z \end{pmatrix}$$

or P (x, y, z). σ_{yz} = P (–x, y, z). So, the matrix for

$$\sigma_{yz} = \begin{pmatrix} -1 & 0 & 0 \\ 0 & 1 & 0 \\ 0 & 0 & 1 \end{pmatrix}$$

Similarly

$$\sigma_{xz}: \begin{pmatrix} x' \\ y' \\ z' \end{pmatrix} \begin{pmatrix} 1 & 0 & 0 \\ 0 & -1 & 0 \\ 0 & 0 & 1 \end{pmatrix} \begin{pmatrix} x \\ y \\ z \end{pmatrix}$$

or P (x, y, z). σ_{xz} = P (x, –y, z) So, the matrix for

$$\sigma_{xz} = \begin{pmatrix} 1 & 0 & 0 \\ 0 & -1 & 0 \\ 0 & 0 & 1 \end{pmatrix}$$

Inversion: During inversion operation it is clear that signs of all the co-ordinates change. So, we need a negative unit matrix to define inversion, or

$$\begin{pmatrix} x' \\ y' \\ z' \end{pmatrix} = \begin{pmatrix} -1 & 0 & 0 \\ 0 & -1 & 0 \\ 0 & 0 & -1 \end{pmatrix} \begin{pmatrix} x \\ y \\ z \end{pmatrix}$$

i.e.

$$\begin{pmatrix} x' \\ y' \\ z' \end{pmatrix} = \begin{pmatrix} -x \\ -y \\ -z \end{pmatrix}$$

or P (x, y, z).i = P (–x, –y, –z).

So, the matrix for 'i' is

$$i = \begin{pmatrix} -1 & 0 & 0 \\ 0 & -1 & 0 \\ 0 & 0 & -1 \end{pmatrix}$$

Proper Rotation: In order to construct a matrix for proper rotation operation, consider the axis of rotation as z-axis. So, it is obvious that z co-ordinate remains unchanged. So, the matrix may be

$$\begin{pmatrix} & & 0 \\ & & 0 \\ 0 & 0 & 1 \end{pmatrix}$$

The missing four elements can be found out by considering the effect of rotation on rest of the two dimensions i.e. x, y dimensions in xy plane. Let P_1 defines by the co-ordinate x, y and after rotation by an angle 8. P_1 changes to P_2 with co-ordinates x', y' can be expressed according to the following equations

$$x' = x \cos \theta - y \sin \theta$$

$$y' = x \sin \theta + y \cos \theta$$

So, the matrix is represented as

$$\begin{pmatrix} x' \\ y' \end{pmatrix} = \begin{pmatrix} \cos \theta & -\sin \theta \\ \sin \theta & \cos \theta \end{pmatrix} \begin{pmatrix} x \\ y \end{pmatrix}$$

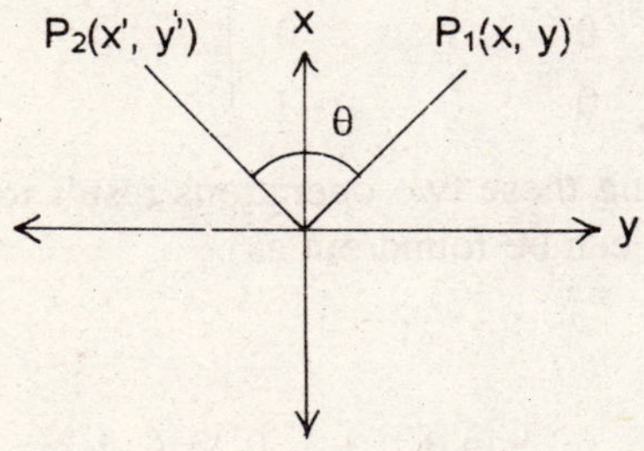

Fig. 3.4

This is the result obtained if anticlockwise rotation is considered, but according to the definition, a rotation operation is clockwise. So, we have to change the quadrants as cos θ = cos (–θ) and – sin θ = sin (–θ). So, the matrix corresponding to rotation changes as

$$\begin{pmatrix} \cos\theta & \sin\theta \\ -\sin\theta & \cos\theta \end{pmatrix}$$

Combining the effect of rotation on x, y co-ordinates with z co-ordinate complete matrix for rotations is

$$C_n = \begin{pmatrix} \cos\theta & \sin\theta & 0 \\ -\sin\theta & \cos\theta & 0 \\ 0 & 0 & 1 \end{pmatrix}$$

Improper rotation: Improper rotation is the combination of proper rotation followed by reflection through plane which is perpendicular to this rotation axis. Matrix corresponding to proper rotation has already constructed as

$$\begin{pmatrix} \cos\theta & \sin\theta & 0 \\ -\sin\theta & \cos\theta & 0 \\ 0 & 0 & 1 \end{pmatrix}$$

considering z axis as axis of rotation. If we consider the reflection through the plane which is perpendicular to this z axis i.e. σ_{xy} whose matrix is

$$\begin{pmatrix} 1 & 0 & 0 \\ 0 & 1 & 0 \\ 0 & 0 & -1 \end{pmatrix}$$

On combining these two operations result for S_n operation or improper rotation can be found out as

$S_n = C_n \cdot \sigma_{xy}$

So,

$$\underset{C_n}{\begin{pmatrix} \cos\theta & \sin\theta & 0 \\ -\sin\theta & \cos\theta & 0 \\ 0 & 0 & 1 \end{pmatrix}} \underset{\sigma_{xy}}{\begin{pmatrix} 1 & 0 & 0 \\ 0 & 1 & 0 \\ 0 & 0 & -1 \end{pmatrix}}$$

$$= \begin{pmatrix} \cos\theta & \sin\theta & 0 \\ -\sin\theta & \cos\theta & 0 \\ 0 & 0 & -1 \end{pmatrix}$$

is the matrix for S_n operation.

Section 3.3

Group multiplication table using matrices for different symmetry operations

Group multiplication tables as discussed in the previous chapter can be reconstructed here using matrices representations for various symmetry operations and this can also be proved that the symmetry elements (and their related operations) form a group. Let us consider them one by one.

C_{2v}: List of symmetry elements (or operations) is E, C_2, σ_v and σ_v'. Let us consider a suitable example i.e. water molecule in the frame of axes with oxygen at origin. So, σ_v may be defined as σ_{xz} and σ_v' as σ_{yz}. The matrices for these operations are

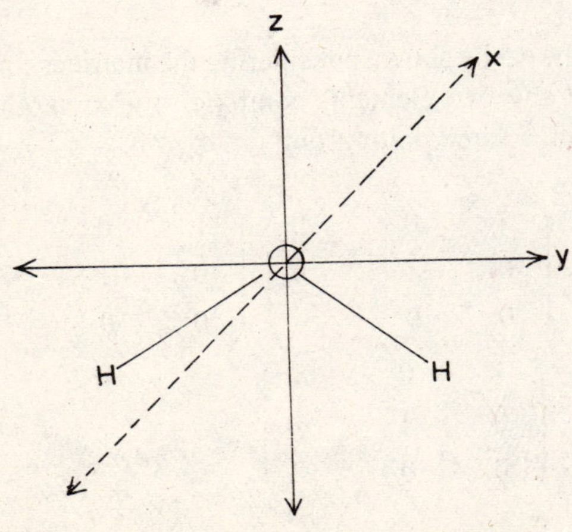

Fig. 3.5

$$E = \begin{pmatrix} 1 & 0 & 0 \\ 0 & 1 & 0 \\ 0 & 0 & 1 \end{pmatrix}$$

$$C_2 = \begin{pmatrix} \cos 180° & \sin 180° & 0 \\ -\sin 180° & \cos 180° & 0 \\ 0 & 0 & 1 \end{pmatrix}$$

$$C_2 = \begin{pmatrix} -1 & 0 & 0 \\ 0 & -1 & 0 \\ 0 & 0 & 1 \end{pmatrix}$$

$$\sigma_{xz} = \begin{pmatrix} 1 & 0 & 0 \\ 0 & -1 & 0 \\ 0 & 0 & 1 \end{pmatrix} \quad \text{and}$$

$$\sigma_{yz} = \begin{pmatrix} -1 & 0 & 0 \\ 0 & 1 & 0 \\ 0 & 0 & 1 \end{pmatrix}$$

Let us try to apply the conditions of formation of a point group

This can be easily shown considering the matrices for symmetry operations that any two elements combine to give another element that belongs to the same point group

$$C_2.\sigma_{xz} = \begin{pmatrix} -1 & 0 & 0 \\ 0 & -1 & 0 \\ 0 & 0 & 1 \end{pmatrix} \begin{pmatrix} 1 & 0 & 0 \\ 0 & -1 & 0 \\ 0 & 0 & 1 \end{pmatrix}$$

$$= \begin{pmatrix} -1 & 0 & 0 \\ 0 & 1 & 0 \\ 0 & 0 & 1 \end{pmatrix}$$

$$= \sigma_{yz}$$

Similarly this can be shown that

D (C_2). D (σ_{yz}) = D (σ_{xz}) and D (σ_{xz}) . D (σ_{yz}) = D (C_2) and so on where D (E), D (C_2), D (σ_{xz}), D (σ_{yz}) etc. are notations for matrices belonging to E,C_2, σ_{xz} and σ_{yz} operations.

There is a matrix i.e. identity matrix exist which can commute with all other matrices of the symmetry operations i.e. this can be shown that D (E). D (C_2) = D (C_2)

$$\begin{pmatrix} 1 & 0 & 0 \\ 0 & 1 & 0 \\ 0 & 0 & 1 \end{pmatrix} \begin{pmatrix} -1 & 0 & 0 \\ 0 & -1 & 0 \\ 0 & 0 & 1 \end{pmatrix}$$

$$= \begin{pmatrix} -1 & 0 & 0 \\ 0 & -1 & 0 \\ 0 & 0 & 1 \end{pmatrix}$$

Similarly this can be proved that D (E). D (σ_{xz}) = D (σ_{xz}); D (E). (σ_{yz}) = D (σ_{yz})

Generally associative law holds good for matrix multiplication. Let us prove D (E). [D (C_2). D (σ_{xz})] = [D (E). D (C_2).] D (σ_{xz})

L.H.S.

D (C_2). D (σ_{xz}) =

$$\begin{pmatrix} -1 & 0 & 0 \\ 0 & -1 & 0 \\ 0 & 0 & 1 \end{pmatrix} \begin{pmatrix} 1 & 0 & 0 \\ 0 & -1 & 0 \\ 0 & 0 & 1 \end{pmatrix} = \begin{pmatrix} -1 & 0 & 0 \\ 0 & 1 & 0 \\ 0 & 0 & 1 \end{pmatrix}$$

and D (E). [D (C_2). D (σ_{xz})] is

$$\begin{pmatrix} 1 & 0 & 0 \\ 0 & 1 & 0 \\ 0 & 0 & 1 \end{pmatrix} \begin{pmatrix} -1 & 0 & 0 \\ 0 & 1 & 0 \\ 0 & 0 & 1 \end{pmatrix} = \begin{pmatrix} -1 & 0 & 0 \\ 0 & 1 & 0 \\ 0 & 0 & 1 \end{pmatrix}$$

R.H.S.

D (E). D (C_2) =

$$\begin{pmatrix} 1 & 0 & 0 \\ 0 & 1 & 0 \\ 0 & 0 & 1 \end{pmatrix} \begin{pmatrix} -1 & 0 & 0 \\ 0 & -1 & 0 \\ 0 & 0 & 1 \end{pmatrix} = \begin{pmatrix} -1 & 0 & 0 \\ 0 & -1 & 0 \\ 0 & 0 & 1 \end{pmatrix}$$

and D (E). [D (C_2). D (σ_{xz})] is

$$\begin{pmatrix} -1 & 0 & 0 \\ 0 & 1 & 0 \\ 0 & 0 & 1 \end{pmatrix}\begin{pmatrix} 1 & 0 & 0 \\ 0 & -1 & 0 \\ 0 & 0 & 1 \end{pmatrix} = \begin{pmatrix} -1 & 0 & 0 \\ 0 & 1 & 0 \\ 0 & 0 & 1 \end{pmatrix}$$

Hence, L.H.S. = R.H.S. So, associative law for the operations E, C_2 and σ_{xz} has been verified. Similarly this can be proved for other symmetry elements also.

Existence of inverse for any one of these matrices for these operations can also be checked in the same way. Therefore, it is proved that E, C_2, σ_{xz} and σ_{xz} constitute a point group i.e. C_{2v}. The group multiplication table can also be constructed using matrix multiplications for various symmetry elements as

C_{2v}	D (E)	D (C_2)	D (σ_{xz})	D (σ_{yz})
D (E)	D (E)	D (C_2)	D (σ_{xz})	D (σ_{yz})
D (C_2)	D (C_2)	D (E)	D (σ_{yz})	D (σ_{xz})
D (σ_{xz})	D (σ_{xz})	D (σ_{yz})	D (E)	D (C_2)
D (σ_{yz})	D (σ_{yz})	D (σ_{xz})	D (C_2)	D (E)

<u>C_{2h}</u> : Consider a suitable example i.e. trans-$C_2H_2Cl_2$ in the frame of axes. List of symmetry elements is E, C_2, σ_h and 'i'. Let z-axis is the proper axis of rotation i.e. the C_2 axis which is coming out of the plane of the molecule and hence the horizontal plane is xy plane i.e. $\sigma_h = \sigma_{xy}$ plane. Therefore, matrices for various symmetry elements are

$$E = \begin{pmatrix} 1 & 0 & 0 \\ 0 & 1 & 0 \\ 0 & 0 & 1 \end{pmatrix}$$

$$C_2 = \begin{pmatrix} \cos 180° & \sin 180° & 0 \\ -\sin 180° & \cos 180° & 0 \\ 0 & 0 & -1 \end{pmatrix} = \begin{pmatrix} -1 & 0 & 0 \\ 0 & -1 & 0 \\ 0 & 0 & 1 \end{pmatrix}$$

$$\sigma_h = \sigma_{xy} \begin{pmatrix} 1 & 0 & 0 \\ 0 & 1 & 0 \\ 0 & 0 & -1 \end{pmatrix} \text{ and}$$

$$\text{'i'} = \begin{pmatrix} -1 & 0 & 0 \\ 0 & -1 & 0 \\ 0 & 0 & -1 \end{pmatrix}$$

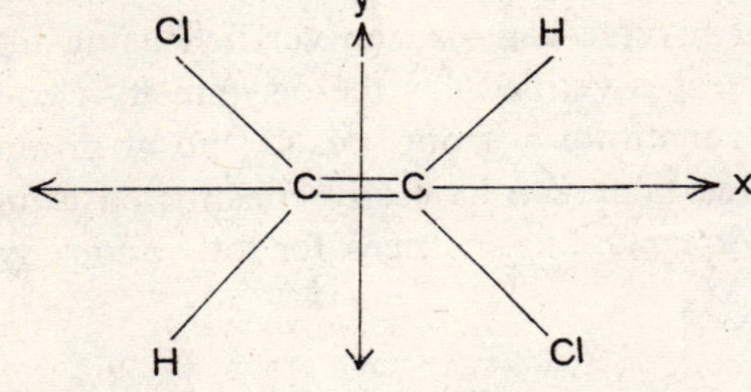

Fig. 3.6

To prove that these elements constitute a group let us try to verify the conditions for formation of a point group. Let us consider D (E), D (C_2), D (σ_{xy}) and D (i) as notations for the E, C_2, σ_{xy} and 'i' matrices.

This can be easily shown using matrices of these symmetry operations that two elements combine to give another element that belongs to same point group

$$\begin{pmatrix} -1 & 0 & 0 \\ 0 & -1 & 0 \\ 0 & 0 & 1 \end{pmatrix} \begin{pmatrix} 1 & 0 & 0 \\ 0 & 1 & 0 \\ 0 & 0 & -1 \end{pmatrix} = \begin{pmatrix} -1 & 0 & 0 \\ 0 & -1 & 0 \\ 0 & 0 & -1 \end{pmatrix}$$

i.e. D (C_2). D (σ_{xy}) = D (i) and 'i' is the symmetry element of this group. Similarly this can be shown for other symmetry operations using their corresponding matrices.

Identity matrix is the matrix that can combine with matrices of other symmetry elements leaving them unchanged.

i.e. $D(E).\ D(C_2) = D(C_2)$

$D(E).\ D(\sigma_{xy}) = D(\sigma_{xy})$ and

$D(E).\ D(i) = D(i)$

As generally associative law holds good for matrices. So, this can be easily shown that this law is valied for the matrices of these symmetry operations, such as

$$D(E).\ [D(C_2).\ D(\sigma_{xy})] = [D(E).\ D(C_2)].\ D(\sigma_{xy}).$$

Similarly this can be proved for other symmetry elements also.

Existence of inverse can also be verified using these four matrices. Therefore, it is verified that these symmetry elements viz. E, C_2, σ_{xy} and i constitute a group i.e. C_{2h} point group whose group multiplication table can be easily constructed using multiplication of the corresponding matrices for the various symmetry operations i.e.

C_{2h}	D (E)	D (C_2)	D (σ_{xy})	D (i)
D (E)	D (E)	D (C_2)	D (σ_{xy})	D (i)
D (C_2)	D (C_2)	D (E)	D (i)	D (σ_{xy})
D (σ_{xz})	D (σ_{xy})	D (i)	D (E)	D (C_2)
D (i)	D (i)	D (σ_{xy})	D (C_2)	D (E)

$\underline{C_{3v}}$: Consider a suitable example i.e. NH_3 whose point group is C_{3v}. This point group has six operations viz. E, C_3^1, C_3^2, σ_v, σ_v' and σ_v'' (ref. previous chapter section 2.6). Consider z-axis as C_3 axis coming out of the plane of paper or xy plane and N-atom is on z-axis. So, NH_3 can be represented as a triangle (Figure 3.7). As C_3^1 is rotation by 120° (or 2π/3) and C_3^2 is the rotation by 240° (or 4π/3) about z-axis and σ_v can be considered as σ_{yz} or plane passing through NH_1, so, matrices for these symmetry elements are:

$$D(E) = \begin{pmatrix} 1 & 0 & 0 \\ 0 & 1 & 0 \\ 0 & 0 & 1 \end{pmatrix}$$

$$D(C_3^1) = \begin{pmatrix} -1/2 & \sqrt{3}/2 & 0 \\ -\sqrt{3}/2 & -1/2 & 0 \\ 0 & 0 & 1 \end{pmatrix}$$

after substituting the value of θ = 120°. Similarly

$$D(C_3^2) = \begin{pmatrix} -1/2 & -\sqrt{3}/2 & 0 \\ \sqrt{3}/2 & -1/2 & 0 \\ 0 & 0 & 1 \end{pmatrix}$$

after substituting the value of θ = 240° and

$$D(\sigma_v) = D(\sigma_{yz}) = \begin{pmatrix} -1 & 0 & 0 \\ 0 & 1 & 0 \\ 0 & 0 & 1 \end{pmatrix}$$

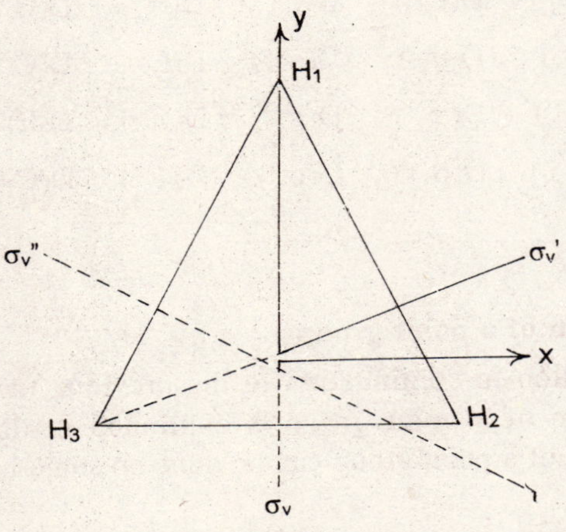

Fig. 3.7

Matrices for other two planes can be obtained according to the combination of two operations i.e. $D(\sigma_v') = D(C_3^2).D(\sigma_v)$

$$= \begin{pmatrix} -1/2 & -\sqrt{3}/2 & 0 \\ \sqrt{3}/2 & -1/2 & 0 \\ 0 & 0 & 1 \end{pmatrix} \begin{pmatrix} -1 & 0 & 0 \\ 0 & 1 & 0 \\ 0 & 0 & 1 \end{pmatrix} = \begin{pmatrix} 1/2 & -\sqrt{3}/2 & 0 \\ -\sqrt{3}/2 & -1/2 & 0 \\ 0 & 0 & 1 \end{pmatrix}$$

and $D(\sigma_v'') = D(C_3^1) . D(\sigma_v)$

$$= \begin{pmatrix} -1/2 & \sqrt{3}/2 & 0 \\ -\sqrt{3}/2 & -1/2 & 0 \\ 0 & 0 & 1 \end{pmatrix} \begin{pmatrix} -1 & 0 & 0 \\ 0 & 1 & 0 \\ 0 & 0 & 1 \end{pmatrix} = \begin{pmatrix} 1/2 & \sqrt{3}/2 & 0 \\ \sqrt{3}/2 & -1/2 & 0 \\ 0 & 0 & 1 \end{pmatrix}$$

These six matrices follow the necessary conditions for the formation of a group which can be shown by considering the appropriate multiplications of the required matrices. Similarly the group multiplication table can also be drawn as

C_{3v}	D(E)	$D(C_3^1)$	$D(C_3^2)$	$D(\sigma_v)$	$D(\sigma_v')$	$D(\sigma_v'')$
D(E)	D(E)	$D(C_3^1)$	$D(C_3^2)$	$D(\sigma_v)$	$D(\sigma_v')$	$D(\sigma_v'')$
$D(C_3^1)$	$D(C_3^1)$	$D(C_3^2)$	D(E)	$D(\sigma_v'')$	$D(\sigma_v)$	$D(\sigma_v')$
$D(C_3^2)$	$D(C_3^2)$	D(E)	$D(C_3')$	$D(\sigma_v')$	$D(\sigma_v'')$	$D(\sigma_v)$
$D(\sigma_v)$	$D(\sigma_v)$	$D(\sigma_v')$	$D(\sigma_v'')$	D(E)	$D(C_3^1)$	$D(C_3^2)$
$D(\sigma_v')$	$D(\sigma_v')$	$D(\sigma_v'')$	$D(\sigma_v)$	$D(C_3^2)$	D(E)	$D(C_3^1)$
$D(\sigma_v'')$	$D(\sigma_v'')$	$D(\sigma_v)$	$D(\sigma_v')$	$D(C_3^1)$	$D(C_3^2)$	D(E)

Section 3.4

Representation of a point group

This section in continuation to the previous two sections. Representation of a point group is explained in this section. Representation of a point group can be done on some basis. These basis are:

(i) Cartesian co-ordinates x, y, z or basis vectors of specific length or position vectors etc.

(ii) Rotational vectors can also be taken as basis of representations.

(iii) Mathematical functions or function space as p_x, p_y and p_z or d-orbital set also form the basis of representation.

Effect of any symmetry operation can be considered and represented using these set of basis. This representation will vary if the basis of representation is changed. Let us try to have some representations for some important point groups as example. Using x, y, z co-ordinates and basic vectors of specific length same matrix representations can be obtained for various symmetry operations belonging to different point groups as it is explained in the section 3.2.

C_{2v} point group

Let us try to have representations for C_{2v} point group on the basis of p-orbital set (i.e. p_x, p_y and p_z), considering +1 for no change in the orbital on any symmetry operation and –1 for change on any operation. This group has 4 symmetry elements. Let us discuss their effect on p_x, p_y & p_z separately.

During identity operation or during E operation p_x remains as p_x. Under the influence of C_2 + ve lobe of p_x attains the position of –ve lobe and –ve lobe attains the position of + ve lobe. So, the net effect on p_x is –1 p_x (i.e. p_x changes from the original position). In σ_{xy} plane there is no change in the p_x orbital w.r.t. its original position but under the influence of σ_{yz}, + ve lobe of p_x becomes –ve and – ve lobe of p_x changes to + ve lobe. So, the net effect can be written as –1 p_x. The effect of all these symmetry elements can be summarised as

$$E(p_x) = (+1)\ p_x$$

$$C_2(p_x) = (-1)\ p_x$$

$$\sigma_{xz}(p_x) = (+1)p_x \text{ and}$$

$$\sigma_{yz}(p_x) = (-1)p_x$$

Similarly effects of these four symmetry operations can be discussed on p_y and p_z orbitals as

$$E(p_y) = (+1)\ p_y$$

$$C_2(p_y) = (-1)\ p_y$$

$$\sigma_{xz}(p_y) = (-1)p_y \text{ and}$$

$$\sigma_{yz}(p_y) = (+1)p_y$$

and

$$E(p_z) = (+1)\ p_z$$
$$C_2(p_z) = (+1)\ p_z$$
$$\sigma_{xz}(p_z) = (+1)p_z \text{ and}$$
$$\sigma_{yz}(p_z) = (+1)p_z$$

These effects can be presented in a table as

C_{2v}	E	C_2	σ_{xz}	σ_{yz}	Basis
	+1	–1	+1	–1	p_x
	+1	–1	–1	+1	p_y
	+1	+1	+1	+1	p_z

From the discussion given above it is clear that p_x, p_y and p_z orbitals give different representations. So, these orbitals are of different energy and are non-degenerate. Now let us try to discuss the effect of the operations of these symmetry elements on the various rotational vectors. Rotational vectors are considered as vectors rotating around x, y or z axes respectively. Effect of these operations on R_x, R_y and R_z rotational vectors may be summarized as

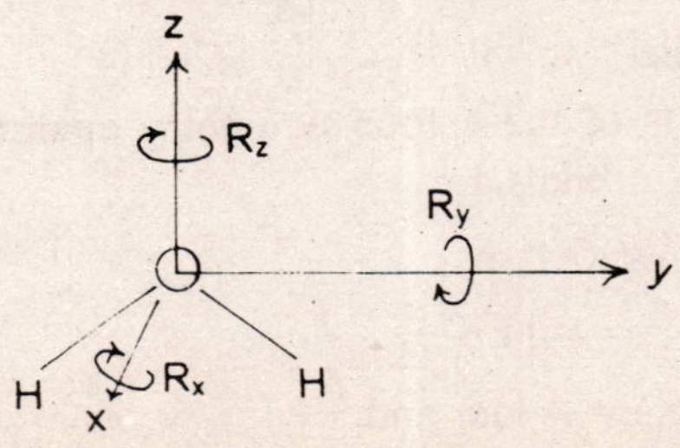

Fig. 3.8

$E(R_x) = +1\ (R_x)$; $E(R_y) = +1\ (R_y)$; $E(R_z) = +1\ (R_z)$

$C_2(R_x) = -1\ (R_x)$; $C_2(R_y) = -1\ (R_y)$; $C_2(R_z) = +1\ (R_z)$

$\sigma_{xz}(R_x) = -1\ (R_x)$; $\sigma_{xz}(R_y) = +1\ (R_y)$; $\sigma_{xz}(R_z) = -1\ (R_z)$

$\sigma_{yz}(R_x) = +1\ (R_x)$; $\sigma_{yz}(R_y) = -1\ (R_y)$; $\sigma_{yz}(R_z) = -1\ (R_z)$

If we tabulate these representations with the representations generated because of p_x, p_y and p_z orbitals, following table is obtained

C_{2v}	E	C_2	σ_{xz}	σ_{yz}	Basis
	+1	−1	+1	−1	p_x, R_y
	+1	−1	−1	+1	p_y, R_x
	+1	+1	−1	−1	R_z
	+1	+1	+1	+1	p_z

Similarly, these representations can be obtained for different function space or multiple mathematical functions i.e. $d_{x^2-y^2}$ (corresponding function is $x^2 - y^2$), d_{z^2} (corresponding function may be written as z^2), d_{yz} (corresponding function is yz) and so on. Let us try to have representations for each of these function space, taking suitable example (SO_2) in the similar manner as it is done for p-orbitals and rotational vectors i.e. considering +1 for no change and −1 for change in the respective orbital on performing operation related to the symmetry element.

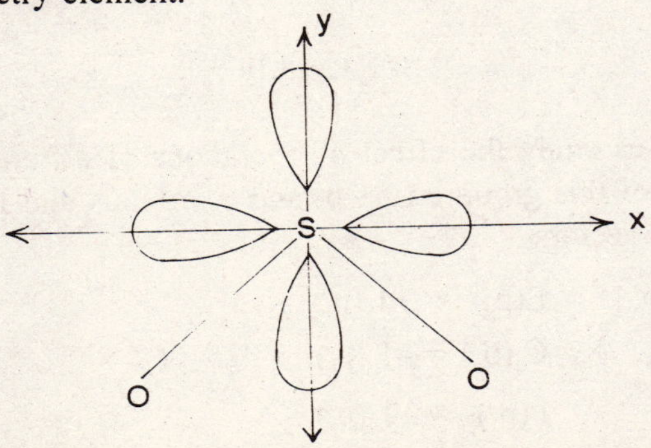

Fig. 3.9

The results so obtained can be summarized as a table along with the representations of p_x, p_y and p_z and the rotational vectors R_x, R_y and R_z as

C_{2v}	E	C_2	σ_{xz}	σ_{yz}	Basis	
	+1	−1	+1	−1	p_x, R_y	xz
	+1	−1	−1	+1	p_y, R_x	yz
	+1	+1	−1	−1	R_z	xy
	+1	+1	+1	+1	p_z	x^2, y^2, z^2

C_{2h} point group

Let us try to discuss the same for C_{2h} point group considering the suitable example i.e. trans-$C_2H_2Cl_2$. List of symmetry elements in this molecule is E, C_2, σ_h and 'i'. If we consider z-axis as proper axis of rotation i.e. C_2 axis then σh will be the xy plane i.e. ($\sigma_h = \sigma_{xy}$).

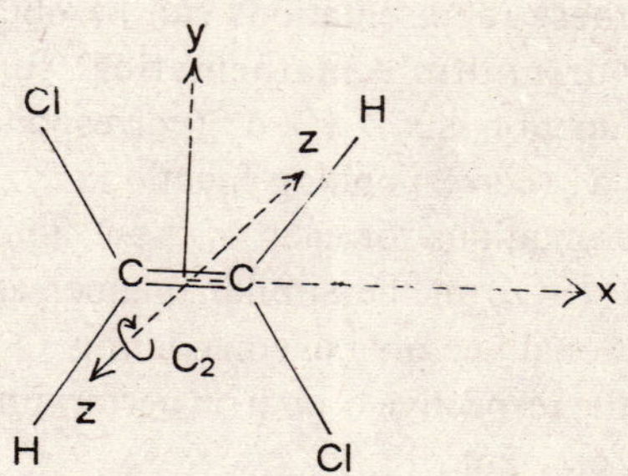

Fig. 3.10

Let us study the effect of operations of different symmetry elements of this group on p_x, p_y and p_z orbitals and R_x, R_y and R_z rotational vectors

$$E(p_x) = +1\ (p_x)$$
$$C_2(p_x) = -1\ (p_x)$$
$$i\ (p_x) = -1\ (p_x)$$
$$\sigma_{xy}(p_x) = +1\ (p_x)$$

$$E(p_y) = +1\ (p_y)$$
$$C_2(p_y) = -1\ (p_y)$$
$$i\ (p_y) = -1\ (p_y)$$
$$\sigma_{xy}(p_y) = +1\ (p_y)$$

$$E(p_z) = +1\ (p_z)$$
$$C_2(p_z) = +1\ (p_z)$$
$$i\ (p_z) = -1\ (p_z)$$
$$\sigma_{xy}(p_z) = -1\ (p_z)$$

and the effect on rotational vectors is summarised as

$$E(R_x) = +1\ (R_x)$$
$$C_2(R_x) = -1\ (R_x)$$
$$i\ (R_x) = +1\ (R_x)$$
$$\sigma_{xy}(R_x) = -1\ (R_x)$$

$$E(R_y) = +1\ (R_y)$$
$$C_2(R_y) = -1\ (R_y)$$
$$i\ (R_y) = +1\ (R_y)$$
$$\sigma_{xy}(R_y) = -1\ (R_y)$$

$$E(R_z) = +1\ (Rz)$$
$$C_2(R_z) = +1\ (R_z)$$
$$i\ (R_z) = +1\ (R_z)$$
$$\sigma_{xy}(R_z) = +1\ (R_z)$$

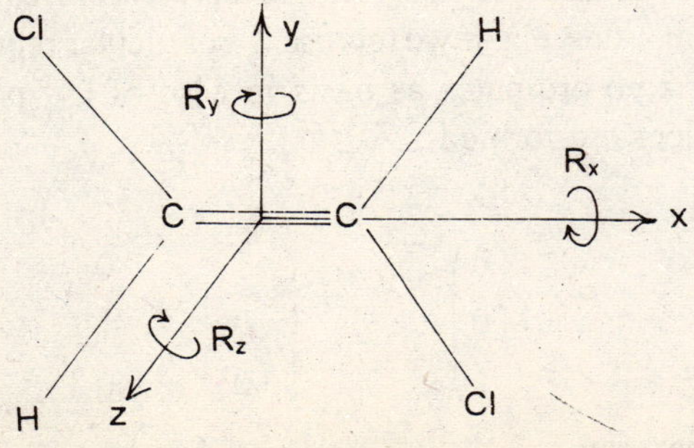

Fig. 3.11

This can be tabulated as follows

C_{2h}	E	C_2	i	σ_{xy}	Basis
	1	1	1	1	R_z
	1	–1	1	–1	R_x, R_y
	1	1	–1	–1	P_z
	1	–1	–1	1	p_x, p_y

Let us try to study the effect of operation of these symmetry operations on multiple function space i.e. on the d-orbital set (viz. $x^2–y^2$, z^2, xy, yz and xz functions) and on summarizing it in the tabulated form following table can be obtained

C_{2h}	E	C_2	i	σ_{xy}	Basis	
	1	1	1	1	R_z	x^2, y^2, z^2, xy
	1	–1	1	–1	R_x, R_y	xz, yz
	1	1	–1	–1	p_z	
	1	–1	–1	1	p_x, p_y	

Section 3.5

Reducible and irreducible representations

A representation is defined as a set of matrices or their characters which represent the operations of a point group. The set of vectors of co-ordinate system or the mathematical function space, with respect to which these matrices are defined are called as the basis of representations, e.g. if we represent a particular representation using x, y and z co-ordinates as basis in case of C_{2v} point group following matrices are formed.

$$E \qquad \begin{pmatrix} 1 & 0 & 0 \\ 0 & 1 & 0 \\ 0 & 0 & 1 \end{pmatrix} \qquad \text{Character} = 3$$

$$C_2 \qquad \begin{pmatrix} -1 & 0 & 0 \\ 0 & -1 & 0 \\ 0 & 0 & 1 \end{pmatrix} \qquad \text{Character} = -1$$

$$\sigma_{xz} \qquad \begin{pmatrix} 1 & 0 & 0 \\ 0 & -1 & 0 \\ 0 & 0 & 1 \end{pmatrix} \qquad \text{Character} = 1$$

$$\sigma_{yz} \qquad \begin{pmatrix} -1 & 0 & 0 \\ 0 & 1 & 0 \\ 0 & 0 & 1 \end{pmatrix} \qquad \text{Character} = 1$$

Character of the identity operation is known as dimension of the representation. So, this particular representation is a three dimensional representation.

Representations with respect to x, y and z axes (or the corresponding p-orbitals) separately are shown in the table of C_{2v} (section 3.4). All these representations are one dimensional representations.

Representations which have high dimensions are reducible representations and these can be reduced to the representations of lower dimensions which are known as irreducible representations.

A reducible representation and its reduction can be obtained by carrying a similarity transformation. Consider P, Q, R and S as a representations of a group in which [Q] [R] = S. On similarity transformation the matrices (diagonal elements) may be obtained as

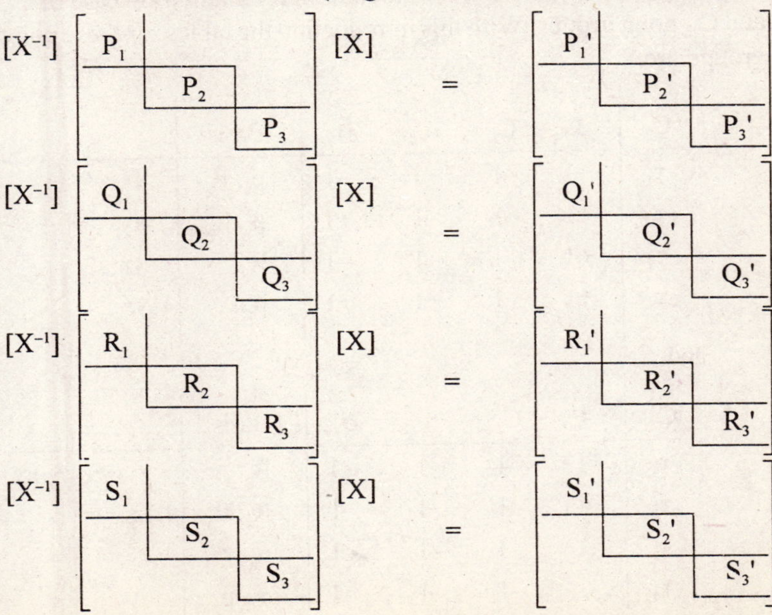

The matrices that are obtained after similarity transformation are block factorized. If in the original representation [Q] [R] = S. So, $Q_1'. R_1' = S_1'$; $Q_2'.R_2' = S_2'$ and $Q_3'.R_3' = S_3'$. Therefore, the original representation i.e. P, Q, R, & S is a reducible representation as

$$P_1\ Q_1\ R_1 \text{ and } S_1,\ P_2\ Q_2\ R_2 \text{ and } S_2 \text{ and } P_3\ Q_3\ R_3 \text{ and } S_3.$$

A representation is symbolised as the symbol τ (tou). We can represent the corresponding irreducible representation in the reducible representation as $\tau_r = \tau_1 \oplus \tau_2 \oplus \tau_3$ ....

which means that the reducible representation i.e. τ_r can be reduced to the irreducible representations i.e. τ_1, τ_2 and τ_3 respectively by similarity transformation. An irreducible representation or its equivalent representation may occur in a reducible representation more than once. Their relation can be expressed as

$$\tau_r = \Sigma\, a_u.\tau_u$$

where τ_r = reducible representation

a_u = number of times an representation (τ_u) occur in the given reducible representation. Therefore, we can conveniently introduce the symbols τ_1, τ_2, τ_3 and τ_4 for the four representations in case of C_{2v} and C_{2h} point groups. With this introduction the tables for these point groups are

C_{2v}	E	C_2	σ_{xz}	σ_{yz}	Basis	
τ_1	+1	−1	+1	−1	p_x, R_y	x^2, y^2, z^2
τ_2	+1	−1	−1	+1	p_y, R_x	xy
τ_3	+1	+1	−1	−1	R_z	xz
τ_4	+1	+1	+1	+1	p_z	yz

and

C_{2h}	E	C_2	i	σ_{xy}	Basis	
τ_1	1	1	1	1	R_z	x^2, y^2, z^2, xy
τ_2	1	−1	1	−1	R_x, R_y	xz, yz
τ_3	1	1	−1	−1	p_z	
τ_4	1	−1	−1	1	p_x, p_y	

For the matrix representation following points are to be noted :

(i) by performing similarity transformations characters of matrices does not change

(ii) conjugate matrices have identical characters

(iii) if, A, B, C and D are matrices for representation and AB = D and BA = C then characters of D and C are equal.

Section 3.6

Great Orthogonality Theorem : (G.O.T.)

The great orthogonality theorem is given as

$$\Sigma\left[\tau_i(R)_{mn}\right]\left[\tau_j(R)_{m'n'}\right]^* = \left(h/\sqrt{\ell_i\ell_j}\right)\left(\delta_{ij}\delta_{mm'}\delta_{nn'}\right)$$

This means that in the set of matrices constituting any one irreducible representation, a set of corresponding matrix elements, one from each matrix, behaves as the components of a vector in h-dimensional space such that all these vectors are mutually orthogonal and each is normalised, so, that the square of its length equals h/ℓ_i.

In this equation of G.O.T., h is the order of the group, ℓ_i and ℓ_j represent the dimensions of two irreducible representations [they also represent the order of the matrices in the reducible representation]. R stands for various symmetry elements (operations) of the group. The element in the nth row and m^{th} column of the matrix correspond to an operation R in the irreducible representation 'i' and $\tau_i R(_{m'n'})$ represents its complex conjugate and can be replaced by the term itself if it does not involved imaginary or complex part in numbers.

This G.O.T. equation can be expressed into three simpler equations as $\Sigma\tau_i\ (R)_{mn}\ \tau_j\ (R)_{mn} = 0$ if $i \neq j$

Which means that the elements of corresponding matrices of different irreducible representations are orthogonal to each other.

$$\Sigma\tau_i\ (R)_{mn}\ \tau_j\ (R)_{m'n'} = 0 \text{ if } m \neq m' \text{ or/ \& } n \neq n'$$

Which means that the elements of different set of matrices of the same irreducible representations are orthogonal to each other and finally

$$\Sigma\tau_i\ (R)_{mn}\ \tau_i\ (R)_{mn} = h / \ell_i$$

Which means that square of the length of any such vector for which $\delta_{ij} = \delta_{mm} = \delta_{nn} = 1$, equals to h / ℓ_i. Five rules about irreducible representation can be framed on the basis of G.O.T. These are

(i) The number irreducible representations in a group is equal to the number of classes of operations of the group.

(ii) The sum of squares of the dimensions of the irreducible representations of a group is equal to the order of the group i.e. $\Sigma l_i^2 = l_1^2 + l_2^2 + \ldots\ldots\ldots = h$

(iii) The sum of squares of the characters in any irreducible representations in a group is equal to h i.e., order of group, $\Sigma [X_i (R)]^2 = h$

(iv) The characters of the two different irreducible representations of a same group are orthogonal to each other i.e. $\Sigma X_i (R) X_j (R) = 0$ when $i \neq j$

(v) In a given representation (whether reducible or irreducible) the characters of all the matrices belonging to the same class are identical.

Let us try to prove these five points taking examples of C_{2v} and C_{2h} point groups.

C_{2v} point group

Table as derived in section 3.5 for this point group is

C_{2v}	E	C_2	σ_{xz}	σ_{yz}
τ_1	+1	+1	+1	+1
τ_2	+1	+1	−1	−1
τ_3	+1	−1	+1	−1
τ_4	+1	−1	−1	+1

It is clear from this table that number of four irreducible representations in this group i.e. τ_1, τ_2, τ_3 and τ_4 are equal to the number of classes i.e. four.

The sum of squares of the dimensions of the irreducible representations should be equal to the order of the group (i.e. = 4)

$$\Sigma l_i^2 = 1^2 + 1^2 + 1^2 + 1^2 = 4$$

The sum of squares of characters in any irreducible representations is equal to h (order of the group). Let us try it for all representations one by one

$$\tau_1 = 1^2 + 1^2 + 1^2 + 1^2 = 4$$

$$\tau_2 = 1^2 + 1^2 + (-1)^2 + (-1)^2 = 4$$

$$\tau_3 = 1^2 + (--1)^2 + 1^2 + (-1)^2 = 4$$

$$\tau_4 = 1^2 + (-1)^2 + (-1)^2 + 1^2 = 4$$

The characters of two different irreducible representation of a same group are orthogonal to each other. Let us try this for different sets of irreducible representations

i.e. $\tau_1 . \tau_2$

$$= 1 \times 1 \times 1 + 1 \times 1 \times 1 + 1 \times 1 \times (-1) + 1 \times 1 \times (-1)$$

$$= 1 + 1 - 1 - 1 = 0$$

So, τ_1 & τ_2 are orthogonal to each other. Similarly we can show it for other sets of irreducible representations, as for. $\tau_1 . \tau_3$

$$= 1 \times 1 \times 1 + 1 \times 1 \times (-1) + 1 \times 1 \times 1 + 1 \times 1 \times (-1)$$

$$= 1 - 1 + 1 - 1 = 0$$

It is obvious that in all the irreducible representations i.e. $\tau_1 \tau_2 \tau_3$ and τ_4 representations the characters of the matrices belonging to the operations of same class are identical i.e. characters of the matrices belonging to E or C_2 or σ_{xz} or σ_{yz} are identical.

C_{2h} point group

Let us try to prove these points for C_{2h} point group. As its table (ref. to section 3.5) is

C_{2h}	E	C_2	i	σ_{xy}
τ_1	1	1	1	1
τ_2	1	–1	1	–1
τ_3	1	1	–1	–1
τ_4	1	–1	–1	1

Number of irreducible representations (four) in this group also is equal to the number of classes (i.e. four).

The sum of squares of dimensions of the irreducible representations is equal to the order of the group.

i.e. $\Sigma 1_i^2 = \Sigma X^2 (E)$

$$= 1^2 + 1^2 + 1^2 + 1^2 = 4$$

The sum of squares of characters in any irreducible representation is equal to h (order of the group).

$$\tau_1 = 1^2 + 1^2 + 1^2 + 1^2 = 4$$

$$\tau_2 = 1^2 + (-1)^2 + 1^2 + (-1)^2 = 4$$

$$\tau_3 = 1^2 + 1^2 + (-1)^2 + (-1)^2 = 4$$

$$\tau_4 = 1^2 + (-1)^2 + (-1)^2 + 1^2 = 4$$

The characters of two different irreducible representations of a group are orthogonal.

$$\tau_1 . \tau_2 = 1 \times 1 \times 1 + 1 \times 1 \times (-1) + 1 \times 1 \times 1 + 1 \times 1 \times (-1)$$

$$= 1 - 1 + 1 - 1 = 0$$

So, these two irreducible representations are orthogonal to each other.

Similarly

$$\tau_1 . \tau_3 = 1 \times 1 \times 1 + 1 \times 1 \times 1 + 1 \times 1 \times (-1) + 1 \times 1 \times (-1)$$

$$= 1 + 1 - 1 - 1 = 0$$

Similarly this can be proved for other representations also.

For all irreducible representations i.e. $\tau_1 \, \tau_2 \, \tau_3$ and τ_4, the characters of the matrices belonging to the operations of the same class are identical i.e. characters belonging to E, C_2, σ_h, and 'i' within same class are identical.

Section 3.7

Reduction Formula :

It is the formula which is used to find out the numbers irreducible representations present in a reducible representation.

As according to Great Orthogonality Theorem

$$\sum_R X^2(R) = h \qquad \text{(iii) rule}$$

and

$$\sum_R X_i(R)X_j(R) = 0 \qquad \text{(iv) rule}$$

This can be expressed into one equation as

$$\sum_R X_i(R)X_j(R) = \delta_{ij}h$$

if $i = j$ then $\delta_{ij} = 1$ and if $i \neq j$ then $\delta_{ij} = 0$ condition of normalisation and orthogonality of irreducible representations.

Let X (R) be the character of operation (R) in reducible representation. Let a_i be the number of times i^{th} irreducible representations represents in the reducible representation, so, that

$$X(R) = \sum_i a_i X_i(R)$$

Multiply both the sides by X_j (R) and sum over all classes

$$\sum_R X(R)X_j(R) = \sum_R \sum_i a_i X_i(R)X_j(R)$$

All other terms in the R.H.S. vanish except

$$a_i X_i(R)X_j(R) = h$$

So, putting $i = j$ $\delta_{ij} = 1$; $\delta_{ij}\,h = h$

R.H.S. = a_j h

$$a_j = \frac{1}{h} \sum_R X(R)X_j(R)$$

This can be further simplified i.e. instead of summing over all R, we can sum over all classes. So,

$$a_j = \frac{1}{h} \sum_{\substack{\text{over}\\ \text{all}\\ \text{classes}}} n_R X(R)X_j(R)$$

This is the reduction formula using which we can obtain that how many times X_i (R) irreducible representation is present in a X (R) reducible representation where n_R is the number of operations in a class.

e.g. To reduce

C_{2v}	E	C_2	i	σ_{xy}
τ_r	9	-1	1	3

we have to consider the table of C_{2v} point group.

Applying the reduction formula for τ_1, τ_2, τ_3, & τ_4

$a\tau_1 = \frac{1}{4}[1 \times 1 \times 9 + 1 \times 1 \times (-1) + 1 \times 1 \times 1 + 1 \times 1 \times 3] = 3$

$a\tau_2 = \frac{1}{4}[1 \times 1 \times 9 + 1 \times 1 \times 1 + 1 \times 1 \times (-1) + 1 \times (-1) \times 3] = 1$

$a\tau_3 = \frac{1}{4}[1 \times 1 \times 9 + 1 \times (-1) \times (-1) + 1 \times 1 \times 1 + 1 \times (-1) \times 3] = 2$

$a\tau_4 = \frac{1}{4}[1 \times 1 \times 9 + 1 \times (-1) \times (-1) + 1 \times (-1) \times 1 + 1 \times 1 \times 3] = 3$

So, $\tau_R = 3\tau_1 \oplus \tau_2 \oplus 2\tau_3 \oplus 3\tau_4$

If we count the dimension for τ_1 τ_2 τ_3 and τ_4 that are contributing to τ_R i.e. $3 + 1 + 2 + 3 = 9$ and total dimension of τ_R is 9. So, it is clear that dimension of reducible representation is equal to the sum of dimensions all irreducible representations contributing to it.

Criteria of irreducibility

Let us suppose that τ_1 τ_2 τ_3 . . . . be the irreducible representations that contribute b, c, d . . . . times in a reducible representation as

$$T_a = b\tau_1 + c\tau_2 + d\tau_3 + \ . \ . \ . \ .$$

Character of the reducible representation can be written as

$$X_a(R) = \sum_{u=1}^{k} a_u . X_u(R) \qquad ...(1)$$

Multiply it with conjugate i.e. $X_a(R)^*$ and sum it over all R

$$\sum_R X_a(R).X_a^*(R) = \sum_R \sum_{u=1}^{k} a_u.X_u(R).X_a^*(R)$$

$$\Rightarrow \qquad \sum_R X_a(R).X_a^*(R) = \sum_{u=1}^{k} \sum_R a_u.X_u(R).X_a^*(R)$$

As complex conjugate is represented as

$$X_a^*(R) = \sum_{v=1}^{k} a_v.X_v(R)$$

$$\Rightarrow \quad \sum_{R} X_a(R).X_a^*(R) = \sum_{u=1}^{k}\sum_{v=1}^{k}\sum_{R} a_u.a_v.X_u(R).X_v(R)$$

According to Great Orthogonality Theorem

$$\sum_{R} X_u(R).X_v(R) = \delta_{uv}h$$

$$\Rightarrow \quad \sum_{R} X_a(R).X_a^*(R) = \sum_{u=1}^{k}\sum_{v=1}^{k} a_u.a_v.\delta_{uv}h$$

If u = v; $\delta_{uv} = 1$

$$\Rightarrow \quad \sum_{R} X_a(R).X_a^*(R) = \sum_{u=1}^{k} a_u^2 h$$

If initial representation is irreducible all $a_u = 0$ except one (i.e. $a_u = 1$) which is the representation itself [i.e. X_a (R) = X_u (R]. So, if we put $a_u = 1$

$$\sum_{R} X_a(R).X_a^*(R) = h$$

This is the criteria for irreducibility. We can convert it into summation over all classes. Therefore, the criteria is

$$\sum_{\substack{\text{all}\\ \text{classes}}} n_R X_a(R) X_a^*(R) = h$$

e.g. let us prove that τ_R is a reducible representation and τ_1 τ_2 τ_3 and τ_4 are irreducible representations in C_{2v} point group. Please refer to the table of C_{2v} point group (section 3.6). As criteria of irreducibility is

$$\sum_{\substack{\text{all}\\ \text{classes}}} n_R X_a(R) X_a^*(R) = h$$

If there is not complex conjugate then criteria is

$$\sum_{\substack{\text{all}\\ \text{classes}}} n_R [X_a(R)]^2 = h$$

As it is proved earlier (section 3.6) for τ_1 τ_2 τ_3 and τ_4 that application, this formula gives values equal to order (i.e. 4). So, it can be concluded that these are irreducible representations of C_{2v} point group, and check for τ_R.

$$\sum_{\substack{\text{all} \\ \text{classes}}} n_R\left[X_a(R)\right]^2$$

$$= 1 \times (9)^2 + 1 \times (-1)^2 + 1 \times 1^2 + 1 \times 1\ (3)^2 \neq h \text{ (i.e. 4).}$$

So, τ_R is the reducible representation.

Section 3.8

Character Table

With the help of G.O.T. and its five rules, we can construct a table for different point groups which may provide the information of number of irreducible representations possible in a group, their dimensions, characters of representations and the vectors and the functions space that are made basis for representations. Such a table is known as character table for that point group. For C_{2v} and C_{2h} point groups character tables are already constructed (ref. section 3.4) but the symbols to different irreducible representations i.e. τ_1 τ_2 τ_3 and τ_4 are to be assigned. These symbols are assigned on the basis of Mulikan's notations.

A character table has six portions in all. Descriptions for various six portions of a character table and Mulikan's notations for used to symbolize the different irreducible representations are given below:

Portion A. It represents the symbols of the point group in Schnoflies system of notations.

Portion B. It has the list of symmetry operations separated in different classes.

Portion C. It has symbols for irreducible representations in the Mulikan's notations.

Portion D. This portion contains characters for the different irreducible representations.

Portion E. This portion has single functions (as p_x, p_y, p_z, R_x, R_y, R_z) as basis for different irreducible representations.

Portion F. This portion multiple functions space (as x^2, y^2, z^2, $x^2 - y^2$, xy, xz, yz etc.) as basis for different irreducible representations.

Ⓐ C_{2v}	E	C_2 Ⓑ	σ_{xz}	σ_{yz}	Basis	
τ_1	+1	+1	+1	+1	z	x^2, y^2, z^2
Ⓒ τ_2	+1	+1 Ⓓ	−1	−1	Ⓔ R_z	xy Ⓕ
τ_3	+1	−1	+1	−1	x, R_y	xz
τ_4	+1	−1	−1	+1	y, R_x	yz

Mulikan's Notations for Irreducible Representations: Mulikan's notations for different irreducible representations are as follows:

(i) One dimensional representations are assigned the symbol A or B and E symbol is assigned to two dimensional representations while three dimensional representations are assigned the symbols T or F (T is used to discuss the electronic transitions and F is used for vibrations). As in the table of C_{2v} point group all the representations are one dimensional [as X (E) = 1], so, symbols A or B are used for these representations.

(ii) If a particular irreducible representation is symmetric with respect to C_n the principal axis, i.e. X (C_n) = 1, then it is symmetric w.r.t. C_n axis. In such a case symbol A is used to represent the one dimensional representation otherwise if it is asymmetric w.r.t. C_n axis i.e. if X (C_n) = −1, then symbol B is used to represent that one dimensional representation. So, in C_{2v} group τ_1 and τ_2 shouls be symbolize with the symbol A but τ_3 and τ_4 with the symbol B.

(iii) If a particular irreducible representation is symmetric with respect to $C_2 \perp C_n$ axis [i.e. if X ($C_2 \perp C_n$) = 1] subscript 1 is given to the representation but if it is asymmetric w.r.t. $C_2 \perp C_n$ axis [i.e. if X ($C_2 \perp C_n$) = − 1] subscript 2 is given to the irreducible representation. If there is no C_2 axis present which is perpendicular to C_n axis in the particular point group then this criteria is adopted w.r.t. σ_v. e.g. in case of C_{2v} point group look for $\sigma_v = \sigma_{xz}$. It is clear that $\tau_1 = A_1$, $\tau_2 = A_2$, $\tau_3 = B_1$ and $\tau_4 = B_2$.

(iv) Primes (') and double primes ('') are given as notation to the irreducible representation if that particular representation is symmetric w.r.t. horizontal plane i.e.

w.r.t. σ_h i.e. if X (σ_h) = 1; symbol single prime (') is given to the representation in its notation but if X (σ_h) = – 1; symbol double prime ('') is given to that representation in its notation, e.g. in case of D_{3h} point group in the representation A_1' X (σ_h) = 1 and in A_1'' X (σ_h) = –1.

(v) If a particular irreducible representation is symmetric w.r.t. inversion centre i.e. X (i) = 1 then symbol 'g' (means gerade) is given to that irreducible representation but if X (i) = –1 i.e. that particular representation is asymmetric w.r.t. inversion centre then symbol 'u' (means ungerade) is given to that particular representation, e.g. in case of C_{2h} point group the representations A_g and B_g are symmetric w.r.t. inversion centre i.e. for these representations X (i) = 1 and for A_u and B_u representations X (i) = –1.

(vi) For point group C_1 C_s and C_i where there is no C_n axis only one dimensional representation symbol i.e. A symbol is used to represent the irreducible representation.

(vii) For the point group D_2 & D_{2h} where there are $3C_2$ axes which are perpendicular to each other and if X (C_2) = 1, symbol A is given to the irreducible representation and if X (C_2) = – 1 then symbol B is given for irreducible representation.

(viii) Regarding two dimensional and other representations symbols are given in the same manner i.e. E', E'', E_g, E_u, E_1 & E_2 etc.

Therefore, now it is clear that character table for C_{2v} point group is

C_{2v}	E	C_2	σ_{xz}	σ_{yz}	Basis	
A_1	+1	+1	+1	+1	z	x^2, y^2, z^2
A_2	+1	+1	-1	–1	R_z	xy
B_1	+1	–1	+1	–1	x, R_y	xz
B_2	+1	–1	–1	+1	y, R_x	yz

Character Table for C_{2h} point group

Table is constructed in the section 3.5 and the Mulikan's

notations for different irreducible representations i.e. τ_1 τ_2 τ_3 and τ_4 according to above rules are $\tau_1 = A_g$, $\tau_2 = B_g$, $\tau_3 = A_u$ and $\tau_4 = B_u$. So, complete character table for C_{2h} point group is given below:

C_{2h}	E	C_2	i	σ_{xy}	Basis	
A_g	1	1	1	1	R_z	x^2, y^2, z^2, xy
B_g	1	–1	1	–1	R_x, R_y	xz, yz
A_u	1	1	–1	–1	p_z	
B_u	1	–1	–1	1	p_x, p_y	

Character Table for C_{3v} point group

In this point group there are six symmetry operations with three classes viz. E, $2C_3$ (C_3^1 & C_3^2) and $3\sigma_v$ (σ_v, σ_v' and σ_v''). As according to G.O.T. (first corollary) the number of classes is equal to number of irreducible representations. So, there should be three irreducible representations (i.e., τ_1 τ_2 and τ_3) present in C_{3v} point group.

As according to G.O.T. (second corollary), the sum of squares of the dimensions should be equal to h (i.e. order of the point group or 6) i.e. $\Sigma X^2 (E) = h$. In this point group this has to be followed if $1^2 + 1^2 + 2^2 = 6$ i.e. dimension of irreducible representations should be 1,1, 2. So, the table upto point should be

C_{3v}	E	$2C_3$	$3\sigma_v$
τ_1	1		
τ_2	1		
τ_3	2		

In any point group it is necessary that there should be a irreducible representation which should be symmetrical to all operations i.e. the characters corresponding to all classes of this representation should be 1. So, the table should be

C_{3v}	E	$2C_3$	$3\sigma_v$
τ_1	1	1	1
τ_2	1		
τ_3	2		

Furthermore for τ_1

i.e. $1 \times 1^2 + 2 \times 1^2 + 3 \times 1^2 = 1 + 2 + 3 = 6.$

In order to find out the characters for other representations i.e. for τ_2 and τ_3, let us apply the criteria of orthogonality i.e. for τ_1 τ_2 and for τ_1 τ_3

τ_1 τ_2	E	$2C_3$	$3\sigma_v$
τ_1	1	1	1
τ_2	1	X_1	X_2

Let X_1 be the character of τ_2 w.r.t. $2C_3$ and X_2 be the character w.r.t. $3\sigma_v$. Applying G.O.T. to both of these representations

$$\sum_R n_R X\tau_1(R).X\tau_2(R) = 0$$

$$= 1 \times 1 \times 1 + 2 \times 1 \times X_1 + 3 \times 1 \times X_2 = 0 \quad .... \quad .... \qquad (1)$$

This equation is true if $X_1 = 1$ and $X_2 = -1$. So, the characters for the irreducible representations τ_2 are

	E	$2C_3$	$3\sigma_v$
τ_2	1	1	-1

Upto this character table for C_{3v} is

C_{3v}	E	$2C_3$	$3\sigma_v$
τ_1	1	1	1
τ_2	1	1	-1
τ_3	2	X_3	X_4

In order to find out X_3 & X_4, consider the orthogonality of τ_1 τ_3.

$1 \times 1 \times 2 + 2 \times 1 \times X_3 + 3 \times 1 \times X_4 = 0$

$= 2 + 2X_3 + 3X_4 = 0$.......... (2)

& considering orthogonality of τ_2 τ_3

i.e. $1 \times 1 \times 2 + 2 \times 1 \times X_3 + 3X\,(-1) \times X_4 = 0$

$= 2 + 2X_3 - 3X_4 = 0$......... (3)

Now equation (2) and (3) are the simultaneous equations and can be solved for X_3 and X_4 with the values obtained as $X_3 = -1$ and $X_4 = 0$ on solution of these equations. So, the character table is

C_{3v}	E	$2C_3$	$3\sigma_v$
τ_1	1	1	1
τ_2	1	1	−1
τ_3	2	−1	0

τ_1 is one dimensional irreducible representation and is symmetrical w.r.t. C_3 i.e. the principal axis and to σ_v, so, the Mulikan notation for this irreducible representation is A_1. τ_2 is also one dimensional representation, symmetrical w.r.t. the principal axis i.e. C_3 axis but is asymmetrical w.r.t. σ_v, so, Mulikan notation A_2 is assigned to this representation. As τ_3 is two dimensional representation, so, symbol E is used to represent this representation. Therefore, the character table for this point group is

C_{3v}	E	$2C_3$	$3\sigma_v$
A_1	1	1	1
A_2	1	1	−1
E	2	−1	0

Basis for representations

Let us consider a base vector of specific length along z axis, this remains unchanged w.r.t. all symmetry operations viz. E, C_3 and σ_v. So, its characters are 1,1,1 and these characters belong to A_1 irreducible representation. The base vectors along x, y axes are not independent but are interdependent on performing these operations. So, these can be considered as two dimensional representation with the matrices

$$E \quad \begin{pmatrix} 1 & 0 \\ 0 & 1 \end{pmatrix} \qquad C_3^1 \quad \begin{pmatrix} -1/2 & \sqrt{3}/2 \\ -\sqrt{3}/2 & -1/2 \end{pmatrix} \qquad C_3^2 \quad \begin{pmatrix} -1/2 & -\sqrt{3}/2 \\ \sqrt{3}/2 & -1/2 \end{pmatrix}$$

Characters

2   − 1   − 1

$$\sigma_v \qquad\qquad \sigma_v' \qquad\qquad \sigma_v''$$

$$\begin{pmatrix} 1 & 0 \\ 0 & -1 \end{pmatrix} \begin{pmatrix} -1/2 & -\sqrt{3}/2 \\ -\sqrt{3}/2 & 1/2 \end{pmatrix} \begin{pmatrix} -1/2 & \sqrt{3}/2 \\ \sqrt{3}/2 & 1/2 \end{pmatrix}$$

Characters

$$0 \qquad\qquad 0 \qquad\qquad 0$$

So, x, y belongs to E representations. Let us consider now the rotational vectors. Consider R_z rotational vector on z axis with the effect of E and C_3 operations character of R_z remains as + 1 (as C_3 axis or principal axis is z axis) but w.r.t. σ_v (or σ_{yz} plane ref. Figure 3.7) character of R_z becomes – 1, so, representation A_2 belongs to R_z. R_x and R_y are not independent but form two dimensional representation, so, R_x and R_y belong to E representations. Similarly on considering square or multiple function space i.e. it can be shown that x^2, y^2 and z^2 form the basis of A_1 representation while $x^2 - y^2$, xy, xz and yz collectively form the basis for two dimensional E representation. So, the complete character table for the C_{3v} point group is

C_{3v}	E	$2C_3$	$3\sigma_v$	Basis	
A_1	1	1	1	z	$x^2 + y^2, z^2$
A_2	1	1	–1	R_z	
E_1	2	–1	0	x, y	$x^2 - y^2$, xy
				R_x, R_y	xz, yz

Problems

Q. 1: Show that the irreducible representations (AB_1, AB_2 & AB_3) in the point group D_2, are orthogonal.

Ans: Refer to character table for D_2 point group. Apply

$$\sum_R n_R X_i(R) X_j(R) = 0$$

Orthogonality of A & B_1

$1 \times 1 \times 1 + 1 \times 1 \times 1 + 1 \times 1 \times (-1) + 1 \times 1 \times (-1) = 0$

Orthogonality of A & B_2

$1 \times 1 \times 1 + 1 \times 1 \times (-1) + 1 \times 1 \times 1 + 1 \times 1 \times (-1) = 0$

Orthogonality A & B_3

$1 \times 1 \times 1 + 1 \times 1 \times (-1) + 1 \times 1 \times (-1) + 1 \times 1 \times 1 = 0$

Q. 2: Find out the point group representation for dx^2-y^2 mathematical function space for D_{4h} point group (e.g. Pt Cl_4^{2-}).

Ans: The $d_{x^2-y^2}$ orbital of Pt Cl_4^{2-} is shown in figure. The operations C_4^1 and σ_h are shown in this Figure, which suggest that $C_4^1(d_{x^2-y^2}) = (-1)\, d_{x^2-y^2}$ & $\sigma_h(d_{x^2-y^2}) = (+1)\, d_{x^2-y^2}$.

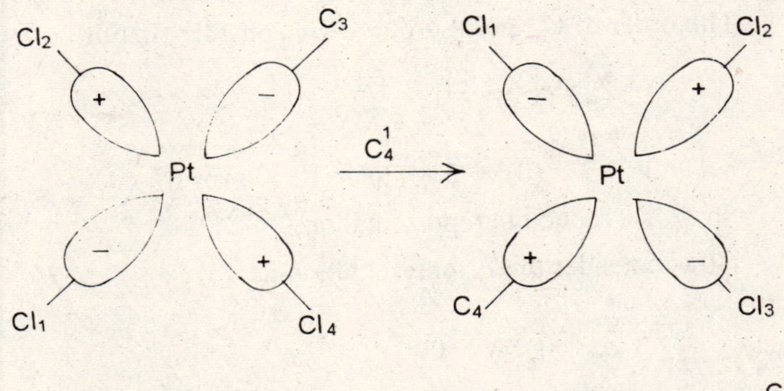

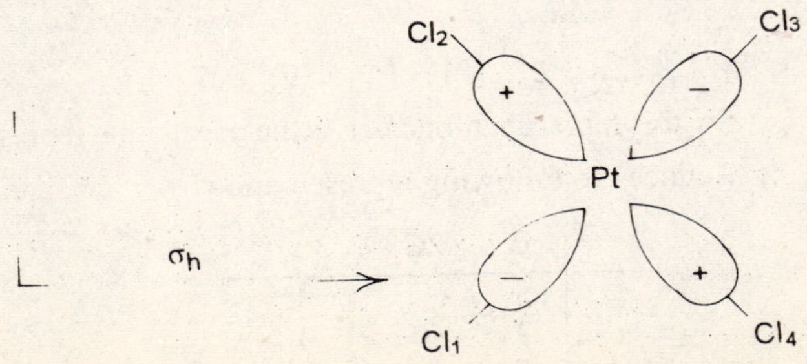

Similarly the effect of other operations can be discussed and can be summarized as

D_{4h}	E	$2C_4$	C_2	$2C_2'$	$2C_2''$	'i'	$2S_4$	σ_h	$2\sigma_v$	$2\sigma_d$
	1	–1	1	1	–1	1	–1	1	2	–1

Q. 3: Prove that for C_{3v} point group τ_p as obtained on the molecule of NH_3 i.e.

	E	$2C_3$	$3\sigma_v$
τ_p	3	0	1

is a reducible representation but the representation E i.e.

	E	$2C_3$	$3\sigma_v$
τ_p	2	–1	01

is a irreducible representation

Ans: As according to the criteria of irreducibility

$$\sum_{\substack{\text{all}\\\text{classes}}} n_R[X_a(R)]^2 = h$$

The order of C_{3v} point group is 6. Consider first the τ_p

$$\sum_{\substack{\text{all}\\\text{classes}}} n_R[X_a(R)]^2$$

$= 1 \times 3^2 + 2 \times (0)^2 + 3 \times (1)^2 = 12 \neq h$ (or 6)

So, it is a reducible representation.

Now, consider the E representation i.e.

$$\sum_{\substack{\text{all}\\\text{classes}}} n_R[X_a(R)]^2$$

$= 1 \times (2)^2 + 2 \times (-1)^2 + 3 \times (0)^2 = 6 = h$

So, this representation i.e. E is the irreducible representation.

Q. 4: Reduce the following representations

	E	$2C_3$	$3\sigma_v$
τ_d	5	–1	1
τ_p	3	0	1

for C_{3v} point group.

	E	C_2	σ_{xz}	σ_{yz}
τ_p	3	–1	1	1

for C_{2v} point group

Ans: Considering the character table for C_{3v} point group and reducing T_d representation

$T_d = A_1 \oplus 2E$

Similarly reduction can be done for T_p, considering C_{3v} point group and this reduced to

$\tau_d = A_1 \oplus E$

On reducing the T_p considering the C_{2v} character table, this representation reduced to

$T_d = A_1 \oplus B_1 \oplus B_2$

Q. 5: Give the irreducible representations for the following reducible representations

D_{3d}	E	$2C_3$	$3C_3$	'i'	$2S_6$	$3\sigma_d$
T_r	5	2	3	–1	2	1

C_{2v}	E	C_2	σ_v	σ_v'
T_r	6	0	2	4

T_d	E	$8C_3$	$3C_2$	$6S_4$	$6\sigma_d$
T_r	8	2	0	2	2

Ans: Considering the respective character tables and applying reduction formula to find out irreducible representations, to the above reducible representation, following results are obtained

D_{3d} : $T_r = 2A_{1g} \oplus A_{1u} \oplus E_u$

C_{2v} : $T_r = 3A_1 \oplus B_1 \oplus 2B_2$

T_d : $T_r = 2A_1 \oplus B_1 \oplus T_2$

Q. 6: Given the matrices :

$$A = \begin{pmatrix} 1 & 1 & 2 \\ 2 & 1 & 1 \\ 3 & 2 & 1 \end{pmatrix} \quad \& \; B = \begin{pmatrix} 2 & 0 & 2 \\ 1 & 1 & 1 \\ 0 & 0 & 1 \end{pmatrix}$$

(i) Find out their characters

(ii) Find C which is equal to BA and D which is equal to AB, also calculate their characters.

Ans: Given

$$A = \begin{pmatrix} 1 & 1 & 2 \\ 2 & 1 & 1 \\ 3 & 2 & 1 \end{pmatrix}$$

Its character is the sum of diagonal elements i.e. 1+1+1 = 3

Similarly

$$B = \begin{pmatrix} 2 & 0 & 2 \\ 1 & 1 & 1 \\ 0 & 0 & 1 \end{pmatrix}$$

Its character is 2 + 1 + 1 = 4

Now, let us find out C = BA

$$\begin{pmatrix} 2 & 0 & 2 \\ 1 & 1 & 1 \\ 0 & 0 & 1 \end{pmatrix} \quad \begin{pmatrix} 1 & 1 & 2 \\ 2 & 1 & 1 \\ 3 & 2 & 1 \end{pmatrix}$$

$$\begin{pmatrix} 8 & 6 & 6 \\ 6 & 4 & 4 \\ 3 & 2 & 1 \end{pmatrix}$$ So, its character is 13

D = AB

$$= \begin{pmatrix} 1 & 1 & 2 \\ 2 & 1 & 1 \\ 3 & 2 & 1 \end{pmatrix} \quad \begin{pmatrix} 2 & 0 & 2 \\ 1 & 1 & 1 \\ 0 & 0 & 1 \end{pmatrix}$$

$$= \begin{pmatrix} 3 & 1 & 5 \\ 3 & 1 & 4 \\ 8 & 2 & 8 \end{pmatrix}$$ So, its character is 12.

4

Group Theory and Quantum Mechanics

Previous chapters explain the concepts of symmetry and group theory i.e. about point groups, representations of symmetry elements and point groups, character tables etc. It has also been mentioned in the previous chapters that group theory provides a link between molecular symmetry and quantum mechanics which provides the basis of solution of molecular structure. In this chapter, some basic concepts of quantum mechanics, such as postulates of quantum mechanics, introduction to wave function, Schrodinger equation and concepts of quantization of energy along with the concept that irreducible representation can be taken as representation of a wave function, concepts of direct product of irreducible representations etc. are discussed.

Section 4.1

Postulates of quantum mechanics

Postulates of quantum mechanics are as follows:

Postulate I: State of a system can be defined by wave function i.e. ψ (q, p, t) where q is the position, p is the momentum and t is the time coordinate. Overall wave function separating the time part becomes

$$\psi (q, p, t) = \psi (q, p). e^{-i\hbar Et} .$$

Probability of finding an electron is given by

$$\psi\psi^{+} = \psi(q,p)\psi^{+}(q,p)\ e^{-i\hbar Et} . e^{+i\hbar Et}$$

Where ψ^* is the complex conjugate of ψ, as ψ may have on imaginary part i.e. i (iota = – 1). Therefore in order to estimate the probability, this wave function is multiplied with its complex conjugate i.e. in $\psi\psi^*$ imaginary term is removed and the probability comes out to be real which has the physical significance. If ψ does not contain any imaginary part then $\psi\psi^*$ may be replaced by ψ^2.

Postulate II: For every observable property of a system, there exists a linear operator and the physical properties of the observable can be treated as free from the mathematical property of the associated operator.

Postulate III: For every observable property of a system, this operator can be operated over the function to get some constant values called eigen values and function back as

$$\hat{A}\psi = a\psi$$

This equation is an eigen equation and a is the eigen value and any result of measurement of a dynamic variable is one of the eigen values of the corresponding operator.

Postulate IV: For any normalized acceptable function 'ψ' the expectation value of energy is given by the variation principle as

$$\langle E \rangle = \int \psi^* \hat{H}\psi \, d\tau$$

or

$$\overline{E} = \langle \psi^* |\hat{H}| \psi \rangle$$

Wave Function: Wave function is a function which expresses the state of any electron (or particle). There are some criteria which are necessary for choosing a well behaved wave function or ψ. These criteria are

(i) It should have single valued solution for Schrodinger equation i.e., for each value of x it should have one or same value. This may be expressed as

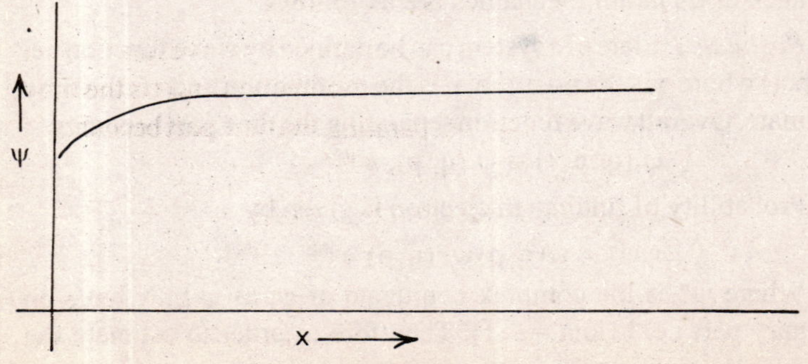

Fig. 4.1

(ii) A well behaved wave function should be a finite function as shown in the figure

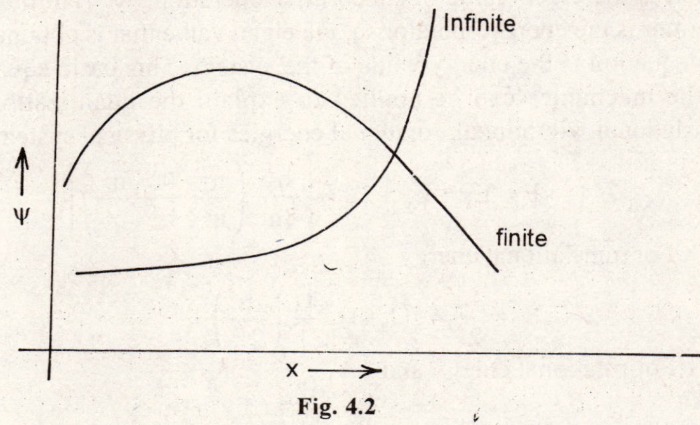

Fig. 4.2

(iii) Third criteria is that the well behaved wave function should be continuous.

Wave function (ψ) may be a mathematical function which has no significance as such but it is ψ^2 or $\psi\psi^*$ (where ψ^* is complex conjugate of ψ) is significant physically in determining the probability of any electron (or particle) in any atom (or space).

Section 4.2

Basic Equation of Quantum mechanics and effect of Symmetry Transformation on it

Section 4.1 has explained the postulates of quantum mechanics. The Schrodinger equation that can be very well explain for any physical system is

$$\nabla^2 \psi (x,y,z) + \frac{8\pi m^2}{h^2}(E - V) \psi (x,y,z) = 0$$

Which can be expressed as

$$\hat{H}\psi = E\psi$$

Where $\hat{H}$ is the Hamiltonian operator which directs some operations to be carried out on the ψ which is a normalised wave function. This equation is an eigen equation as on operation of $\hat{H}$ on the normalised wave function ψ we get back the function ψ and E which is the eigen value obtained after operation. As Hamiltonian operator is the energy operator so, the eigen value that is obtained in this equation is the energy value of the system. This basic equation of the mechanics can be applied to explain the quantization of translational, vibrational, rotational energies for physical system as

$$E = E_x + E_y + E_z = \left[\frac{h^2}{8m}\left(\frac{n_x^2}{a^2}\ \frac{n_y^2}{b^2}\ \frac{n_z^2}{c^2}\right)\right]$$

For translational energy

$$E_j = \frac{L^2}{2I} = J(J+1) \cdot \frac{1}{2I}\left(\frac{h}{2\pi}\right)^2$$

For rotational energy and

$$E_v = (v + ½)\ \frac{h}{2\pi}\sqrt{\frac{k_y}{m}}$$

For vibrational energy

This detail is beyond the scope of this book. The only property of the Hamiltonian which shall be used in explicit concern is the symmetry with respect to interchange of like particles in the system to which it applied. In an atomic system particles are electrons and atomic nuclei. In any symmetry operation, if any of the indistinguishable particles are changed, the energy of the system remains the same that is Hamiltonian must remains unchanged or we can say that symmetry operator must commutes with the Hamiltonian, so,

$$\hat{O}_R.\hat{H} = \hat{H}.\hat{O}_R$$

Where O_r is the transformation operator. If Hamiltonian commutes with a constant value, so,

$$\hat{H}C\psi = C\hat{H}\psi = CE\psi$$

In a system where there is no change in energy values, we can say that the eigen values are degenerate i.e.

$$\hat{H}\psi_{in} = E_i\psi_{in}$$

So, E_i is n-fold degenerate. This can also be true in case of linear combination, as

$$\hat{H}\sum_j a_{ij}\ \psi_{ij} = E_i \sum_j a_{ij}\ \psi_{ij}$$

We can express the orthogonality and the normalisation of ψ, the wave function as an integral

$$\int \psi . \psi^* d\tau = \delta_{ij}$$

Now, if $\delta_{ij} = 0$ the eigen functions are said to be orthogonal and if $\delta_{ij} = 1$ these functions are said to be normalised. This equation can also be written as

$$\int \sum_j a_{ij} \psi_{ij} a_{ij} \psi^*_{ij} d\tau = \sum_j a^2_{ij} = 1$$

for the linear combination. With the effect of transformation operator under any symmetry operation the eigen equation can be rewritten as

$$\hat{H} \hat{O}_R \psi_i = E_i \hat{O}_R \psi_i$$

So, $O_r \psi$ is explained as the eigen function. ψ is normalised, so,

$$\hat{O}_R \psi = +1 \psi_i$$
$$\text{or} \quad -1 \psi_i$$

Therefore on applying each and every operation in the group to an eigen function ψ_i belonging to non-degenerate eigen value, there must generate a representation of the group with matrix equal to + 1 or –1. As the generated representations are one dimensional, so, they are irreducible. Therefore for degenerate set of eigen values.

$$\hat{H} \hat{O}_R \psi_{in} = E_i \hat{O}_R \psi_{in}$$

The eigen function R ψ_{in} may be written in a linear combination form as

$$\hat{O}_R \psi_{in} = \sum_{j=1}^{k} O_{in} \psi_{ij}$$

Similar equations can be written for different symmetry operations and for the combinations of symmetry operations.

In case of a free atom, 's' orbital is non-degenerate, 'p' is triply degenerate and 'd' is penta degenerate. Thus,

$$\hat{H} \psi_{px} = E_i \psi_{px}$$

$$\hat{H} \psi_{py} = E_i \psi_{py}$$

$$\hat{H} \psi_{pz} = E_i \psi_{pz}$$

So, on operation of $\hat{H}$ Hamiltonian on ψ's of p-orbital same eigen values are obtained, (in a free atom system). This can also be confirmed with the matrix that describes the transformation of triply degenerate p-orbitals as it is three dimensional representation matrix i.e.

$$E = \begin{pmatrix} 1 & 0 & 0 \\ 0 & 1 & 0 \\ 0 & 0 & 1 \end{pmatrix} \quad \chi = 3 \text{ (see chapter 3, section 3.2)}$$

Similarly, it can be shown that 'd' orbital set is penta-degenerate in free atom state as the corresponding matrix representation is five dimensional representation.

These orbitals i.e. p or d-orbitals if considered in a molecule, this can be shown easily that the representations that describe the transformations of p and d set of orbitals are reducible one and can be reduced to irreducible representations, one or two dimensional in case of p-orbitals and one two or three dimensional in case of d-orbitals. So, we can say that the degeneracy of the eigen values corresponding to the different orbitals set of the free atom, is spilt-up in case of the molecule. Let us discuss it with some examples (section 4.4) before that shapes of the orbitals are explained here in the section 4.3

Section 4.3

Shape of Orbitals

The wave equation that can be written on expanding the Hamiltonian in terms of polar co-ordinates is

$$\frac{1}{r^2}\frac{d}{dr}\left(r^2\frac{d\psi}{dr}\right)+\frac{1}{r^2\sin^2\theta}\frac{d^2\psi}{d\phi^2}+\frac{1}{r^2\sin\theta}\frac{d}{d\theta}\left(\sin\theta\frac{d\psi}{d\theta}\right)$$
$$+\frac{8\pi^2 m}{h}E\psi = 0$$

Where $\psi = R(r)\,\Theta(\theta)\,\Phi(\phi)$ in terms of polar co-ordinates.

This equation is separable into three equations in terms of, ϕ, θ and r, whose solutions are

$$\phi_m = \frac{1}{\sqrt{2m}}e^{\pm im\phi}$$

Where m = 0, 1, 2, 3 ... ... etc. and solutions of θ is

$$\theta_{l,m} = \sqrt{\frac{(2\ell+1)(\ell-|m|)!}{2(\ell-|m|)!}} - p_\ell^{|m|}\cos\theta$$

This can be concluded that the θ part of the wave function is dependent on the two numbers that are l and m as

$$m = 0, \pm 1, \pm 2, \pm 3 \ldots \ldots \pm l$$

The solution of r part of the equation is in turn dependent on two numbers that are n, l with n = 1, 2, 3 ... ... and l = 0, 1, 2, 3 ... ... (n – 1)

Thus it can be inferred that the three parts of the wave function are dependent upon n, 1, m numbers which are principal, azumithal and magnetic quantum numbers. The shapes of different orbitals are dependent only upon the angular part of the wave function, it is independent of the radial part. Therefore the shapes of different orbitals depends on only l & m numbers i.e. on azumithal and magnetic quantum numbers. These different shapes are basically spherical harmonics (or the contour maps if we compare them in Geography). These different spherical harmonics or the shapes of the orbitals are given below:

s-orbital

For this orbital $l = 0$ & m = 0 and

$$\phi = \frac{1}{2\sqrt{\pi}}$$

Therefore the shape of s-orbital is spherical or it is spherically symmetrical as it is the orbital for which the angular wave function is independent of φ and θ. The shape is given in the figure 4.3. In this orbital the region showing the maximum probability is equidistant from the nucleus.

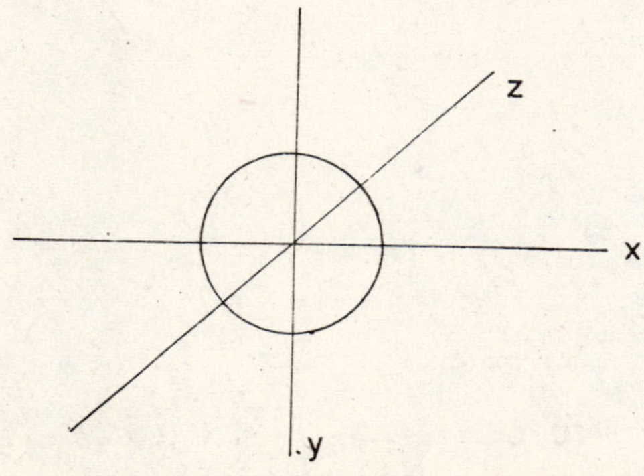

Fig. 4.3

p-orbitals

For p-orbital $l = 1$ and m = 0, + 1 or – 1, so, three solutions are possible for angular parts of wave function which are

$$\phi = \frac{1}{\sqrt{2\pi}} \quad \text{and} \quad \theta = \sqrt{\frac{6}{2}}\cos\theta$$

So,

$$\psi(\theta, \phi) = \frac{1}{\sqrt{2\pi}}\sqrt{\frac{6}{2}}\cos\theta$$

When $l = 0$ and m = 0. The corresponding shape is dump-bell and ψ is dependent upon θ. It is given in figure 4.4

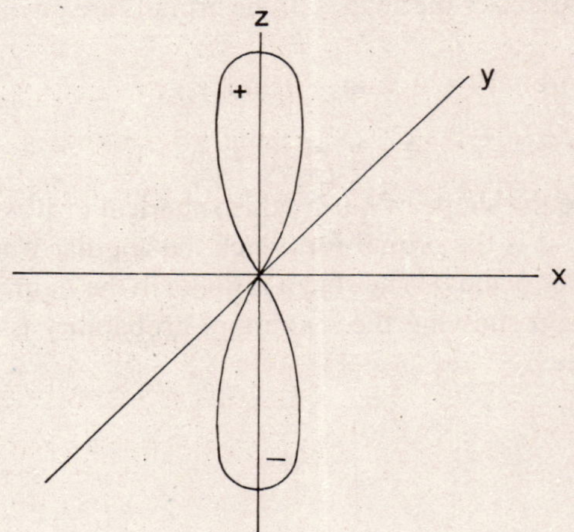

Fig. 4.4

Similarly

$$\psi(\theta, \phi) = \frac{1}{\sqrt{2\pi}}\sqrt{\frac{3}{2}}\sin\theta\,(\cos\phi + i\sin\phi)$$

When $l = 1$, m = + 1 and

$$\psi(\theta, \phi) = \frac{1}{\sqrt{2\pi}} \sqrt{\frac{3}{2}} \sin\theta \, (\cos\phi - i \sin\phi)$$

When 1 = 1, m = – 1

To avoid the imaginary part i.e. 'i', we can take the result of these two wave functions as obtained by linear combination as

$$\frac{1}{\sqrt{2\pi}} (\psi_{1,1} + \psi_{1,-1}) = \sqrt{\frac{3}{4\pi}} \sin\theta \cos\phi$$

$$\frac{1}{\sqrt{2\pi}} (\psi_{1,1} - \psi_{1,-1}) = \sqrt{\frac{3}{4\pi}} \sin\theta \sin\phi$$

So, these functions are dependent on θ and ϕ. Therefore, on plotting them we get shapes of different p-orbitals which are again dump-bell along x-axis and y-axis (Figure 4.5).

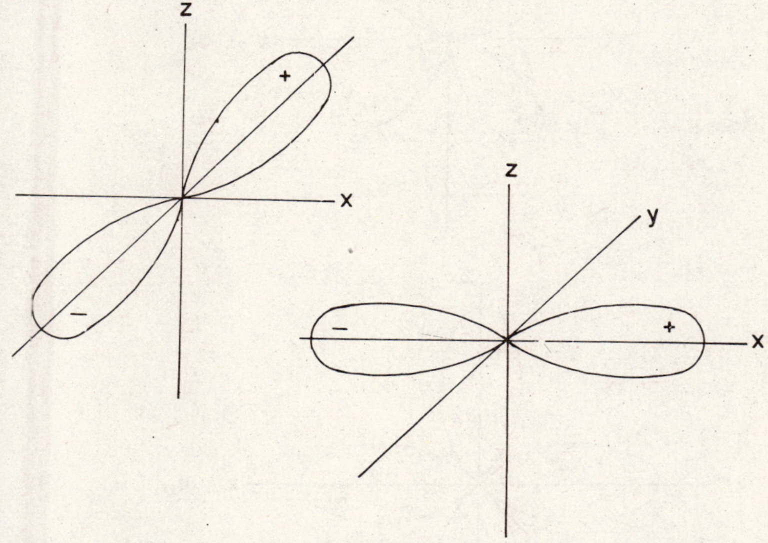

Fig. 4.5

Similarly from the spherical harmonics of 'd' and 'f' orbitals the shapes obtained are given in the figure 4.6 and angular functions for these orbitals are given in Table 4.1.

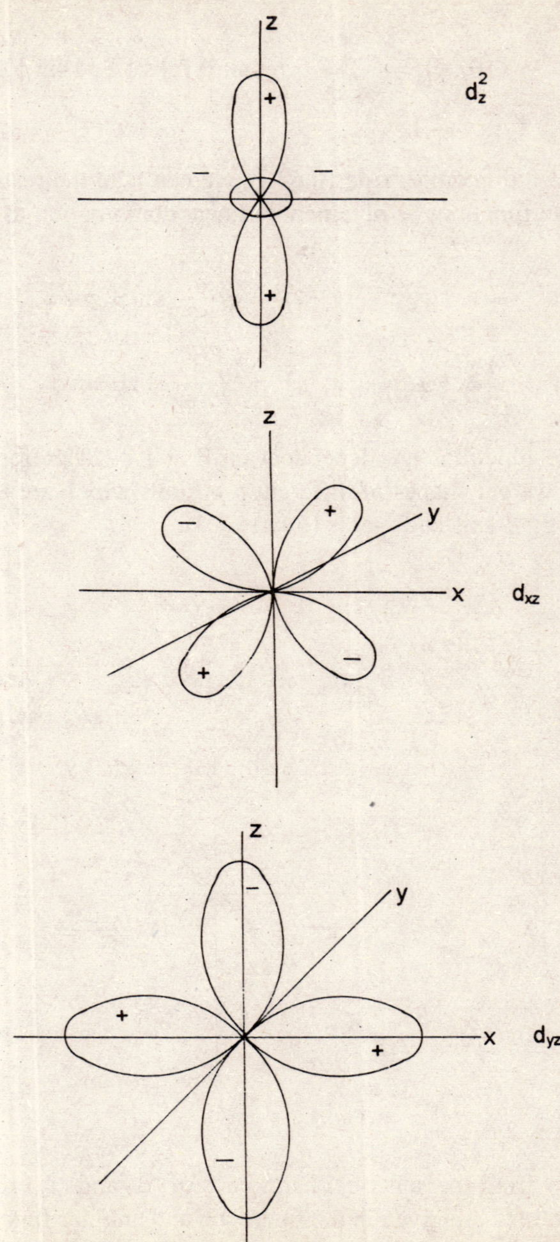
z
+
d_z^2
–
–
+
z
–
+
y
x
d_{xz}
+
–
z
–
y
+
x
d_{yz}
+
–

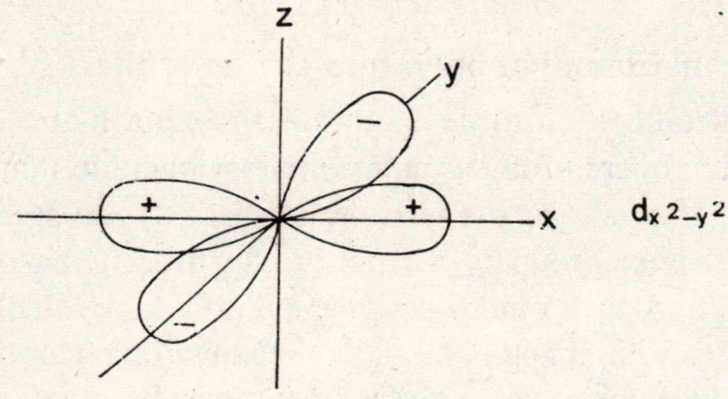

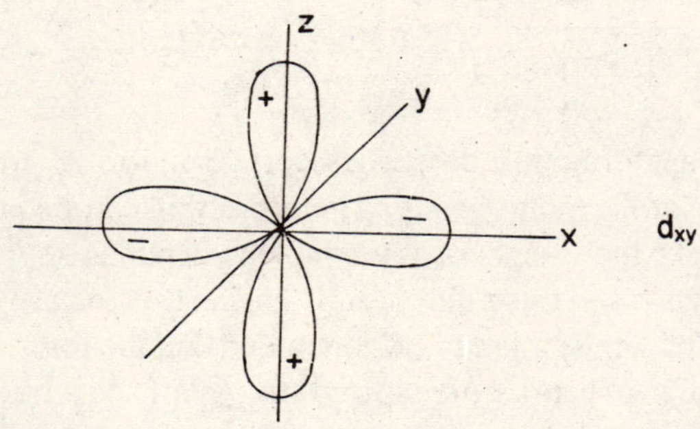

Fig. 4.6

Table 4.1 : Angular functions for d and f orbitals

Orbital	*Angular Function*
dz^2	$3\cos^2\theta - 1$
$d_{x^2-y^2}$	$\sin^2\theta \cos 2\phi$
d_{yz}	$\sin^2\theta \cos 2\phi$
d_{xz}	$\sin\theta \cos\theta \cos\phi$
d_{yz}	$\sin\theta \cos\theta \sin\phi$
f_{x^3}	$\sin\theta \cos\phi\ (5 \sin^2\theta \cos^2\phi - 3)$
f_{y^3}	$\sin\theta \sin\phi\ (5 \sin^2\theta \cos^2\phi - 3)$
f_{z^3}	$\cos\theta\ (5 \sin^2\theta - 3)$
$f_{z(x^2-y^2)}$	$\sin^2\theta \cos\theta \cos 2\phi$
f_{xyz}	$\sin^2\theta \cos\theta \sin 2\phi$
$f_{y(z^2-x^2)}$	$\sin\theta \sin\phi\ (\cos^2\theta - \sin^2\theta \cos^2\phi)$
$f_{x(z^2-y^2)}$	$\sin\theta \cos\phi\ (\cos^2\theta - \sin^2\theta \sin^2\phi)$

Section 4.4

Effect of transfromation operators on the orbitals

Let us take a suitable example to explain the effect of transformation operator under any symmetry operation on different orbitals. For this take s, p and d orbitals of oxygen atom in H_2O molecule with point group C_{2v}. On performing the different operations i.e. E, C_2 σ_{xz} and σ_{yz} on these orbitals and no noting the effect with character +1 on no change and – 1 on change, this can be seen s-orbital remains unchanged in all the operations. So,

E	C_2	σ_{xz}	σ_{yz}
1	1	1	1

These transformation properties correspond to A_1 irreducible representation of C_{2v} point group. Therefore this can be concluded that the s-orbital which is spherically symmetrical remains symmetrical to all the operations in this point group. This is true in case of all the point groups and hence this can be concluded that the transformation properties of s-orbital correspond to A_1 or A_{1g} irreducible representations.

Now, let us consider p-orbitals one by one. p-orbitals can be written as

$$p_z = R \cos \theta$$

$$p_x = R \sin \theta \cos \phi \text{ and}$$

$$p_y = R \sin \theta \sin \phi$$

Considering only the angular part of wave functions, in C_{2v} point group in different operations θ part remains unchanged while only φ changes. So, the resulting wave functions can be expressed as

$$O_E (p_z) = E (R \cos \theta) = R \cos \theta = p_z$$

$$O_{C_2} (p_z) = C_2 (R \cos \theta) = R \cos \theta = p_z$$

$$O_{\sigma_{xz}} (p_z) = \sigma_{xz} (R \cos \theta) = R \cos \theta = p_z$$

$$O_{\sigma_{yz}} (p_z) = \sigma_{yz} (R \cos \theta) = R \cos \theta = p_z$$

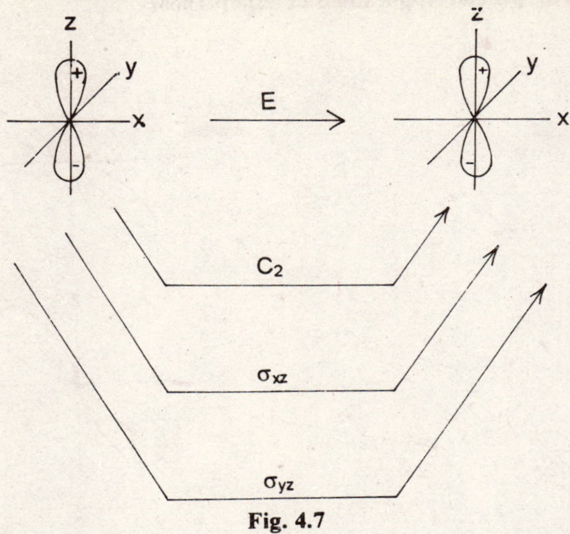

Fig. 4.7

This can be written as

	E	C_2	σ_{xz}	σ_{yz}	
A_1	+1	+1	+1	+1	p_z

Which corresponds to A_1 irreducible representation. Similarly for p_z and p_y orbitals the resulting wave functions can be shown as follows :

$$O_E\,(p_x) = E\,(R \sin \theta \cos \phi) = (R \sin \theta \cos \phi) = p_x$$

$$O_{C_2}\,(p_x) = C_2\,(R \sin \theta \cos \phi) = [R \sin \theta \cos (\phi + \pi)] = -\,p_x$$

$$O_{\sigma_{xz}}\,(p_x) = \sigma_{xz}\,(R \sin \theta \cos \phi) = [R \sin \theta \cos (\phi)] = p_x$$

$$O_{\sigma_{yz}}\,(p_x) = \sigma_{yz}\,(R \sin \theta \cos \phi) = [R \sin \theta \cos (\phi + \pi)] = -\,p_x$$

So, the characters are

	E	C_2	σ_{xz}	σ_{yz}	
B_1	1	−1	1	−1	px

Which corresponds to B_1 irreducible representation.

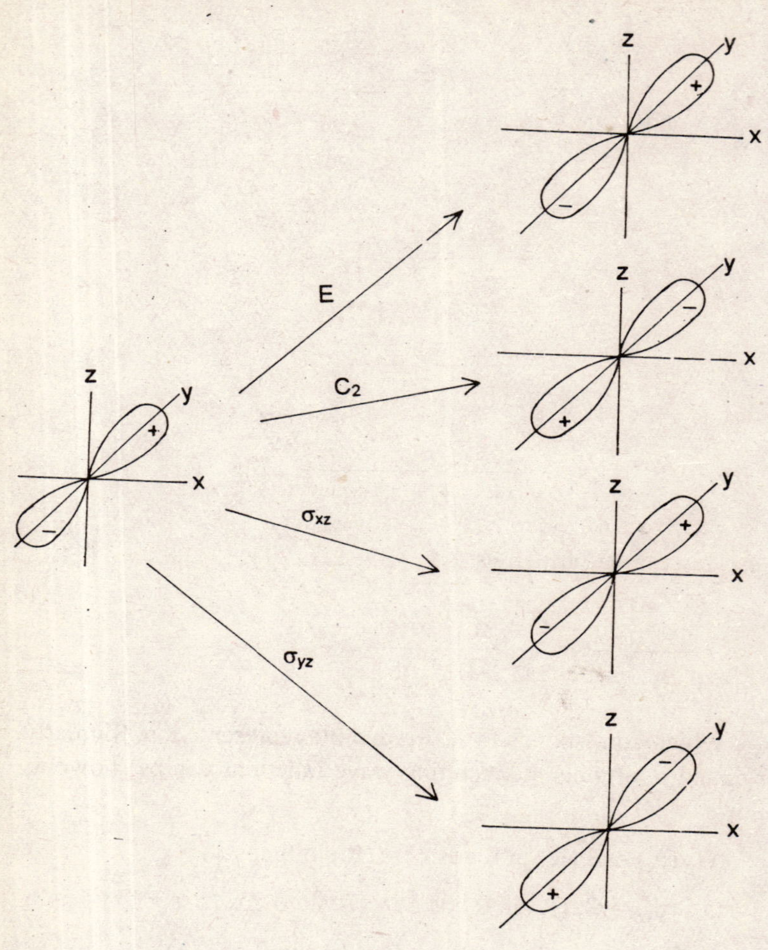

Fig. 4.8

Similarly for p_y the resulting wave functions are

$$O_E (p_y) = E (R \sin \theta \sin \phi) = (R \sin \theta \sin \phi) = p_y$$

$$O_{C_2} (p_y) = C_2 (R \sin \theta \sin \phi) = [R \sin \theta \sin (\phi + \pi)] = -p_y$$

$$O_{\sigma_{xz}} (p_y) = \sigma_{xz} (R \sin \theta \sin \phi) = [R \sin \theta \sin (\phi + \pi)] = -p_y$$

$$O_{\sigma_{yz}} (p_y) = \sigma_{yz} (R \sin \theta \sin \phi) = [R \sin \theta \sin (\phi)] = +p_y$$

Therefore, the characters are

	E	C_2	σ_{xz}	σ_{yz}	
B_2	1	–1	–1	1	p_y

So, p_y belongs to B_2 irreducible representation. So, degenerate p-orbitals with corresponding reducible representation

	E	C_2	σ_{xz}	σ_{yz}	
T_p	1	–1	1	1	p

Transformed to non-degenerate p_x, p_y and p_z orbitals with corresponding irreducible representations viz. A_1, B_1 and B_2. This can also be shown using reduction formula that

$$T_p = A_1 \oplus B_1 \oplus B_2$$

Similarly the irreducible representations corresponding to d-orbitals can be obtained. A simple way to know the transformation properties of an orbital in a point group is to look at the irreducible representation corresponding to the same co-ordinate or binary product functions. The transformation properties of d-orbitals in C_{2v} point group can be seen from the corresponding character table as

d_{xy}	A_2
d_{xz}	B_1
d_{yz}	B_2
$d_z{}^2$	A_1
$d_{x^2-y^2}$	A_1

Let us consider another example to illustrate the effect of transformations operator under different symmetry operations on p_x & p_y orbitals of N-atom in NH_3 (molecular point group C_{3v}). These orbitals are represented by the following eigen functions

$$p_x = R \sin\theta \cos\phi$$

$$p_y = R \sin\theta \sin\phi$$

θ remain same as it is the angle measured from the reference axis (z axis) or the principal axis but ϕ changes in rotation in C_3 operation is by the angle $2\pi/3$. So, the effect of transformation operator on p_x and p_y are

$O_E = E\ (R \sin\theta \cos\phi) = R \sin\theta \cos\phi = p_x$

$O_E = E\ (R \sin\theta \sin\phi) = R \sin\theta \sin\phi = p_y$

In the matrix form this result can be shown as

$$\begin{pmatrix} 1 & 0 \\ 0 & 1 \end{pmatrix} \begin{pmatrix} p_x \\ p_y \end{pmatrix} = E \begin{pmatrix} p_x \\ p_y \end{pmatrix}$$

$\Rightarrow \qquad X\ (E) = 2$

$$\begin{aligned} O_{C_3} p_x &= C_3(R\sin\theta\cos\phi) = R(\sin\theta)\left[\cos\left(\phi + \tfrac{2\pi}{3}\right)\right] \\ &= R(\sin\theta)\left[\cos\theta\cos\tfrac{2\pi}{3} - \sin\phi\sin\tfrac{2\pi}{3}\right] \\ &= R(\sin\theta)\left[-1/2\cos\phi - \sqrt{3}/2\sin\phi\right] \\ &= -1/2R\sin\theta\cos\phi - \sqrt{3}/2R\sin\theta\sin\phi \\ &= -1/2p_x - \sqrt{3}/2p_y \end{aligned}$$

$$\begin{aligned} O_{C_3} p_y &= C_3(R\sin\theta\sin\phi) = R(\sin\theta)\left[\sin\left(\phi + \tfrac{2\pi}{3}\right)\right] \\ &= R(\sin\theta)\left[\sin\theta\cos\tfrac{2\pi}{3} + \cos\phi\sin\tfrac{2\pi}{3}\right] \\ &= R(\sin\theta)\left[-1/2\sin\phi + \sqrt{3}/2\cos\phi\right] \\ &= -1/2R\sin\theta\sin\phi + \sqrt{3}/2R\sin\theta\cos\phi \\ &= -1/2p_y + \sqrt{3}/2p_x \end{aligned}$$

in matrix this can be represented as

$$\begin{pmatrix} -1/2 & -\sqrt{3}/2 \\ \sqrt{3}/2 & -1/2 \end{pmatrix} \begin{pmatrix} p_x \\ p_y \end{pmatrix} = C_z \begin{pmatrix} p_x \\ p_y \end{pmatrix}$$

$\Rightarrow \qquad X\ (C_3) = -1$

Similarly for σ_v (consider σ_v, see chapter 3, section 3.3, figure 3.7.)

$$O\sigma_v(R\sin\theta\cos\phi) = O\sigma_v(p_x) = p_x$$

$$O\sigma_v(R\sin\theta\sin\phi) = O\sigma_v(p_y) = -R\sin\theta\sin\phi = -p_y$$

And in the matrix as

$$\begin{pmatrix} 1 & 0 \\ 0 & -1 \end{pmatrix} \begin{pmatrix} p_x \\ p_y \end{pmatrix} = \sigma_v \begin{pmatrix} p_x \\ p_y \end{pmatrix}$$

$$\Rightarrow \qquad X(\sigma_v) = 0$$

Therefore, p_x, p_y are the degenerate set of orbitals and the irreducible representation corresponding to these orbitals is

	E	C_3	σ_v	
E	2	−1	0	p_x, p_y

This irreducible representation is E which is two dimensional representation.

Section 4.5

The direct product representation

Let us suppose that R is any operation in a symmetry group of a molecule with x_1, x_2, x_3 ... ... x_m and y_1, y_2 ... ... y_n be the set of functions, which are the basis of representations of the group. On operation on any of the basis function, we can write

$$RX_i = \sum_{j=1}^{m} x_{ij} X_j$$

$$RY_k = \sum_{l=1}^{n} y_{lk} Y_l$$

So, this equation is also true

$$RX_iY_k = \sum_{j=1}^{m} \sum_{l=1}^{n} x_{ij}\, y_{lk}\, X_j\, Y_l = \sum_{j} \sum_{k} z_{jlik} X_i Y_l$$

In this equation the set of functions X_i Y_k is called direct product of functions x_i and y_k which also forms a basis of representation of the same group and the $z_{il,\,ik}$ are the elements of a matrix Z of order (mn) × (nm).

In case of a direct product representation the characters of the direct product representation are equal to the products of the characters of the representations based on the individual function, e.g. direct product of some irreducible representations of a group i.e. C_{4v} are as follows:

C_{4v}	E	C_2	$2C_4$	$2\sigma_v$	$2\sigma_d$
A_1	1	1	1	1	1
A_2	1	1	1	–1	–1
B_1	1	1	–1	1	–1
B_2	1	1	–1	–1	1
E	2	–2	0	0	0
A_1A_2	1	1	1	–1	–1
A_1EB_2	2	–2	0	0	0

Similarly for C_{3v}.

C_{3v}	E	$2C_3$	$3\sigma_v$
A_1	1	1	1
A_2	1	1	–1
E	2	–1	0
A_1A_2	1	1	–1
A_2E	2	–1	0
A_1E	2	–1	0

Secondly, the representation of a direct product T_{AB} will be totally symmetric if and if only irreducible representation T_A = the irreducible representation T_B. Direct product can be represented by the matrix multiplication as $D^i R \otimes D^j (R)$.

Let us consider an example of direct product in the matrix notation.

$$\begin{pmatrix} A_{11} & A_{12} \\ A_{21} & A_{22} \end{pmatrix} \begin{pmatrix} B_{11} & B_{12} & B_{13} \\ B_{21} & B_{22} & B_{23} \\ B_{31} & B_{32} & B_{33} \end{pmatrix}$$

$$\begin{pmatrix} A_{11}B_{11} & A_{11}B_{12} & A_{11}B_{13} & A_{12}B_{11} & A_{12}B_{12} & A_{12}B_{13} \\ A_{11}B_{21} & A_{11}B_{22} & A_{11}B_{23} & A_{12}B_{21} & A_{12}B_{22} & A_{12}B_{23} \\ A_{11}B_{31} & A_{11}B_{32} & A_{11}B_{33} & A_{12}B_{31} & A_{12}B_{32} & A_{12}B_{33} \\ A_{21}B_{11} & A_{21}B_{12} & A_{21}B_{13} & A_{22}B_{11} & A_{22}B_{12} & A_{22}B_{13} \\ A_{21}B_{21} & A_{21}B_{22} & A_{21}B_{23} & A_{22}B_{21} & A_{22}B_{22} & A_{22}B_{23} \\ A_{21}B_{31} & A_{21}B_{32} & A_{21}B_{33} & A_{22}B_{31} & A_{22}B_{32} & A_{22}B_{33} \end{pmatrix}$$

This is a direct product representation in the form of matrices and as it is in this case the matrices are 2×2 and 3×3, so, the result is 6×6 as shown.

Problems

Q. 1: Perform the operations of C_{3v} and show with the effect of transformation operator on p_z orbital, that this orbital belongs to one dimensional representation.

Ans: As the spherical harmonics and including the 'r' part the p_z orbital is R cos θ. On performing operations i.e. on applying the transformation operator on p_z, following resulting wave functions can be obtained

$$O_E\,(p_z) = E\,(R\cos\theta) = R\cos\theta = p_z$$

$$O_{C_3}\,(p_z) = C_3\,(R\cos\theta) = R\cos\theta = p_z$$

$$O_{\sigma_v}\,(p_z) = \sigma_v\,(R\cos\theta) = R\cos\theta = p_z$$

This can be written as the irreducible representation

	E	C_3	σ_v
A_1	1	1	1

Therefore, p_z belongs to the A_1 irreducible representation.

Q. 2: Show that in T_d point group the p-orbital belongs to a three dimensional irreducible representation.

Ans: As set of p-orbitals can be represented as

$$p_z = R\cos\theta$$

$$p_x = R\sin\theta\cos\phi \text{ and}$$

$$p_y = R\sin\theta\sin\phi$$

On performing different operations in T_d molecular point group i.e. with the effect of transformation operator following wave functions can be obtained

$$O_E\,(p_z) = E\,(R\cos\theta) = R\cos\theta = p_z$$

$$O_E\,(p_x) = E\,(R\sin\theta\cos\phi) = R\sin\theta\cos\phi = p_x$$

$$O_E\,(p_y) = E\,(R\sin\theta\sin\phi) = R\sin\theta\sin\phi = p_y$$

Which can be shown in the matrix as

$$\begin{pmatrix} 1 & 0 & 0 \\ 0 & 1 & 0 \\ 0 & 0 & 1 \end{pmatrix} \begin{pmatrix} p_x \\ p_y \\ p_z \end{pmatrix} = E \begin{pmatrix} p_x \\ p_y \\ p_z \end{pmatrix}$$

$\Rightarrow$ X (E) = 3

$$O_{C_3} p_x = C_3(R\cos\theta) = R\cos\theta = p_z$$

$$\begin{aligned} O_{C_3} p_x &= C_3(R\sin\theta\cos\phi) = R(\sin\theta)\left[\cos\left(\phi + \tfrac{2\pi}{3}\right)\right] \\ &= R(\sin\theta)\left[\cos\phi\cos\tfrac{2\pi}{3} - \sin\phi\sin\tfrac{2\pi}{3}\right] \\ &= R(\sin\theta)\left[-1/2\cos\phi - \sqrt{3}/2\sin\phi\right] \\ &= -1/2R\sin\theta\cos\phi - \sqrt{3}/2R\sin\theta\sin\phi \\ &= -1/2p_x -- \sqrt{3}/2p_y \end{aligned}$$

$$\begin{aligned} O_{C_3} p_y &= C_3(R\sin\theta\sin\phi) = R(\sin\theta)\left[\sin\left(\phi + \tfrac{2\pi}{3}\right)\right] \\ &= R(\sin\theta)\left[\sin\theta\cos\tfrac{2\pi}{3} + \cos\phi\sin\tfrac{2\pi}{3}\right] \\ &= R(\sin\theta)\left[-1/2\sin\phi + \sqrt{3}/2\cos\phi\right] \\ &= -1/2R\sin\theta\sin\phi + \sqrt{3}/2R\sin\theta\cos\phi \\ &= -1/2p_y + \sqrt{3}/2p_x \end{aligned}$$

which can be represented in the matrix as

$$\begin{pmatrix} -1/2 & -\sqrt{3}/2 & 0 \\ \sqrt{3}/2 & -1/2 & 0 \\ 0 & 0 & 1 \end{pmatrix} \begin{pmatrix} p_x \\ p_y \\ p_z \end{pmatrix} = C_3 \begin{pmatrix} p_x \\ p_y \\ p_z \end{pmatrix}$$

$\Rightarrow$ X $(C_3) = 0$

O_{C_2} $(P_z) = C_2$ (R cos θ) = R cos θ = p_z

O_{C_2} $(P_x) = C_2$ (R sin θ cos ϕ) = [R sin θ cos (ϕ + π)] = – p_x

O_{C_2} $(P_y) = C_2$ (R sin θ sin ϕ) = [R sin θ sin (ϕ + π)] = – p_y

And the corresponding matrix is

$$\begin{pmatrix} -1 & 0 & 0 \\ 0 & -1 & 0 \\ 0 & 0 & 1 \end{pmatrix} \begin{pmatrix} p_x \\ p_y \\ p_z \end{pmatrix} = C_2 \begin{pmatrix} p_x \\ p_y \\ p_z \end{pmatrix}$$

$\Rightarrow$ X $(C_2) = -1$

Similarly on operations this can be shown that matrix for σ_d operation is

$$\begin{pmatrix} -1 & 0 & 0 \\ 0 & 1 & 0 \\ 0 & 0 & 1 \end{pmatrix} \begin{pmatrix} p_x \\ p_y \\ p_z \end{pmatrix} = \sigma_d \begin{pmatrix} p_x \\ p_y \\ p_z \end{pmatrix}$$

$\Rightarrow$ X $(\sigma_d) = 1$

and for the S_4 operation X $(S_4) = -1$

Therefore, this can be expressed in a irreducible representation as

	E	$8C_3$	$3C_2$	$6S_4$	$6\sigma_d$	
T_2	3	0	−1	−1	1	p_x, p_y, p_z

The corresponding irreducible representation is T_2 which can't be reduced. So, it is T_2 representation in T_d point group to which p-orbitals belong and which is a three dimensional irreducible representation.

Q. 3: Write down the characters of the representations for the following direct products and determine the irreducible representation which comparise them for the group.

(i) $D_{6h} : A_{1g} \otimes B_{1g}$

(ii) $C_{3v} : A_1 \otimes A_2$

(iii) $D_{6h} : B_{2u} \otimes E_{1g}$

(iv) $C_{2v} : A_1 \otimes B_1$

(v) $D_{3h} : A_1' \otimes A_2'$

Ans: These direct product representations can be directly obtained by multiplying the corresponding characters of the irreducible representations from their respective character tables

D_{6h}	E	$2C_6$	$2C_3$	$3C_2$	$3C_2'$	$3C_2''$	'i'	$2S_3$	$2S_6$	σ_h	$3\sigma_d$	$3\sigma_v$
$A_{1g} \otimes B_{1g}$	1	–1	1	–1	1	–1	1	–1	1	–1	1	–1

C_{3v}	E	$2C_3$	$3\sigma_v$
$A_1 \otimes A_2$	1	1	–1

D_{6h}	E	$2C_6$	$2C_3$	$3C_2$	$3C_2'$	$3C_2''$	'i'	$2S_3$	$2S_6$	σ_h	$3\sigma_d$	$3\sigma_v$
$B_{2u} \otimes E_{1g}$	2	–1	–1	2	0	0	–1	1	1	–2	0	0

C_{2v}	E	C_2	σ_{xz}	σ_{yz}
$A_1 \otimes B_1$	1	–1	1	–1

D_{3h}	E	$2C_3$	$3C_2$	σ_h	$2S_3$	$3\sigma_v$
$A_1' \otimes A_2''$	1	1	–1	1	1	–1

5

Symmetry and Linear Combination

Linear combination of atomic orbitals is very important concept that should be explained well for the formation of molecular orbitals or during the formation of molecules. In the previous chapter some hints are given for this concept i.e. linear combination of orbitals. Role of symmetry in the linear combination of atomic orbitals is discussed in this chapter. This concept is termed as symmetry adapted linear combination (SALC). In order to get symmetry adapted linearly combined set of A.O's, the first step is to introduce the concept of projection operator and its use in the construction of SALC.

Section 5.1

Concept of Projection Operators

Let us define P^u as operator, operated on the function f^u, the operation of this operator is defined as

$$P^u f^u = (g / n_u) f^u \text{or}$$
$$= (h / n_u) f^u$$

where h is the order of the group and n is dimension of the function space, and P^u and f^u both belong to the same representation. If instead of f^u this projection operator is operated on f^v i.e. if projection operator of one representation is operated on the function belonging to another representation the result will be zero.

i.e. $P^u f^v = 0$

or if function does not belong to the particular u[th] irreducible representation that will vanish. The projection operator p^u is defined as

$$P^u = \Sigma X_u(R)^* O_R$$

Where O_R is the transformation operator. In the next section let us try to apply this projection operator to construct SALC's

Section 5.2

Use of Projection Operator to Construct SALC's

Example 1: Let us consider NH_3. Its point group is C_{3v}. Let us try to construct the SALC's for this case. Let us consider g_1, g_2, g_3 functions (or vectors) as basis of representation.

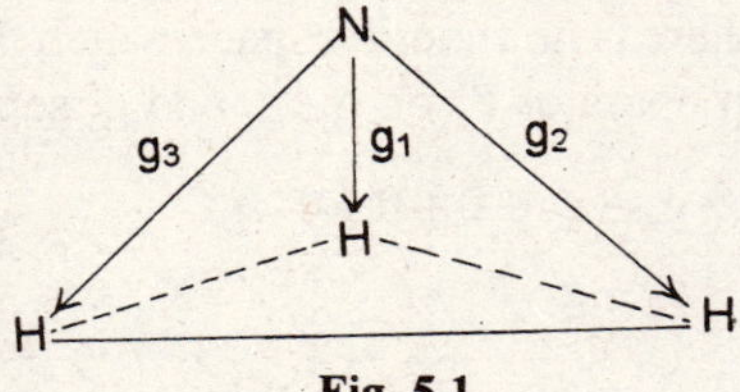

Fig. 5.1

Its character table is

C_{3v}	E	C_3^1	C_3^2	σ_v	σ_v'	σ_v''
A_1	1	1	1	1	1	1
A_2	1	1	1	–1	–1	–1
E	2	–1	–1	0	0	0

Effect of transformation operator O_R on g_1, g_2 and g_3 is given by

g_1	g_1	g_2	g_3	g_1	g_3	g_2
g_2	g_2	g_3	g_1	g_3	g_2	g_1
g_3	g_3	g_1	g_2	g_2	g_1	g_3

Now let us apply the projection operator P^{A1} on g_1, g_2 and g_3 respectively

$P^{A1} = \Sigma X(R) O_R$

P^{A1} on g_1 i.e. $P^{A1}(g_1)$

$= 1 \times g_1 + 1 \times g_2 + 1 \times g_3 + 1 \times g_1 + 1 \times g_3 + 1 \times g_2$

$= 2(g_1 + g_2 + g_3)$ i.e. $(g_1 + g_2 + g_3)$

Similarly on applying projection operator P^{A1} on g_2 we get

$$P^{A1}(g_2) \alpha (g_1 + g_2 + g_3)$$

and similarly

$$P^{A1}(g_3) \alpha (g_1 + g_2 + g_3)$$

Now let us apply projection operator P^{A2} on g_1, g_2 and g_3 respectively

$$P^{A2}(g_1) = 1 \times g_1 + 1 \times g_2 + 1 \times g_3 - 1 \times g_1 - 1 \times g_2 - 1 \times g_3 = 0$$

Similarly this can be shown that $P^{A2}(g_2)$ and $P^{A2}(g_3)$ are also equal to zero. So, there is no function space which belongs to A_2. Let us apply projection operator P^E on g_1, g_2 and g_3 separately

$$P^E(g_1) = 2g_1 - g_2 - g_3 + 0 + 0 + 0$$

$$= 2g_1 - g_2 - g_3$$

Similarly

$$P^E(g_2) = 2g_2 - g_3 - g_1$$

and $P^E(g_3) = 2g_3 - g_2 - g_1$

For linearly independent functions or the linearly independent combination of orbitals sum of the functions must be equal to zero but here it is not equal to zero (if we add all the three functions belonging to the projection operator P^E). So, we can say that these three i.e. $2g_1 - g_2 - g_3, 2g_2 - g_3 - g_1$, and $2g_3 - g_2 - g_1$ functions are linearly dependent functions. In order to get a set of linearly independent functions out of these three, let us perform the following operation i.e.

$$P^E(g_2) - P^E(g_3)$$

$$= (2g_2 - g_3 - g_1) - (2g_3 - g_2 - g_1)$$

$$= 3(g_2 - g_3)$$

This new function i.e. $3(g_2 - g_3)$ is orthogonal to the function obtained by the projection operation on g_1 i.e. function belonging to

$P^E(g_1)$ i.e. scalar product of these two functions viz. $3(g_2 - g_3)$ and $2g_1 - g_2 - g_3$ is equal to zero. Therefore, in all we have three functions which are

$$P^{AI}(g) \; \alpha \; (g_1 + g_2 + g_3)$$
$$P^E(g) \; \alpha \; (2g_2 - g_3 - g_1) \text{ and}$$
$$P^E(g) \; \alpha \; (g_2 - g_3)$$

These three functions are the symmetry adapted linearly combined (SALC's) A.O's for NH_3. These functions can be normalised by considering their coefficients and taking their sqaures and the under-roots. Therefore the functions become

$$P^{AI}(g) \; \alpha \; 1/\sqrt{3}(g_1 + g_2 + g_3)$$
$$P^E(g) \; \alpha \; 1/\sqrt{6}(2g_2 - g_3 - g_1) \text{ and}$$
$$P^E(g) \; \alpha \; 1/\sqrt{2}(g_2 - g_3)$$

Which are the normalised set of functions.

Example 2: Let us try to construct the SALC's of H_2O molecule. Its point group is C_{2v}. Consider g_1 and g_2 vectors (or function space) as basis of representations.

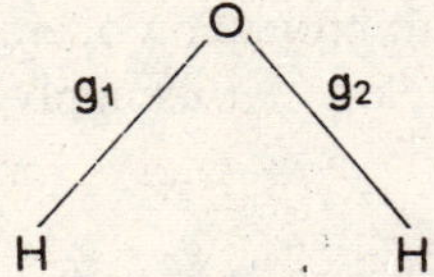

Fig. 5.2

Its character table is given below with effect of transformation operator on g_1 and g_2.

C_{2v}	E	C_2	σ_{xz}	σ_{yz}
A_1	1	1	1	1
A_2	1	1	−1	−1
B_1	1	−1	1	−1
B_2	1	−1	−1	1

Effect of O_R

g_1	g_1	g_2	g_2	g_1
g_2	g_2	g_1	g_1	g_2

Now, on applying the projection operator P^{A1} on g_1 and g_2, we get

$$2(g_1 + g_2) = P^{A1}(g_1 \text{ or } g_2) \alpha (g_1 + g_2)$$

Similarly on applying projection operator P^{A2} on g_1 and g_2

$$P^{A2}(g_1) = g_1 + g_2 - g_2 - g_1 = 0$$

Similarly

$$P^{A2}(g_2) = g_2 + g_1 - g_1 - g_2 = 0$$

So, there is no function space corresponding to A_2. Let us apply projection operator P^{B1} on g_1 and g_2

$$P^{B1}(g_1) = g_1 - g_2 + g_2 - g_1 = 0$$

and

$$P^{B1}(g_2) = g_2 - g_1 + g_1 - g_2 = 0$$

Same is the case with projection operator P^{B1} i.e. no function space corresponds to B_1. Now, let us apply the operator P^{B2} on g_1 and g_2

$$P^{B2}(g_1) = g_1 - g_2 - g_2 + g_1 = 2(g_1 - g_2)$$

Similarly

$$P^{B2}(g_2) = g_2 - g_1 - g_1 + g_2 = 2(g_2 - g_1)$$

So, on two functions SALC's can be written for H_2O molecule and these are $g_1 + g_2$ and $g_1 - g_2$ and their normalisation yields the normalised SALC's as $1/\sqrt{2}(g_1 + g_2)$ and $1/\sqrt{2}(g_1 - g_2)$, as normalisation constant for any SALC is $1/\sqrt{n}$ where n is the sum of squares of individual coefficients.

Example 3: Let us consider example of cyclopropenyl ion with 3π orbitals as basis viz. g_1, g_2 and g_3 vectors.

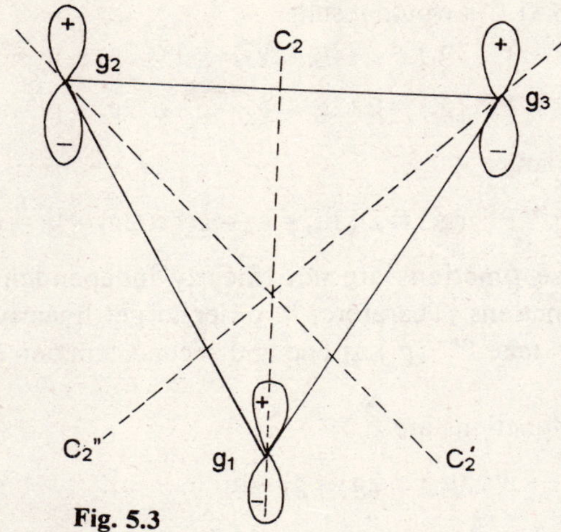

Fig. 5.3

Its point group is D_{3h} and the corresponding T_π or T_{bond} is

D_{3h}	E	C_3^1	C_3^2	C_2	C_2'	C_2''	σ_h	S_3^1	S_3^2	σ_v	σ_v'	σ_v''
T_π	3	0	0	−1	−1	−1	−3	0	0	1	1	1

On reduction of T_π, we get $T_\pi = A_2'' + E''$. So, let us see the effect of projection operators $P^{A2''}$ and $P^{E''}$ on g_1, g_2 and g_3 to get SALC's

D_{3h}	E	C_3^1	C_3^2	C_2	C_2'	C_2''	σ_h	S_3^1	S_3^2	σ_v	σ_v'	σ_v''
T_π	3	0	0	−1	−1	−1	−3	0	0	1	1	1

Effect of Or

g_1	g_1	g_2	g_3	$-g_1$	$-g_3$	$-g_2$	$-g_1$	$-g_2$	$-g_3$	g_1	g_3	g_2
g_2	g_2	g_3	g_1	$-g_3$	$-g_2$	$-g_1$	$-g_2$	$-g_3$	$-g_1$	g_3	g_2	g_1
g_3	g_3	g_1	g_2	$-g_2$	$-g_1$	$-g_3$	$-g_3$	$-g_1$	$-g_2$	g_2	g_1	g_3

On applying the projection operator $P^{A2''}$ on g_1, g_2 and g_3 i.e.

$$P^{A}{}_2''(g) = 4(g_1 + g_2 + g_3) \alpha\, g_1 + g_2 + g_3$$

Similarly on applying the projection operator $P^{E''}$ on g_1, g_2 and

g_3 following SALC's would results

$$P^{E''}(g_1) = 2(2g_1 - g_2 - g_3) \propto 2g_1 - g_2 - g_3$$

$$P^{E''}(g_2) = 2(2g_2 - g_3 - g_1) \propto 2g_2 - g_3 - g_1$$

and similarly

$$P^{E''}(g_3) = 2(2g_3 - g_1 - g_2) \propto 2g_3 - g_1 - g_2$$

So, these functions are not linearly independent they are dependent functions. Therefore, in order to get linearly independent functions take $P^{E''}(g_1)$ as one and second function as $P^{E''}(g_2) - P^{E''}(g_3)$

So, the functions are

$$P^{E''}(g_1) \propto 2g_1 - g_2 - g_3$$

and

$$P^{E''}(g_2) - P^{E''}(g_3) = (2g_2 - g_3 - g_1) - (2g_3 - g_1 - g_2)$$
$$= 3(g_2 - g_3)$$

For linearly independent functions sum of functions must be equal to zero, therefore, we have to carried out this operation to get linearly independent function. This new function is orthogonal to the function $P^{E''}(g_1)$. This can be shown by taking the scalar product of these two functions viz. $P^{E''}(g_1)$ and $P^{E''}(g_2) - P^{E''}(g_3)$ which is equal to zero. So, in all there are three SALC's in this case using π orbitals as basis functions and these are

$$P^{A_2''}(g) \propto g_1 + g_2 + g_3$$

$$P^{E''}(g_1) \propto 2g_1 - g_2 - g_3 \text{ and}$$

$$P^{E''}(g_2) - P^{E''}(g_3) \propto g_2 - g_3$$

which on normalisation give normalised SALC's (or functions) as

$$P^{A2''}(g) \propto 1/\sqrt{3}(g_1 + g_2 + g_3)$$

$$P^{E''}(g_1) \propto 1/\sqrt{6}(2g_1 - g_2 - g_3) \text{ and}$$

$$P^{E''}(g_2) - P^{E''}(g_3) \propto 1/\sqrt{2}(g_2 - g_3)$$

Example 4: Let us consider NO_3^- ion. Its point group is D_{3h}. Consider NO σ bonds as vectors (or functions)

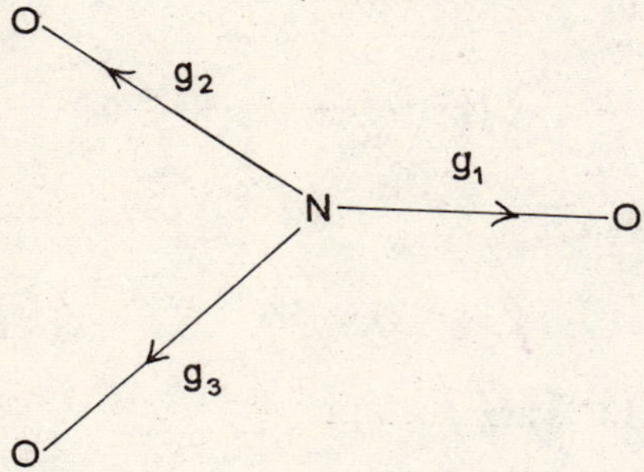

Fig. 5.4

Character table for D_{3h} and the effect of transformation operator is given below

D_{3h}	E	$2C_3$	$3C_2$	σ_h	$2S_3$	$3\sigma_v$
A_1'	1	1	1	1	1	1
A_2'	1	1	–1	1	1	–1
E'	2	–1	0	2	–1	0
A_1''	1	1	1	– 1	–1	–1
A_2''	1	1	– 1	–1	–1	1
E''	2	–1	0	–2	1	0

on forming T_σ and on reducing it we get

$T_\sigma = A_1' \oplus E'$ and the corresponding SALC's are

$P^{A1}\ \alpha\ 1/\sqrt{3}(g_1 + g_2 + g_3)$

$P^{E'}\ \alpha\ 1/\sqrt{6}(2g_1 - g_2 - g_3)$ and

$1/\sqrt{2}(g_2 - g_3)$

Example 5: Using set of π-orbitals in case of NO_3^- ion we get the same functions as in case of σ bonds as vectors (or functions) (ref. example-4).

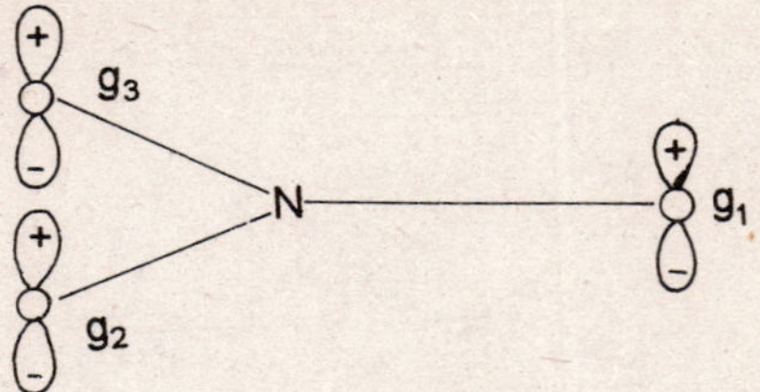

Fig. 5.5

So, the SALC's are

$$P^{A1}\ \ 1/\sqrt{3}(g_1 + g_2 + g_3)$$
$$P^{E'}\ \ 1/\sqrt{6}(2g_1 - g_2 - g_3) \text{ and}$$
$$1/\sqrt{2}(g_2 - g_3)$$

Example 6: For an octahedral AB_6 molecule, the bond vectors may be considered as shown in Figure 5.6. The representation T_{bond} reduces to $A_{1g} + E_g + T_{1u}$ and the corresponding SALC's are

$$A_{1g}:\ \ g_1 + g_2 + g_3 + g_4 + g_5 + g_6$$
$$E_g:\ \ 2g_1 - g_2 - g_3 - g_4 - g_5 - 2g_6$$
$$g_2 - g_3 - g_4 - g_5$$
$$T_{1u}:\ \ g_1 - g_6$$
$$g_2 - g_4$$
$$g_3 - g_5$$

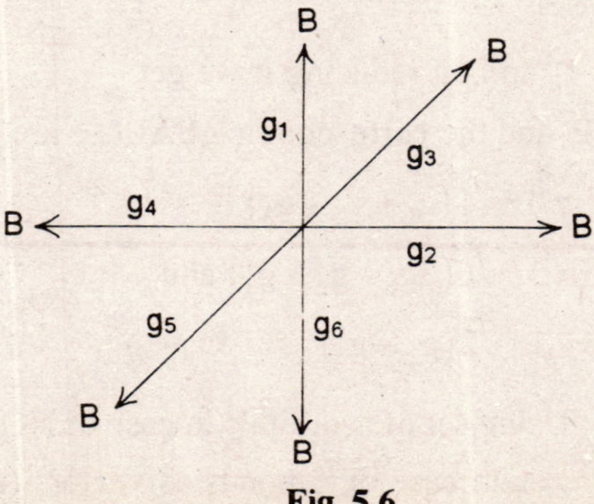

Fig. 5.6

6

Group Theory and Concepts of Bonding–I

There are two approaches viz. (1) Valance bond theory and (2) Molecular Orbital theory, which are used to explain the formation of covalent bond. Valance bond theory was proposed by Heitler and London in 1927 while the alternative molecular orbital was proposed by Hund and Mulikan.

In this chapter, features of valance bond theory as explained by Heitler and London and its modification as suggested by Pauling and Slater, with the concepts of formation of concept σ (sigma) π (pi) bonds and concept of hybridization in light of group theory is explained.

Section 6.1

Introduction to valance bond theory

This theory was explained by Heilter and London in 1927. According to this theory, two hydrogen atoms which are far apart *i.e.,* H_A and H_B with electrons 1 and 2 can be represented by the individual wave functions Ψ_{A1} and Ψ_{B2} *i.e.,* electron 1 is with H_A and electron 2 is around the H_B. When they come close to each other, interaction starts between them and because of interactions between their atomic orbitals during the formation of molecule *i.e.,* Ψ_{A1} and Ψ_{B2} interact with each other. There are two possible structures can be given for hydrogen molecule under these conditions which may be

$$\underset{\text{I}}{H_A^1 H_B^2} \longleftrightarrow \underset{\text{II}}{H_A^2 H_B^1}$$

Wave functions for the corresponding molecular structures can be given by

$$\phi_I = \Psi_{A1}\,\Psi_{B2} \qquad \text{and} \qquad \phi_{II} = \Psi_{A2}\,\Psi_{B1}$$

And the combination for these two orbitals is the Ψ_m (or molecular orbital) which can be written as

$$\Psi_m = C_I\phi_I + C_{II}\,\phi_{II}$$

or

$$\Psi_m = C_I\Psi_{A1}\,\Psi_{B2} + C_{II}\,\Psi_{A2}\,\Psi_{B1}$$

where C_I and C_{II} are the coefficients for ϕ_I & ϕ_{II}. The values of C_I and C_{II} should be such that the resulting wave function has minimum energy and for this

$$C_I^2 = C_{II}^2$$

or

$$C_I = \pm\, C_{II}$$

i.e., if $C_I = 1$; C_{II} should have the values either +1 or –1. If C_{II} is +1, then the symmetric combination of wave functions may be obtained *i.e.*,

$$\Psi_s = \Psi_{A1}\,\Psi_{B2} + \Psi_{A2}\,\Psi_{B1}$$

or

$$\Psi_B = C_I\,(\phi_I + \phi_{II})$$

Similarly for $C_{II} = -1$ antisymmetric combination may be obtained *i.e.*,

$$\Psi_a = \Psi_{A1}\,\Psi_{B2} - \Psi_{A2}\,\Psi_{B1}$$

or

$$\Psi_a = C_I\,(\phi_I - \phi_{II})$$

If the interaction is symmetrical *i.e.*, electronic waves are in the same phase in between the nuclei *i.e.*, the spin of two interacting electrons are opposite or This results in the lower energy state than the isolated H_A & H_B atoms. If symmetrical interaction is there, formation of a covalent bond takes place, with the corresponding energy

$$E_s = \frac{<\phi_I|\hat{H}|\phi_{II}> + <\phi_I|\hat{H}|\phi_I>}{<\phi_I|\phi_{II}> + <\phi_I|\phi_{II}>}$$

The corresponding bond distance where the energy is minimum can be worked out on substituting various parameters in the above energy equation. This distance corresponds to bond distance of H_2 molecule *i.e.* (r_s = 0.87 Å).

In case of antisymmetric combination of orbitals as explained by the equation of Ψ_a, spin of two combining electrons are parallel *i.e.,* the two electronic waves results in are state of higher energy which is given by E_a.

$$E_a = \frac{<\phi_I|\hat{H}|\phi_{II}> - <\phi_I|\hat{H}|\phi_I>}{<\phi_I|\phi_{II}> - <\phi_I|\phi_{II}>}$$

E_a is higher than the energy of the corresponding interacting atoms because of this antisymmetric combination formation of molecule can't takes place. Pauling and Sater suggested two more ionic structures which can contribute to the formation of H_2 molecule as

$$H_{A_{III}}^{-1,2}\, H_B^{+} \longleftrightarrow H_{A_{IV}}^{+}\, H_B^{-1,2}$$

According to them actually during symmetric interaction, the formation of symmetric molecular orbital *i.e.,* Ψ_s is given by

$$\Psi_s = \Psi_{cov.} + \lambda\Psi_{ion}$$

Where λ is the degree of ionic character in bond *i.e.,* 0.17 for H_2 molecule and

$$\Psi_{ion} = \Psi_{III} + \Psi_{IV}$$

and

$$\Psi_{III} = \Psi_{A1}\,\Psi_{A2}$$

and

$$\Psi_{IV} = \Psi_{B1}\,\Psi_{B2}$$

By this modification r_s is calculated as 0.86 Å for H_2 molecule. Thus the assumptions on which Valance bond theory is based are:

(i) Atomic orbitals retained their identity but only electrons form covalent bonds and lose their identity.

(ii) In the interaction or during bond formation only valance electrons are disturbed but other remain undisturbed.

(iii) There should be unpaired electrons required on every atom to form a valance bond. If it is not so than energy is required to unpair the electrons.

(iv) Strength of a valance bond depends upon the extent of overlapping *i.e.,* on the overlap integral which is given by $S = \int \phi_I \phi_{II} dr$. If $S > 0$, stable bond is formed and if $S < 0$, no bond is formed and if $S = 0$ there is neither repulsion nor attraction between the interacting nuclei.

Some features of valance bond theory are:

(i) If there is symmetrical interaction *i.e.,* in case of formation of Ψ_s orbital, σ bond is formed (as σ bond is symmetrical) and it also indicate the direction of bond formation. In case of s-s interaction or p_z–p_z interaction σ bonds are formed.

(ii) If interaction is less, *i.e.,* in case of sideways interaction of orbitals as in case of p_x-p_x and p_y-p_y interaction Π(pi) bonds are formed. Sigma (σ) bond is stronger than pi (π) bond as extent of overlapping or interaction is higher in case of σ bond as compare to π bond.

(iii) Rotation around a sigma (σ) bond is possible, and

(iv) Isomerism can be shown by sigma (σ) bond.

Various orbital interactions and sigma (σ) bond and pi (π) bond formation can be shown as in Fig. 6.1.

Formation of simple molecules *i.e.,* H_2, O_2 and N_2 can be explained on the basis of this concept. In case of H_2 molecule one σ bond is formed, in case of O_2 there is one σ bond and one π bond is formed while in case of N_2 one σ and two π bonds are formed. In case of poly-atomic molecules, in order to explain the directional nature of covalent bonds, Pauling and Slater suggested that the formation of covalent bond takes place in the direction where there is maximum overlap. It was considered that in case of H_2O and NH_3 type molecules there is overlapping of atomic orbitals of hydrogen atoms with p-orbitals of O-atom and N-atom. Therefore the bond angles should be 90° as p-orbitals are at 90° to each other. The greater bond angle in H_2O (~105°) or in NH_3 (~108°) is due to repulsion between the bond pairs and lone pairs of electrons. In order to explain the cases like BeH_2, BF_3 & CH_4 etc. concept of hybridization was introduced.

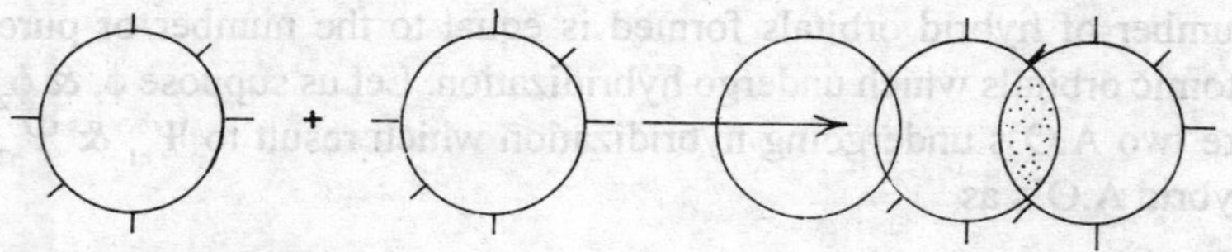

$\sigma_{s\text{-}s}$ bond formation

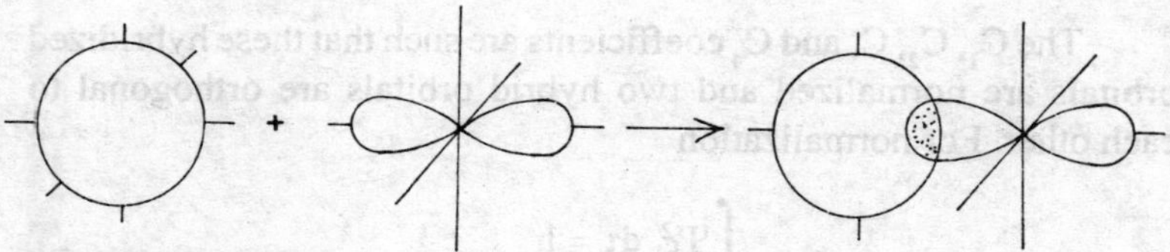

$\sigma_{s\text{-}px}$ bond formation

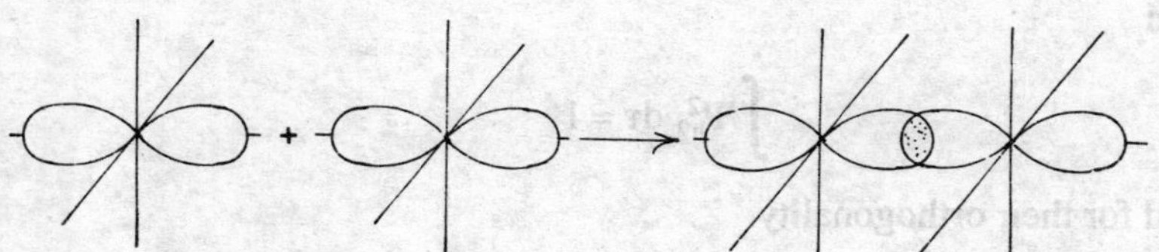

$\sigma_{pz\text{-}pz}$ bond formation

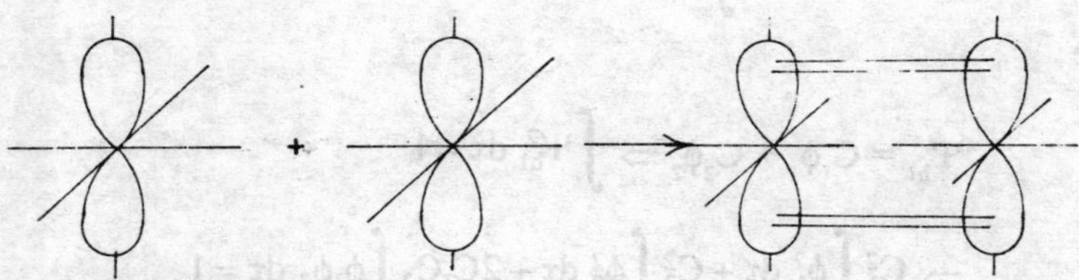

$\pi_{py\text{-}py}$ bond formation

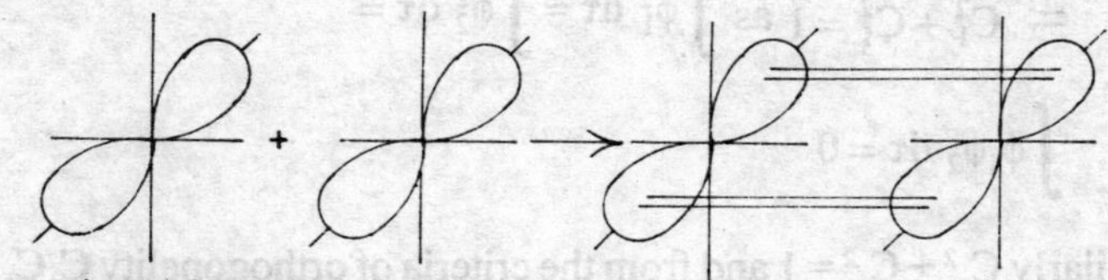

$\pi_{px\text{-}px}$ bond formation

Fig. 6.1

According to this concept, combination of atomic orbitals having same or similar energy inter mix to form hybrid orbitals which can

be effectively overlap to give a definite shape to the molecule. The number of hybrid orbitals formed is equal to the number of pure atomic orbitals which undergo hybridization. Let us suppose ϕ_1 & ϕ_2 are two A.O's undergoing hybridization which result to Ψ_{h1} & Ψ_{h2} hybrid A.O's as

$$\Psi_{h1} = C_1\phi_1 + C_2\phi_2$$
$$\Psi_{h2} = C_3\phi_1 + C_4\phi_2$$

The C_1, C_2, C_3 and C_4 coefficients are such that these hybridized orbitals are normalized and two hybrid orbitals are orthogonal to each other. For normalization

$$\int \Psi_{h1}^2 \, d\tau = 1$$

and

$$\int \Psi_{h2}^2 \, d\tau = 1$$

and for their orthogonality

$$\int \Psi_{h1} \Psi_{h2} \, d\tau = 0$$

since

$$\Psi_{h1} = C_1\phi_1 + C_2\phi_2 \Rightarrow \int \Psi_{h1}^2 \, d\tau = 1$$

$$\Rightarrow C_1^2 \int \phi_1^2 \, d\tau + C_2^2 \int \phi_2^2 \, d\tau + 2C_1C_2 \int \phi_1\phi_2 \, d\tau = 1$$

$$\Rightarrow C_1^2 + C_1^2 = 1 \text{ as } \int \phi_1^2 \, d\tau = \int \phi_2^2 \, d\tau =$$

and $\int \phi_1\phi_2 \, d\tau = 0$

Similarly $C_3^2 + C_4^2 = 1$ and from the criteria of orthogonality $C_1C_3 + C_2C_4 = 0$. So, this can be said that hybrid orbitals are orthogonal to each other. Hybridized orbitals give the formation of stable bonds as they are in the state of lower energy as compared to the energy of combining A.O's and this energy is known as hybridization energy. further more hybridization is not a real physical phenomenon but it is a concept to explain the shape and structure of molecules. In working out the pure atomic orbitals contributing to the formation of hybrid orbitals, concept of group theory can be applied. Construction of hybrid orbitals on the basis of group theory is discussed in section 6.2

Section 6.2

Hybrid Orbitals with σ-bonding

Following points must be taken into account in order to construct the hybrid orbitals with σ-bonding:

(i) Find out the point group for the molecule for which hybrid orbitals are to be constructed.

(ii) Using σ bonds as vectors for representation, find out the reducible representation which may be termed as T_{hybrid}.

(iii) Reduce this reducible representation to get irreducible representations belonging to that particular point group.

(iv) Match the irreducible representations to get the A.O's that form the basis for those irreducible representations (for this see the character table of that point group).

(v) Using these A.O's functions, that form basis for irreducible representations, construct the possible hybrid orbitals.

Let us try to apply these set of rules and construct the suitable examples.

Example 1: Let us take the example of regular AB_4 tetrahedral molecule *e.g.*, CCl_4 or CH_4. This can be represented as in Fig. 6.2. Let v_1, v_2, v_3 and v_4 be the vectors along σ C-H bonds. This molecule belongs to T_d point group. Characters of T_{hybrid} can be obtained by considering the number of vectors that remain un-shifted during any symmetry operation and the T_{hybrid} for this case is given below

T_d	E_3	$8C_2$	$3C_2$	$6s_4$	$6\sigma_d$
T_{hybrid}	4	1	0	0	2

Using character table of T_d point group and on reducing it,

$$T_{hybrid} = A_1 \oplus T_2$$

On matching the irreducible representations *i.e.,* A_1 and T_2 this can be written that A_1 corresponds to totally symmetric *i.e.,* s-orbital and T_2 corresponds to p_x, p_y and p_z single basis functions and d_{xy}, d_{yz}, d_{xz} the multiple basis functions. As there is 1:1 contribution of A_1 & T_{hyb}. So, one can write that

A_1	T_2
s	p_x, p_y, p_z
	d_{xz}, d_{yz}, d_{xy}

Therefore the symmetry arguments alone will take us to a general solution to the problem as set of hybrid orbitals which are linear combination of both as

$$\Psi_{hybrid} = a\,(sp^3) + b(sd^3)$$

but as the energies of s & d orbitals are not comparable, so, we can conclude that the contribution of d is negligible in the case of CH_4 or CCl_4, so, Ψ_{hybrid} can be written as

$\Psi_{hybrid} = sp^3$ *i.e.,* coefficient a = 1 and b is taken as zero.

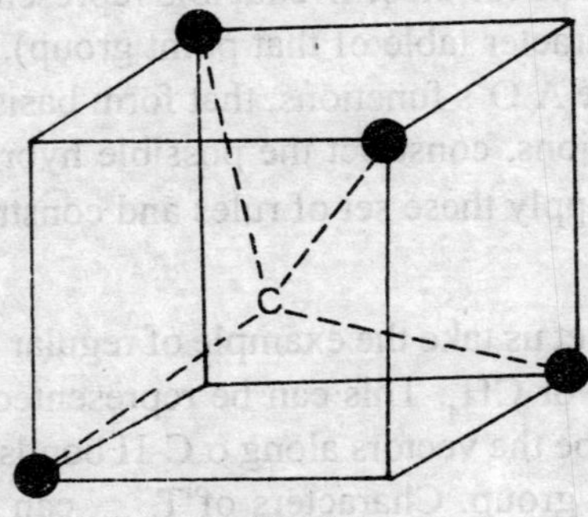

Fig. 6.2

Example 2 : Hybrid orbitals for Mn in MnO_4^- (permanganate ion can also be constructed in the same manner as explained above for carbon in case of CH_4 or CCl_4 (example 1) but in the case of MnO_4^- ion we can't neglect sd^3 hybrid orbitals part as in this case (*i.e.,* in Mn) d-orbitals are of nearly comparable energy. So,

$$\Psi_{hybrid} = a(sp^3) + b(sd^3)$$

in this case.

Example 3: Let us take another example of PCl_5. Point group of this molecule is D_{3h}. Considering all the five σ bonds (*i.e.,* P-Cl bonds) as base vectors, T^{hyb} can be written as

D_{3h}	E	$2C_3$	$3C_2$	σ_h	$2S_3$	$3\sigma_v$
T_{hybrid}	5	2	1	3	0	3

T_{hybrid} can be obtained considering the number of vectors that remain un-shifted during any symmetry operation. Using character table for D_{3h} point group and reducing this representation *i.e.,* T_{hybrid}, we get

$$T_{hybrid} = 2A_1' \oplus A_2'' \oplus E'$$

on matching the irreducible representations *i.e.,* A_1', A_2'' and E' following result can be obtained

D_{3h}	E	$2C_3$	$3C_2$	σ_h	$2S_3$	$3\sigma_v$	
A_1'	1	1	1	1	1	1	x^2+y^2, z^2
A_2''	1	1	−1	−1	−1	1	z
E'	2	−1	0	2	−1	0	x, y, x^2-y^2, xy

So, the corresponding orbital functions to these irreducible representations can written as

A_1'	A_2''	E'
s	p_z	p_x, p_y
(totally summetirc)		
d_{z^2}	$d_{x^2-y^2}$,	d_{xy}

Possible combinations for hybridized orbitals can be written by considering the contribution of A_1' two times as

1. ns (n + 1) s, p_z, p_x, p_y
2. ns (n + 1) s, p_z, $d_{x^2-y^2}$, d_{xy}
3. nd_z^2 (n+1) d_{z^2}, p_z, $d_{x^2-y^2}$, d_{xy}
4. nd_z^2 (n+1) d_{z^2}, p_z, p_x, p_y

5. ns, nd_{z^2}, p_z, p_x, p_y and

6. ns, nd_{z^2}, pz, $d_{x^2-y^2}$, d_{xy}

The first four combinations are irrelevant and not important for consideration as in these combinations we are considering s (or d_{z^2}) from one quantum number and same s (or d_{z^2})from the other quantum number which is not feasible on the basis of energy considerations. So, the combinations number 5 & 6 are significant that are

5. n_s,nd_{z^2}, p_z, p_x, p_y and

6. n_s, nd_{z^2}, p_z, $d_{x^2-y^2}$, d_{xy}

Therefore hybrid orbitals can be written as

$$\Psi_{hybrid} = a(sdp^3) + b(spd^3)$$

Again on the energy consideration d-orbitals are not so, important as p-orbitals in the case of phosphorous (P) in the construction of hybrid orbitals. So, the coefficient a = 1 and b = 0 and

$$\Psi_{hybrid} = sdp^3$$

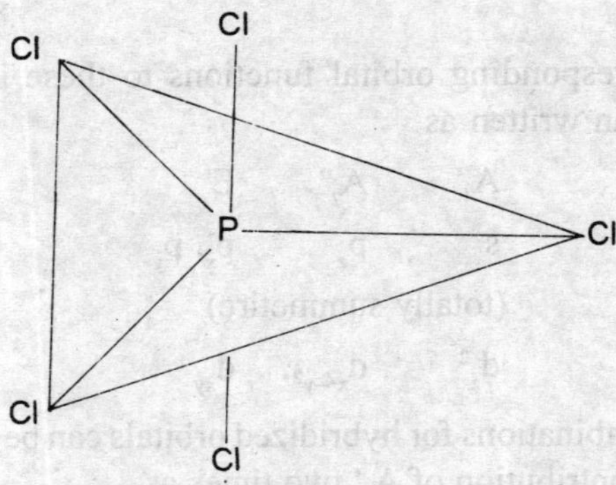

Fig. 6.3

Example 4: Hybrid orbitals for Mo in $MoCl_5$ can also be constructed in the same manner as explained for PCl_5 but in the case of Mo, we can't neglect importance of d-orbitals in the construction of hybrid orbitals. So,

$$\Psi_{hybrid} = a(sdp^3) + b(spd^3)$$

as energy for d-orbitals are comparable.

Example 5: Let us consider another example of regular AB_6 octahedral molecule (*e.g.*, SF_6 or any transition metal complex in regular octahedral geometry as $CoCl_6^{3-}$, CoF_6^{3-} etc.) Molecular point group for this is O_h. Considering all the six σ-bonds (*i.e.*, A-B bonds) as base vectors.

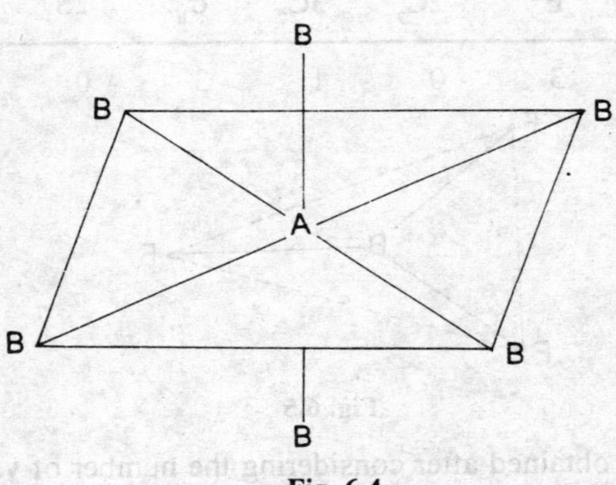

Fig. 6.4

T_{hybrid} can be written as

O_h	E	$8C_3$	$3C_2$	$6C_4$	$6C_2''$	'i'	$8S_6$	$3\sigma_h$	$6S_4$	$6\sigma_d$
T_{hybrid}	6	0	2	2	0	0	0	4	0	2

In the same way as it was discussed in the previous example, T_{hybrid} is obtained after considering the number of vectors that remain un-shifted during any symmetry operation. Using character table of O_h point group and on reducing the T_{hybrid}, we get

$$T_{hybrid} = A_{1g} \oplus E_g \oplus T_{1u}$$

The corresponding atomic are (on matching from the character table for corresponding irreducible representations)

A_{1g}	E_g	T_{1u}
s	d_{z^2}	p_x, p_y, p_z
(totally symmetric)	$d_{x^2-y^2}$	

Therefore, the hybrid orbital is written as

$$\Psi_{hybrid} = sp^3d^2 \text{ or}$$

$$\Psi_{hybrid} = d^2sp^3$$

Depending upon the fact that whether inner orbital (or outer orbital) complex is formed by the central metal ion.

Example 6: Let us consider another example *i.e.,* BF_3. Point group of this molecule is D_{3h}. Considering each σ B-F bond as a base vector the T_{hybrid} can be written as

D_{3h}	E	$2C_3$	$3C_2$	σ_h	$2S_3$	$3\sigma_v$
T_{hybrid}	3	0	1	3	0	1

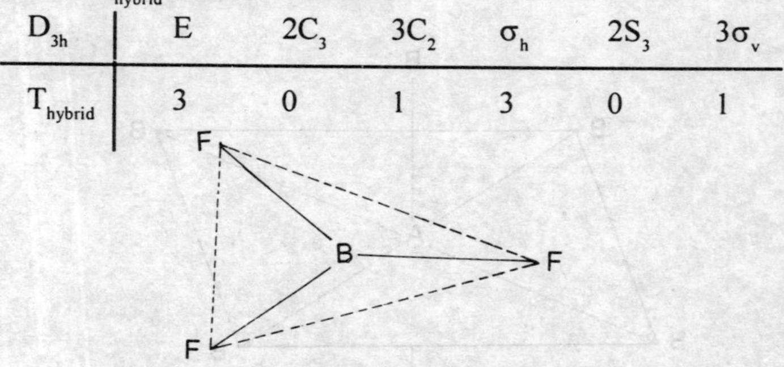

Fig. 6.5

T_{hybrid} is obtained after considering the number of vectors that remain un-shifted during any symmetry operation. On reducing the T_{hyb}, using the character table for D_{3h} point group, we get

$$T_{hybrid} = A_1' \oplus E'$$

on matching these irreducible representations following basis functions can be obtained.

D_{3h}	E	$2C_3$	$3C_2$	σ_n	$2S_3$	$3\sigma_v$	
A_1	1	1	1	1	1		x^2+y^2, z^2
E'	2	−1	0	2	−1	0	x, y, x^2-y^2, xy

So, the corresponding orbitals are

	A_1	E'
	s	p_x, p_y
	(totally symmetric)	$d_{x^2-y^2}$, d_{xy}

Therefore the Ψ_{hybrid} may be written as

$$\Psi_{hybrid} = a(sp^2) + b(sd^2)$$

In case of BF_3 (*i.e.*, B-atom) d-orbitals are not at all significant on the basis of energy consideration, so,

$$\Psi_{hybrid} = a(sp^2) \textit{ i.e.}, a = 1 \text{ and } b = 0$$

or

$$\Psi_{hybrid} = sp^2$$

Section 6.3

Hybrid Orbitals with π-bonding

Group theory can also be applied to explain the construction of hybrid π-orbitals.In this section this is explained considering suitable examples.

Example 7: Let us consider AB_3 triangular planar molecule *e.g.*, BF_3 or BCl_3. The process of construction of hybrid orbitals with Π-bonding is same as it is explained in the previous section. The hybrid orbitals considering σ-bonds only for this molecule is $\Psi_{hybrid} = a(sp^2) + b(sd^2)$ or more precisely $\Psi_{hybrid} = sp^2$ taking the energy consideration of orbitals into account.

Now let us consider π-orbitals. The π-bonds can be formed by the interaction of central B-atom orbitals with the X(Cl or F) atom. π-orbitals which are perpendicular to the plane of the molecule or which are in the plane of the molecule are shown in the fig. 6.6. Let us assign the perpendicular orbitals as π_{perp} and planar orbitals as π_{plane}. Considering these π-orbitals as base vectors T_{hybrid} can be written as

D_{3h}	E	$2C_3$	$3C_2$	σ_h	$2S_3$	$3\sigma_v$
$T_{hyb}^{\pi\,perp}$	3	0	–1	–3	0	1
$T_{hyb}^{\pi\,plane}$	3	0	–1	3	0	–1

T^{hyb} is obtained considering that if vector does not change its position the character is 1 and if its shifts from the original position the character is zero and if the vector shifts and changes in the opposite direction the character assigned as –1.

These reducible representations can be reduced using the

character table of D_{3h} point group as

$$T_{hyb}^{\pi(perp)} = A_2'' \oplus E'' \text{ and}$$

$$T_{hyb}^{\pi(plane)} = A_2' \oplus E'$$

From the character table, the corresponding orbitals can be find out for these representation as

A_2''	E'
p_z	d_{xz}, d_{yz}

and

A_2'	E'
None	p_x, p_y
	$d_{x^2-y^2}$, d_{xy}

Therefore, the Π-orbitals in the perpendicular direction can be formed by three orbitals that are p_z, d_{xz} and d_{yz} of the central atom. As there is no atomic orbitals corresponding to A_2' irreducible representation, the π hybrid orbitals in the planar direction can be considered to form of two orbitals that are p_x, p_y or $d_{x^2-y^2}$ and d_{xy}. As p_x, p_y have already been used in the making of σ hybrid orbitals (example 6, section 6.2), therefore only $d_{x^2-y^2}$ and d_{xy} can be considered to form π-hybrid orbitals in planar direction. In conclusion we can say that its pπ-dπ bonding in this case.

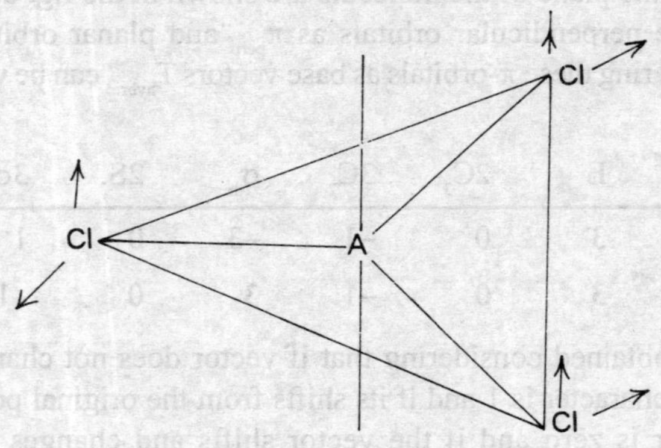

Fig. 6.6

Example 8: Now let us consider a case of an AB_6 octahedral molecule (*e.g.*, transition metal complexes in regular geometry). If we consider two mutually perpendicular π base vectors on each B-atom as shown in fig. 6.7, we get the following T_{hybrid}

O_h	E	$8C_3$	$3C_2$	$6C_4$	$6C_2''$	'i'	$8S_6$	$3\sigma_h$	$6S_4$	$6\sigma_d$
T_{hybrid}	12	0	–4	0	0	0	0	0	0	0

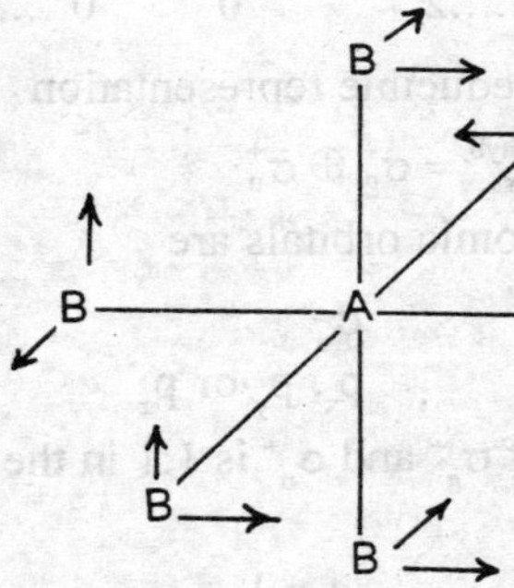

Fig. 6.7

On reduction of this reducible representation we get the following irreducible representations

$$T_{hybrid} = T_{1g} \oplus T_{2g} \oplus T_{1u} \oplus T_{2u}$$

From the corresponding character table of the O_h point group, on matching these irreducible representations, we get the following orbitals as bases functions

T_{1g}	T_{2g}	T_{1u}	T_{2u}
None	d_{xy}, d_{xz}, d_{yz}	p_x, p_y, p_z	none

As it has been explained earlier that all the three orbitals *i.e.*, p_x, p_y, p_z are used up in the construction of σ orbital (example 5 section 6.2). Therefore, only three d-orbitals corresponding to T_{2g} irreducible representation are involved in π-bonding. So, there can be at most three π-bonding. So, there can be at most three π-bonds equally amongst the six A-B bond pairs.

Problems

Q. 1. Work out hybridization for linear AB_2 (e.g., $BeCl_2$) molecule.

Ans. For AB_6 ($BeCl_2$) molecule, molecular point group is $D_{\infty h}$. T^{hyb}_{σ} is given by

	E	$2C_\infty$.....$\infty\sigma_v$	'i'	$2S_\infty$.....C_2
T^{σ}_{hyb}	2	2.........2	0	0.........0

On reduction of this reducible representation

$$T^{hyb}_{\sigma} = \sigma_g^+ \oplus \sigma_u^+$$

The corresponding atomic orbitals are

σ_g^+	σ_u^+
s (totally symmetric)	p_x, p_y or p_z

As the contribution of σ_g^+ and σ_u^+ is 1:1 in the construction of hybrid orbital, so,

Ψ_{hybrid} = a (sp) a = 1

or

Ψ_{hybrid} = sp

Q. 2. Work out the hybridization for σ bonds in case of AB_4 square planar complex.

Ans. Considering all AB bonds as base vectors T_{hyb} can be written as

D_{4h}	E	$2C_4$	C_2	$2C_2'$	$2C_2''$	'i'	$2S_4$	σ_h	$2\sigma_v$	$2\sigma_d$
T_{hyb}	4	0	0	2	0	0	0	4	2	0

T_{hyb} can be reduced to

$T_{hyb} = A_{1g} \oplus B_{1g} \oplus E_u$ and the corresponding orbitals are

A_{1g}	B_{1g}	E_u
s, d_{z^2}	$d_{x^2-y^2}$	p_x, p_y

So, the possible combination for hybrid orbitals are

$$\Psi_{hybrid} = a(dsp^2) + b(d^2p^2)$$

This can be shown that b = 0 and a = 1, so,

$$\Psi_{hybrid} = dsp^2$$

So, the most common hybridization in case of square planar complex AB_4 is dsp^2.

Q. 3. Obtain T_π (reducible) for square planar molecule and reduce this representation.

Ans. Consider four π vectors in perpendicular and four in parallel direction as shown in the fig.

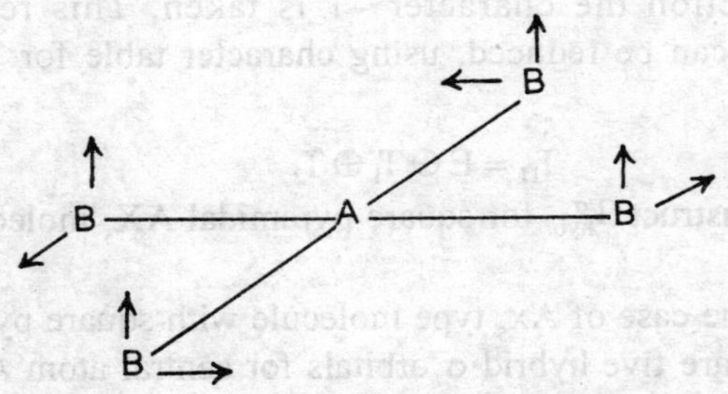

Fig.

They have to be considered considered as a separate sets because the vectros in one direction can be converted to one another only by rotation through 90° along x and y axes and this rotation is not an operation in D_{4h} point group. $T_{\pi perp.}$ and $T_{\pi parallel}$ can be obtained as

D_{4h}	E	$2C_4$	C_2'	$2C_2'$	$2C_2''$	'i'	$2S_4$	σ_h	$2\sigma_v$	$2\sigma_d$
$T_{\pi perp.}$	4	0	0	–2	0	0	0	–4	2	0
$T_{\pi parallel}$	4	0	0	–2	0	0	0	4	–2	0

T_Π (reducible) representations are obtained on the basis of the following fact that if vector does not change in its position then character is 1, if it shifts for the original position then character assigned is zero and if the vectro shifts and changes in opposite direction the character –1 is assigned.

Q. 4. Find out T_π in case of a tetrahedral molecule.

Ans. In case of a tetrahedral molecule (molecular point group

T_d) the π bonds can be formed in the direction perpendicular to the σ bond. These π bonds can be shown as vectors on four B atoms. All these eight vectors are equivalent as they can be interchanged by the σ_v symmetry operation of the T_d point group in a tetrahedral molecule, so T_π can be written as

T_d	E	$8C_3$	$3C_2$	$6S_4$	$6\sigma_d$
T_π	8	–1	0	0	0

This reducible representation can be obtained on the basis of same fact that if vector does not change its position then the character is taken as 1 otherwise zero and if the vector changes its position in opposite direction the character –1 is taken. This reducible representation can be reduced, using character table for T_d point group, as

$$T_\Pi = E \oplus T_1 \oplus T_2$$

Q. 5. Construct T^{σ}_{hyb} for square pyramidal AX_5 molecule and reduce it.

Ans. In the case of Ax_5 type molecule with square pyramidal structure there are five hybrid σ orbitals for central atom A which are directed to the corners of a square pyramid. Molecular point group for this type of molecules is C_{4v}. Considering the five σ bonds as base vectors T^{σ}_{hyb} (reducible) representation can be obtained on the basis of a fact that un-shifted vector can be assigned character 1 and the shifted vector as zero. So, T_{hyb} is

C_{4v}	E	$2C_4$	C_2	$2\sigma_v$	$2\sigma_d$
T_{hyb}	5	1	1	3	1

On reducing T_{hyb} we get

$$T_{hyb} = 2A_1 \oplus B_1 \oplus E.$$

7

Group Theory and Concepts of Bonding–II

The concepts of symmetry and group theory can also be applied for determination of electronic wave function. For this purpose certain approximations are helpful which are:

(i) The electronic wave function are made-up of molecular orbitals.

(ii) The molecular orbitals are formed by the linear combination of atomic orbitals.

(iii) Separate out the part of wave function concerned with σ bond formation ane while dealing with π bond formation deal only with π electronic part.

(iv) Solution of appropriate equations by making assumption about certain integral over the π electronic M.O.'s. This approximation brings us to the Huckel Molecular Orbital (HMO) concept.

Actually, though valance bond theory very well explains the structure of molecules but it has some shortcomings also. It considers only the valance electrons or the outermost orbitals of the atoms, which according to this theory take part in the formation of bonds. This theory was not able to explain paramagnetic character of O_2 molecule. This theory was not able to explain the electronic spectra of the molecules and interpret the magnetic properties of the molecules qualitatively.

Therefore, an alternate theory was proposed by Hund and Mulikan which is known as molecular orbtial theory. According to this theory energy is also qantized in molecules which represent the region along which there is maximum probability of electrons. Analogous to atomic orbtials (A.O.'s) these are termed as molecular orbitals (M.O's).

Section 7.1

Introduction To Molecular Orbital Theory

Hund and Mulikan introduced the formation of molecular orbtials i.e., the molecular orbital theory which is based upon the following points.

(i) Electrons and atoms loses their identities during the formation of molecular orbitals (or molecules).

(ii) Electron remains under the influence of more than one nucleus i.e., a molecular orbital is a polycentric orbital.

(iii) Number of combining (A.O's) atomic orbitals is equal to the number of (M.O's) molecular orbitals formed. If the M.O. is of higher energy than the combining A.O.'s anti bonding M.O is formed but if M.O. has lower energy than the combining A.O.'s the bonding M.O. is formed.

For any molecular electronic problem, the Schrodinger equation can be written as $\hat{H}\ \psi\ (1,2,3,.....,n) = E\ \psi\ (1,2,3.....,n)$.

Where $\hat{H}$ is the Hamiltonian and ψ (1,2,3....,n) is the electronic wave function and explicity is a function of the co-ordinates of all n-electrons and E is the total energy. In Hartree Fock (HF) method or self consistent (SCF) method ψ (1,2,3....,n) functions each of which contains the co-ordinates of just one electron are called molecular orbtial. This is an approximation as in reality position of one electron is always correlated with the position of others, so, the function that describes the one electron can't be independent of the functions describing the other electrons. Therefore some error appears in the HF approximation which is known as correlation energy. So, to express the M.O.'s two fundamental laws must be taken into account.

(i) An electronic wave function must not be distinguish, by treating in a different manner, one electron from another. This law is known as law of indistinguishability.

(ii) The electronics wave function must change its sign when two electrons are interchanged. This is antisymmetry law. so, the Slater determinant can be witten as

$$\psi(1,2,3,....,n)=\frac{1}{\sqrt{n!}}\begin{vmatrix}\phi_1(1) & \phi_2(1)......\phi_n(1)\\ \phi_1(2) & \phi_2(2).....\phi_n(2)\\ . & . \quad\quad .\\ . & . \quad\quad .\\ \phi_1(n) & \phi_2(n).....\phi_n(n)\end{vmatrix}$$

This determinant symbolizes Φ(j) as i^th^ M.O. as well defined function of the co-ordinates of electrons 'j'. The original Hamiltonian $\hat{H}$ commutes with the transformation operators O_R for all operations R of the point group to which the molecule belongs, the M.O.'s associated with a given orbtial energy will form a function space whose basis generates a definite irreducible representations of the point group. This is exactly similar to the situation for the exact total electronic wave functions. The exact M.O.'s may be classified according to the irreducible representation to which they belong and usually the symbol for irreducible representation is used to label the M.O. with which it is associated.

It is common practice to replace HF equation by particular mathematical functions of a parameters which can then be optimized in accordance with HF equation. Since these M.O.'s are not completely flexible, so another approximation is introduced which expresses the space part of the M.O's as various linear combinations of atomic orbitals centred as on the same or different nuclei of the molecule. These Φ (j) are called as linear combination of atomic orbitals (L.C.A.O.) and if they are introduced into the Hartree Fock equations, a simpler set of equations is obtained which can be solved and can be used to determined the optimum coefficients.

According to the LCAO–MO concept a wave function corresponding to any M.O. is explained as

$$\Psi_{MO}=C_A\Psi_A+C_B\Psi_B$$

Where Ψ_A and Ψ_B are the wave functions of A and B respectively and C_A and C_B are their corresponding coefficients.

As

$$\hat{H}\Psi_{MO} = E\Psi_{MO}$$
$$= <\Psi_{MO}|\hat{H}|\Psi_{MO}> = \Psi_{MO}|E|\Psi_{MO}$$
$$= <\Psi_{MO}|\hat{H}|\Psi_{MO}> = E<\Psi_{MO}|\Psi_{MO}>$$

or

$$E = \frac{<C_A\psi_A + C_B\psi_B|\hat{H}|C_A\psi_A + C_B\psi_B>}{<C_A\psi_A + C_B\psi_B|C_A\psi_A + C_B\psi_B>}$$

$$= \frac{C_A^2 <\psi_A|\hat{H}|\psi_A> + C_B^2 <\psi_B <\psi_B|\hat{H}|\psi_B> + C_A C_B <\psi_A|\hat{H}|\psi_B> + C_A C_B <\psi_B|\hat{H}|\psi_A>}{C_A^2 <\psi_A^2> + C_B^2 <\psi_B^2> + 2C_A C_B <\psi_A|\psi_B>}$$

Defining $<\Psi_A|\hat{H}|\Psi_A>$ and $<\Psi_B|\hat{H}|\Psi_B>$ as coloumb integrals i.e., α and $<\Psi_A|\hat{H}|\Psi_B>$ and $<\Psi_B|\hat{H}|\Psi_A>$ as exchange integrals i.e., β and S_{AB} as overlap integral i.e., $<\Psi_A|\Psi_B>$ and since Ψ_A and Ψ_B are normalized, so, $<\Psi_A{}^2> = <\Psi_B{}^2> = 1$.

$$\Rightarrow E = \frac{C_A^2\alpha_A + C_B^2\alpha_B + C_A C_B\beta_{AB} + C_A C_B\beta_{BA}}{C_A^2 + C_B^2 + 2C_A C_B S_{AB}}$$

$$\Rightarrow (C_A{}^2 + C_B{}^2 + 2C_A C_B S_{AB})\,E = C_A{}^2\alpha_A + C_B{}^2\alpha_B + C_A C_B\beta_{AB} + C_A C_B\,\beta_{AB}$$

for minima $\partial E / \partial C_A = 0$

$$\Rightarrow 2C_A\,(\alpha_A - E) + 2C_B(\beta_{AB} - ES_{AB}) = 0$$

similarly for $\partial E / \partial C_B = 0$

$$\Rightarrow 2C_B(\alpha_B - E) + 2C_A(\beta_{AB} - ES_{AB}) = 0$$

These two equations can be expressed as a determinant

$$\begin{vmatrix} \alpha_A - E & \beta_{AB} - ES_{AB} \\ \beta_{AB} - ES_{AB} & \alpha_B - E \end{vmatrix} \begin{vmatrix} C_A \\ C_B \end{vmatrix} = 0$$

Physically C_A and C_B are not equal to zero as if C_A or $C_B = 0$ the probability of finding electrons diminishes

$$= \begin{vmatrix} \alpha_A - E & \beta_{AB} - ES_{AB} \\ \beta_{AB} - ES_{AB} & \alpha_B - E \end{vmatrix} = 0$$

$= (\alpha_A - E)\,(\alpha_B - E) - (\beta_{AB} - ES_{AB})^2 = 0$

In the case of H_2 molecule $\alpha_A = \alpha_B = \alpha$ i.e., α's are same

$= (\alpha - E)(\alpha - E) - (\beta_{AB} - ES_{AB})^2 = 0$

$= \{(\alpha - E) - (\beta_{AB} - ES_{AB})][(\alpha - E) + (\beta_{AB} - ES_{AB})] = 0$

This implies that either

$[(\alpha - E) - (\beta_{AB} - ES_{AB})] = 0$

$= E_2 = (\alpha - \beta_{AB}/1 - S_{AB})$ formation of anti bonding M.O. with energy higher than the combining atomic orbitals, takes place and on considering

$$[(\alpha - E) + (\beta_{AB} - ES_{AB})] = 0$$

$= E_1 = (\alpha + \beta_{AB}/1 + S_{AB})$ i.e. formation of bonding M.O. with lower energy than the combining atomic orbitals, takes place. This can be expressed as

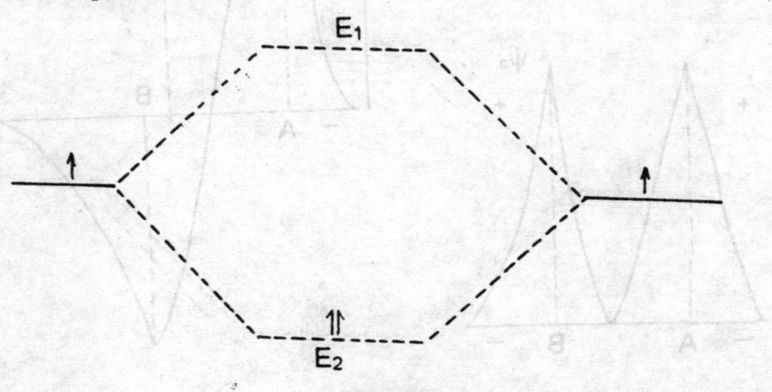

Fig. 7.1

The corresponding wave functions are given by

$$\Psi_{\text{bonding MO}} = \Psi_A + \Psi_B \text{ and}$$

$$\Psi_{\text{antibonding MO}} = \Psi_A - \Psi_B$$

On squaring these $\Psi_{\text{M.O's}}$

$$\Psi^2_{\text{bonding MO}} = \Psi^2_A + \Psi^2_B + 2\Psi_A\Psi_B$$

And

$$\Psi^2_{\text{antibonding MO}} = \Psi^2_A + \Psi^2_B - 2\Psi_A\Psi_B$$

Which clearly shows that in $\psi^2_{\text{bonding MO}}$ electron density is increased by the factor $2\Psi_A\Psi_B$ and in case of $\psi^2_{\text{antibonding MO}}$ it is decreased by $2\Psi_A\Psi_B$. This can be further clear from the Fig. 7.2 which contains plottings for $\Psi_{\text{bonding MO}}$ and $\Psi_{\text{antibonding MO}}$ and their sqaure functions.

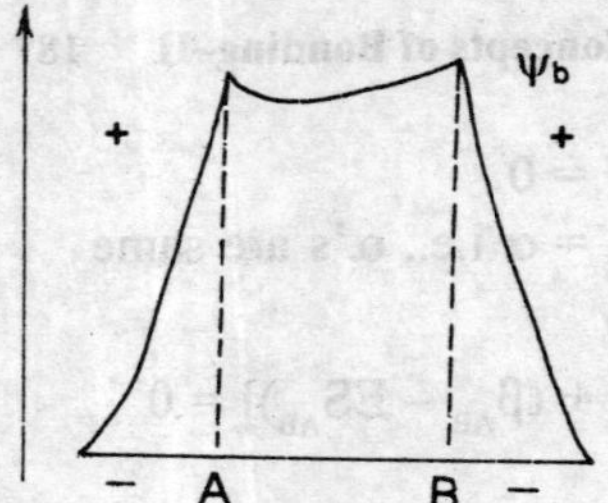

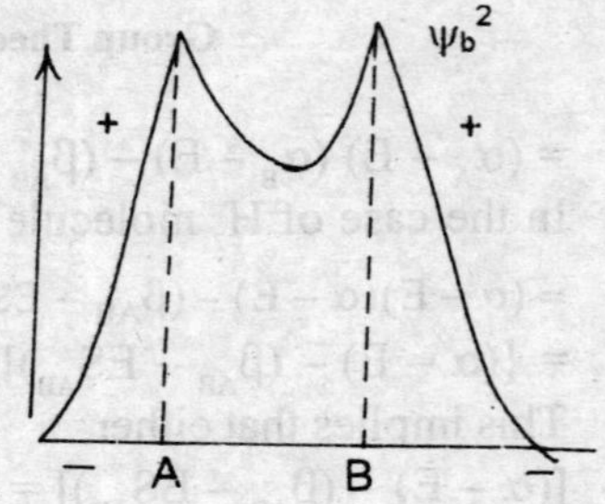

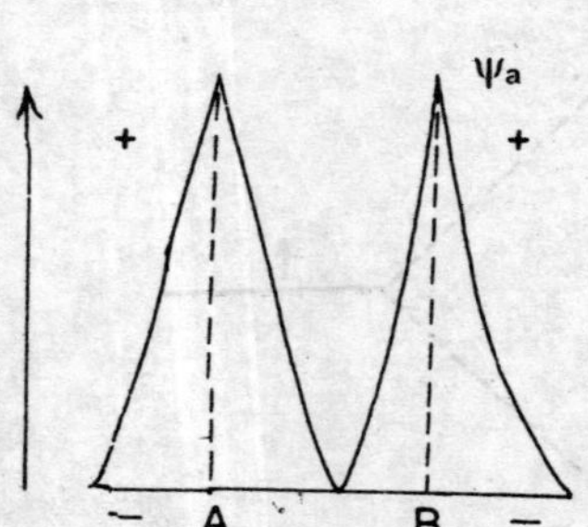

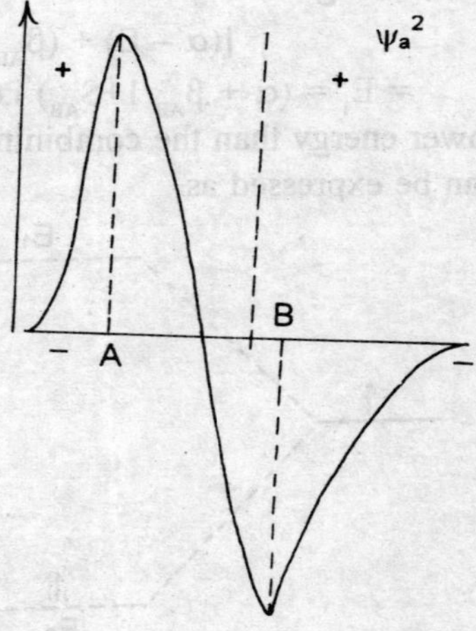

Fig. 7.2

Various M.O's formed by the combination of different A.O.'are given in the fig. 7.3 with the corresponding symmetry of the M.O.'s given in the table 7.1.

Table 7.1

M.O.	*Symmetry*
σ_{ss}	a_{1g}
σ_{pp}, $\pi_{px\,px}$, or π_{pypy}	t_{1u}
M.O. with $d_{x^2-y^2}$, $d_z{}^2$	e_g
d_{xz}, d_{yz}, d_{xy}	t_{2g}

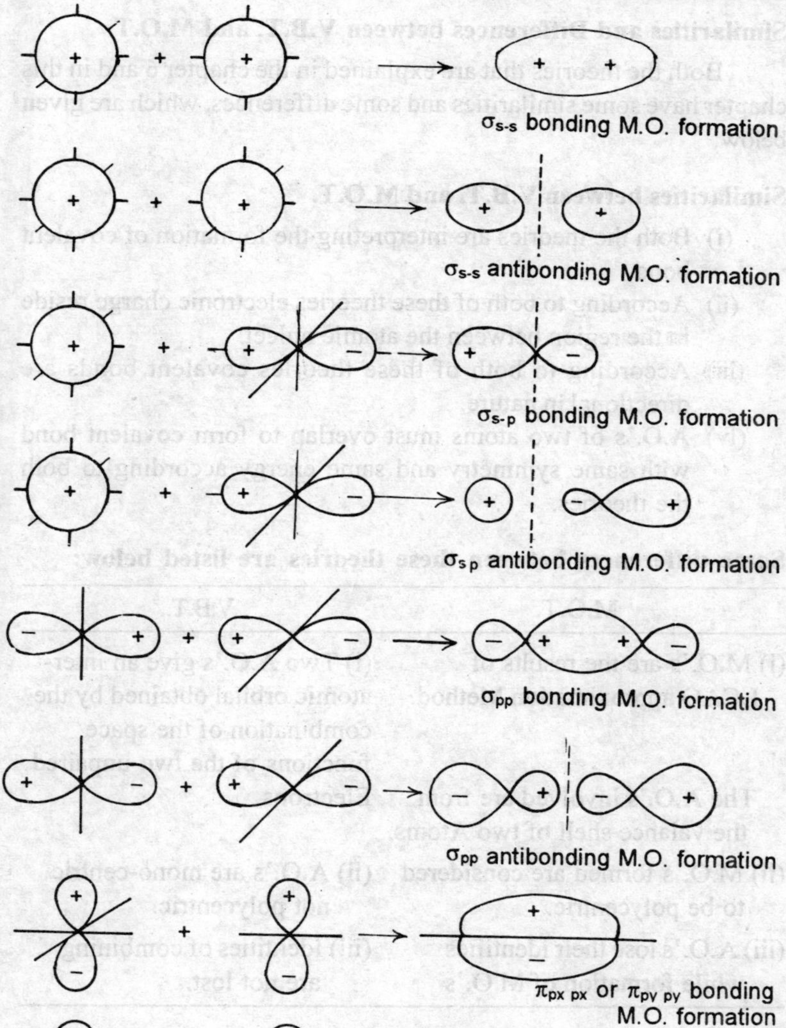

$\sigma_{s\text{-}s}$ bonding M.O. formation

$\sigma_{s\text{-}s}$ antibonding M.O. formation

$\sigma_{s\text{-}p}$ bonding M.O. formation

σ_{sp} antibonding M.O. formation

σ_{pp} bonding M.O. formation

σ_{pp} antibonding M.O. formation

$\pi_{px\ px}$ or $\pi_{py\ py}$ bonding M.O. formation

$\pi_{px\ px}$ or $\pi_{py\ py}$ antibonding M.O. formation

Fig. 7.3

Section 7.2

Similarities and Differences between V.B.T. and M.O.T

Both the theories that are explained in the chapter 6 and in this chapter have some similarities and some differences, which are given below:

Similarities between V.B.T. and M.O.T.

(i) Both the theories are interpreting the formation of covalent bond.

(ii) According to both of these theories electronic charge reside in the region between the atomic nulcei.

(iii) According to both of these theories covalent bonds are directional in nature.

(iv) A.O.'s of two atoms must overlap to form covalent bond with same symmetry and same energy according to both the theories.

Some differences between these theories are listed below:

M.O.T.	V.B.T.
(i) M.O.'s are the results of LCAO approximation Method. The A.O.'s involved are from the valance shell of two Atoms.	(i) Two A.O.'s give an inter-atomic orbital obtained by the combination of the space functions of the two unpaired Electrons.
(ii) M.O.'s formed are considered to be polycentric.	(ii) A.O.'s are mono-centric not polycentric.
(iii) A.O.'s lose their identifies while formation of M.O.'s	(iii) Identities of combining are not lost.

Section 7.3

Molecular orbital energy level diagrams

When a central metal atom A is surrounded by n other atoms or ligands, the construction of a qualitative molecular orbital energy level diagram follows a similar pattern, if only σ-bonding is significant.

Whenever there is a match in symmetry between a central metal atom's atomic orbital and a possible linear combination of other atoms or ligands orbitals, interaction can occur with the formation of bonding and anitbonding molecular orbtials as expalined above. Orbitals of either central atom A or ligand B or other ligands, for which such a match does not occur must necessarily remain non-bonding. Using projection operators, it is possible to construct the explicit forms of the symmetry-adapted linear combinations of the ligand's orbitals. In order to construct an accurate molecular orbital energy level diagram it is necessary to know the exact energies of all the participating atomic orbtials and the extent of mixing of atomic orbitals on the central metal atom. Let us try to explain the simple cases of AX_n system where A is central atom which has s– and p– orbitals available for bonding.

AB_2 Molecule

There are two possible geometries for AB_2 molecules: bent (C_{2v} symmetry) *e.g.*, H_2O etc. or linear ($D_\infty h$ symmetry) *e.g.*, BeH_2 etc. Taking the example of bent molecule *i.e.*, H_2O, first; from the character table it is clear that 2s and $2p_z$ (using the co-ordinate axes shown in Fig. 7.4) atomic orbtials of oxygen are each bases for the A_1 irreducible representation; the $2p_x$ for B_1 and $2p_y$ for B_2 irreducible representations respectively. To determine the symmetries of the possible combinations of hydrogen 1s orbitals, we use ϕ_1 & ϕ_2 to generate a reducible representation for σ bonding.

C_{2v}	E	C_2	σ_{xz}	σ_{yz}
T_σ	2	0	0	2

On reduction

$$T_\sigma = A_1 \oplus B_2$$

So, it is possible to construct one linear combination of A_1 and on e of B_2 symmetry. Using ϕ_1 as a generating function, the projection operator method gives the forms of the functions as

C_{2v}	E	C_2	σ_{xz}	σ_{yz}
ϕ_1	ϕ_1	ϕ_2	ϕ_2	ϕ_1
A_1	1	1	1	1
B_2	1	–1	–1	1

This shows that the A_1 combination, when normalized is $\Psi_{A1} = \frac{1}{\sqrt{2}}(\phi_1 + \phi_2)$ and B_2 combination is $\Psi_{B2} = \frac{1}{\sqrt{2}}(\phi_1 - \phi_2)$ (Fig. 7.4).

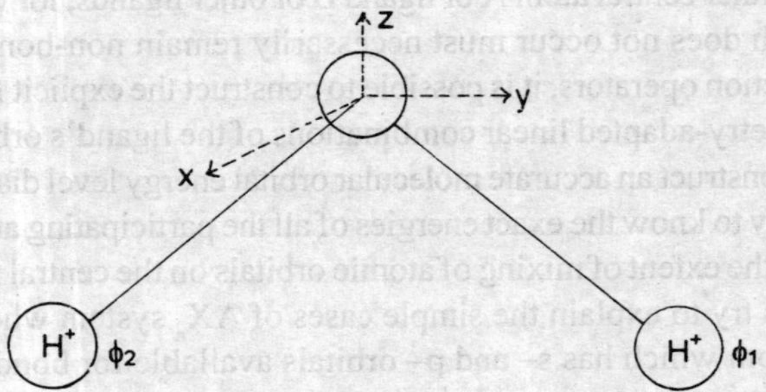

Fig. 7.4

Let us assume that the oxygen 2s and 2p orbitals are independent and the 2s orbitals are too tightly bound (low in energy) to take part in molecular orbital formation. We than see that Ψ_{A1} and $2p_z$ form bonding and antibonding molecular orbitals as do Ψ_{B2} and $2p_y$ leaving $2p_z$ non-bonding (B_1 symmetry) as shown in Fig. 7.5

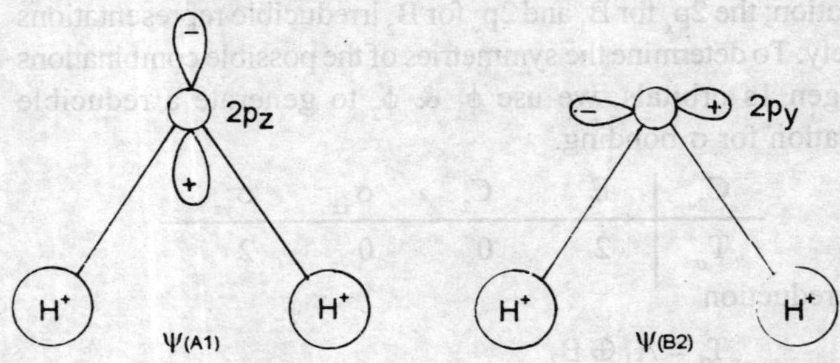

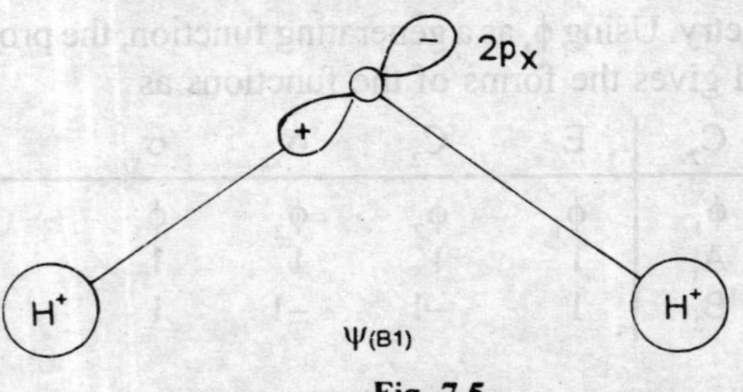

Fig. 7.5

When we have an approximate molecular orbital energy level diagram as shown in Fig. 7.6, the available electrons fill at the bonding and non-bonding molecular orbitals to give a stable electronic configuration.

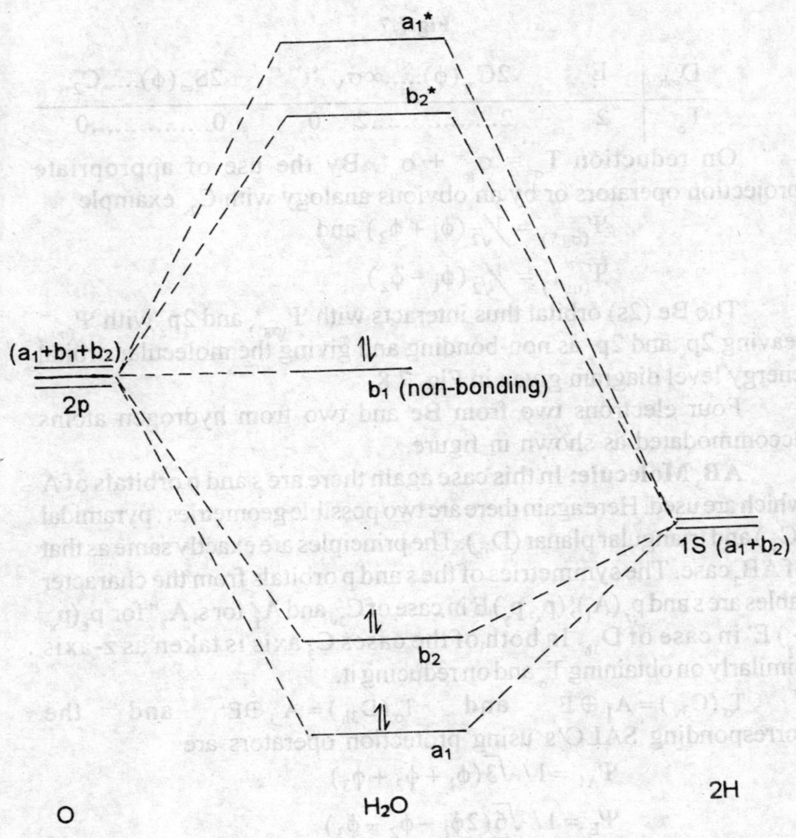

Fig. 7.6

In case of linear AB_2 molecule, the process is analogous, using the character table for $D_{\infty h}$ point group. This shows that the central atom's atomic orbitals have the following symmetries: σ_g^+ and σ_u^+ where the z axis is the molecular axis (Fig. 7.7). Using the hydrogen's

1s atomic orbitals ϕ_1 and ϕ_2 bases for reducible representation we have

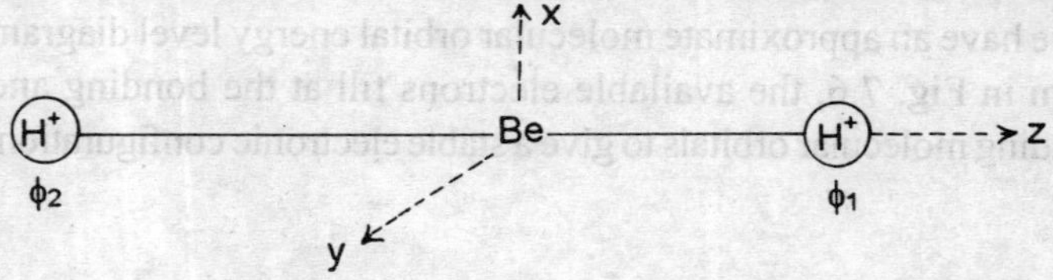

Fig.7.7

$D_{\infty h}$	E	$2C_\infty(\phi)$.....$\infty\sigma_V$	'i'	$2S_\infty(\phi)$.....C_2
T_σ	2	2..............2	0	0...............0

On reduction $T_\sigma = \sigma_g^+ + \sigma_u^+$. By the use of appropriate projection operators or by an obvious analogy with C_{2v} example

$$\Psi_{(\sigma g^+)} = 1/\sqrt{2}(\phi_1 + \phi_2) \text{ and}$$

$$\Psi_{(\sigma u^+)} = 1/\sqrt{2}(\phi_1 - \phi_2)$$

The Be (2s) orbital thus interacts with $\Psi_{(\sigma g^+)}$ and $2p_z$ with $\Psi_{(\sigma u^+)}$ leaving $2p_x$ and $2p_y$ as non-bonding and giving the molecular orbital energy level diagram given in Fig. 7.8.

Four electrons two from Be and two from hydrogen atcms accommodated as shown in figure.

AB_3 Molecule: In this case again there are s and p orbitals of A which are used. Here again there are two possible geometries : pyramidal (C_{3v}) and triangular planar (D_{3h}). The principles are exactly same as that of AB_2 case. The symmetries of the s and p orbitals from the character tables are s and p_z (A_1); (p_x, p_y) E in case of C_{3v} and A_1' for s, A_2'' for p_z (p_x, p_y) E' in case of D_{3h}. In both of the cases C_3 axis is taken as z-axis Similarly on obtaining T_σ and on reducing it.

$T_\sigma(C_{3v}) = A_1 \oplus E$ and $T_\sigma(D_{3h}) = A'_1 \oplus E'$ and the corresponding SALC's using protection operators are

$$\Psi_{A1} = 1/\sqrt{3}(\phi_1 + \phi_2 + \phi_3)$$

$$\Psi_E = 1/\sqrt{6}(2\phi_1 - \phi_2 - \phi_3)$$

& $\Psi_E = 1/\sqrt{2}(\phi_2 - \phi_3)$ for C_{3v} point group and

$$\Psi_{A1'} = 1/\sqrt{3}(\phi_1 + \phi_2 + \phi_3)$$

$$\Psi_{E'} = 1/\sqrt{6}(2\phi_1 - \phi_2 - \phi_3)$$

$$\Psi_{E'} = \frac{1}{\sqrt{2}}(\phi_2 - \phi_3) \text{ for } D_{3h} \text{ po int group}$$

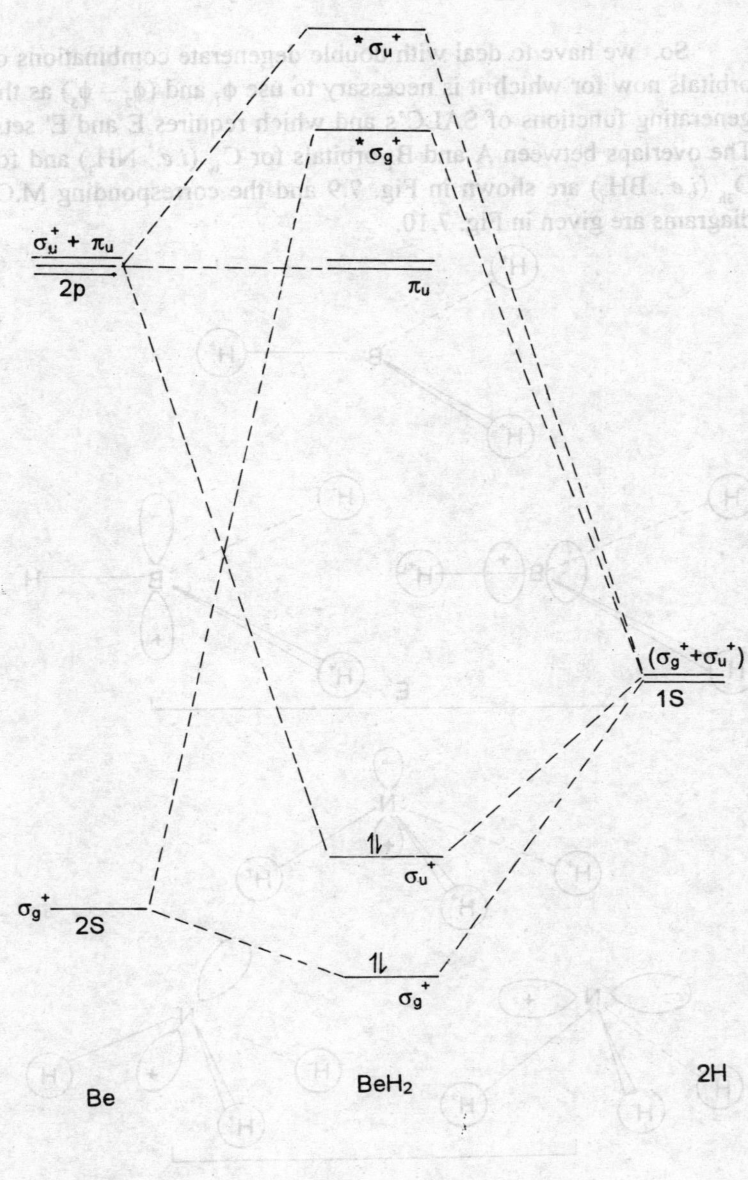

Fig. 7.8

So, we have to deal with double degenerate combinations of orbitals now for which it is necessary to use ϕ_1 and $(\phi_2 - \phi_3)$ as the generating functions of SALC's and which requires E and E' sets. The overlaps between A and B_3 orbitals for C_{3v} (*i.e.*, NH_3) and for D_{3h} (*i.e.*, BH_3) are shown in Fig. 7.9 and the corresponding M.O. diagrams are given in Fig. 7.10.

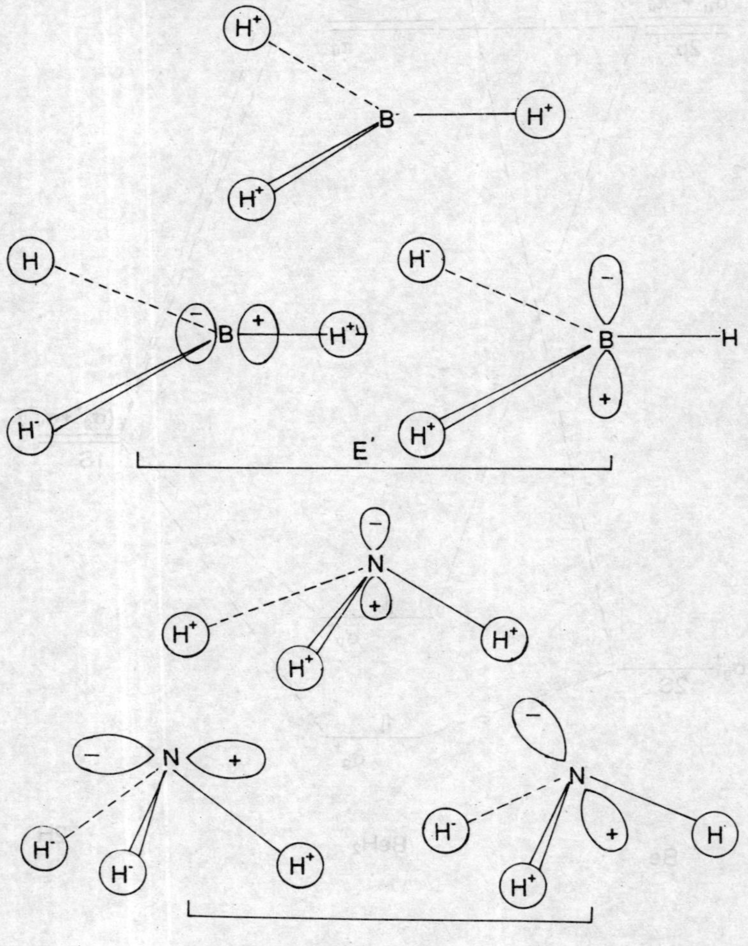

Fig. 7.9

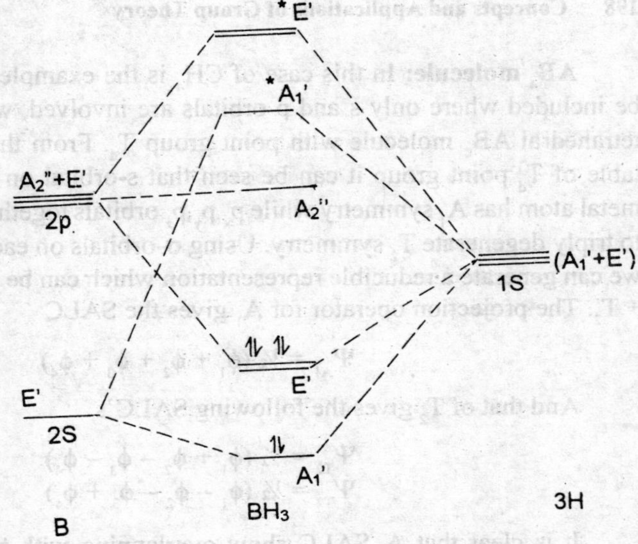
*E'
*A₁'
A₂"+E'
2p
A₂"
(A₁'+E')
1S
E'
E'
2S
A₁'
BH₃
B
3H

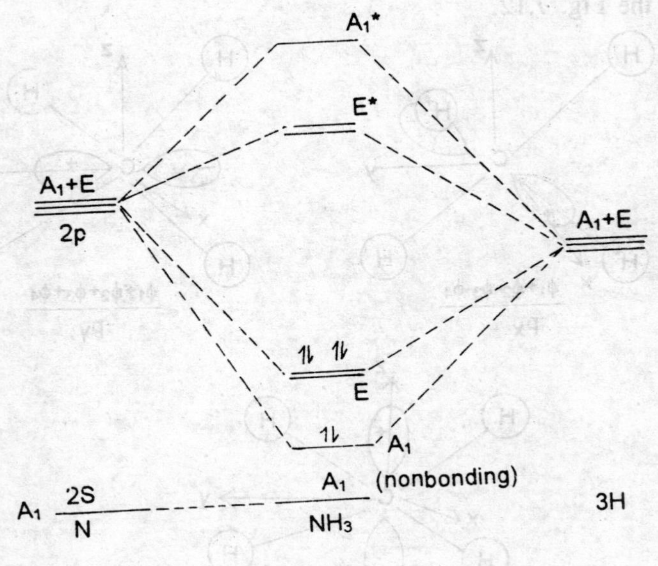
A₁*
E*
A₁+E
2p
A₁+E
E
A₁
A₁ (nonbonding)
A₁ 2S
N
NH₃
3H

Fig. 7.10

AB_4 molecule: In this case of CH_4 is the example which can be included where only s and p orbitals are involved, with regular tetrahedral AB_4 molecule with point group T_d. From the character table of T_d point group it can be seen that s-orbital on the central metal atom has A_1 symmetry while p_x, p_y, p_z orbitals together give rise to triply degenerate T_2 symmetry. Using σ-orbitals on each B-atoms we can generate a reducible representation which can be reduced $A_1 + T_2$. The projection operator for A_1 gives the SALC

$$\Psi_{A1} = \tfrac{1}{2}(\phi_1 + \phi_2 + \phi_3 + \phi_4)$$

And that of T_2 gives the following SALC's

$$\Psi_{T2} = \tfrac{1}{2}(\phi_1 + \phi_2 - \phi_3 - \phi_4)$$
$$\Psi_{T2} = \tfrac{1}{2}(\phi_1 - \phi_2 - \phi_3 + \phi_4)$$

It is clear that A_1 SALC show overlapping with s-orbital on the central metal A as shown in Fig. 7.11 and interaction of T_2 SALC's set with p_x, p_y, p_z on A-atom. The M.O. diagram is given in the Fig. 7.12.

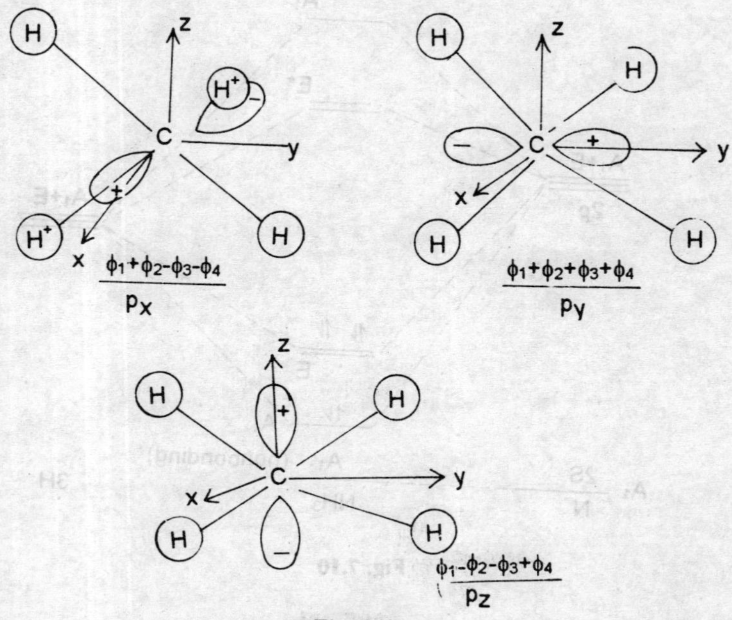

Fig. 7.11

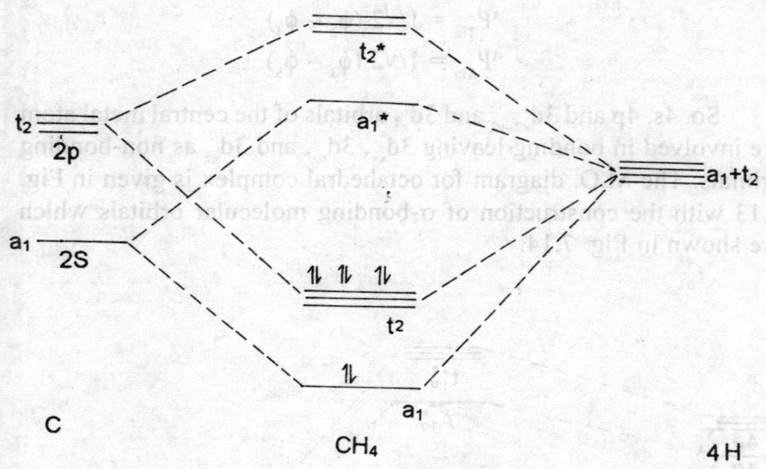

Fig. 7.12

AB$_n$ molecules with d-orbitals including transition metal complexes: In order to study the transition metal (d-block) complexes or compounds of the heavier p-block elements with geometries other than these described above, it is essential to include d-orbitals in the bonding scheme.

The most common geometry for transition metal complexes is regular octahedral ML_6 with O_h point group. As 3d, 4s and 4p orbitals are available for bonding in transition metals, so, on matching the character table of O_h point group it reveals that 4s orbital has A_{1g} symmetry; $3d_{x^2-y^2}$, $3d_{z^2}$ have E_g and $3d_{xy}$, $3d_{yz}$, and $3d_{xz}$ have T_{2g} and $4p_x$, $4p_y$, $4p_z$ have T_{1u} symmetries. Using σ-orbitals as bases functions T_σ can be generated which is reduced to

$$T_\sigma = A_{1g} \oplus E_g \oplus T_{1u}$$

Using pojection operator SALC's can be obtained for octahedral complex as

$$\Psi_{A1g} = 1/\sqrt{6}\,(\phi_1 + \phi_2 + \phi_3 + \phi_4 + \phi_5 + \phi_6)$$
$$\Psi_{Eg} = 1/\sqrt{12}\,(2\phi_1 - \phi_2 - \phi_3 - \phi_4 - \phi_5 + 2\phi_6)$$

$$\Psi_{Eg} = 1/2\ (\phi_2 - \phi_3 + \phi_4 - \phi_5)$$
$$Y_{Tiu} = 1/\sqrt{2}\ (\phi_1 - \phi_6)$$
$$\Psi_{Tiu} = 1/\sqrt{2}\ (\phi_2 - \phi_4)$$
$$\Psi_{Tiu} = 1/\sqrt{2}\ (\phi_3 - \phi_5)$$

So, 4s, 4p and $3d_{x2-y2}$ and $3d_{z^2}$ orbitals of the central metal atom are involved in bonding leaving $3d_{xy}$, $3d_{yz}$, and $3d_{xz}$ as non-bonding orbitals. The M.O. diagram for octahedral complex is given in Fig. 7.13 with the construction of σ-bonding molecular orbitals which are shown in Fig. 7.14.

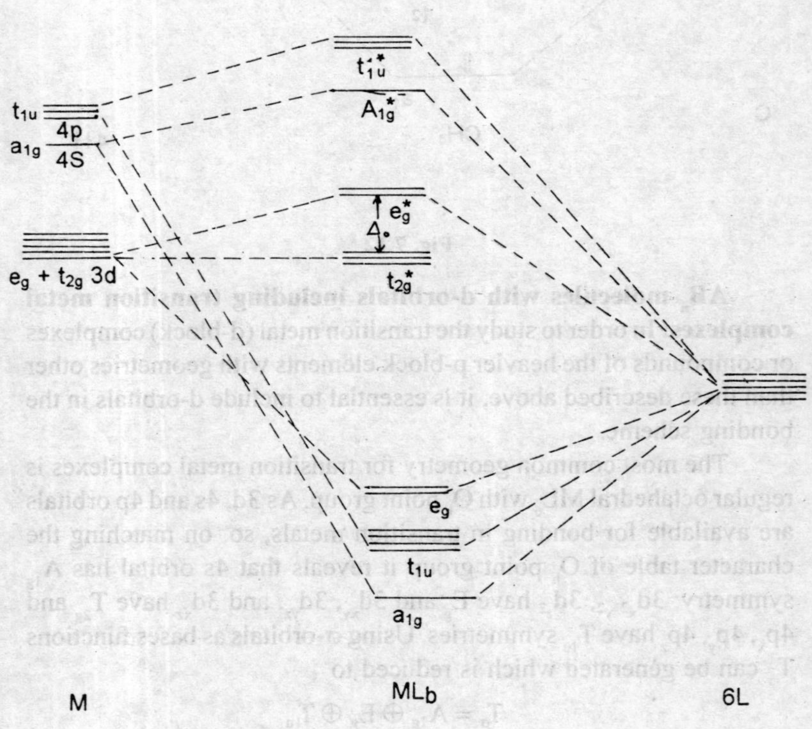

Fig. 7.13

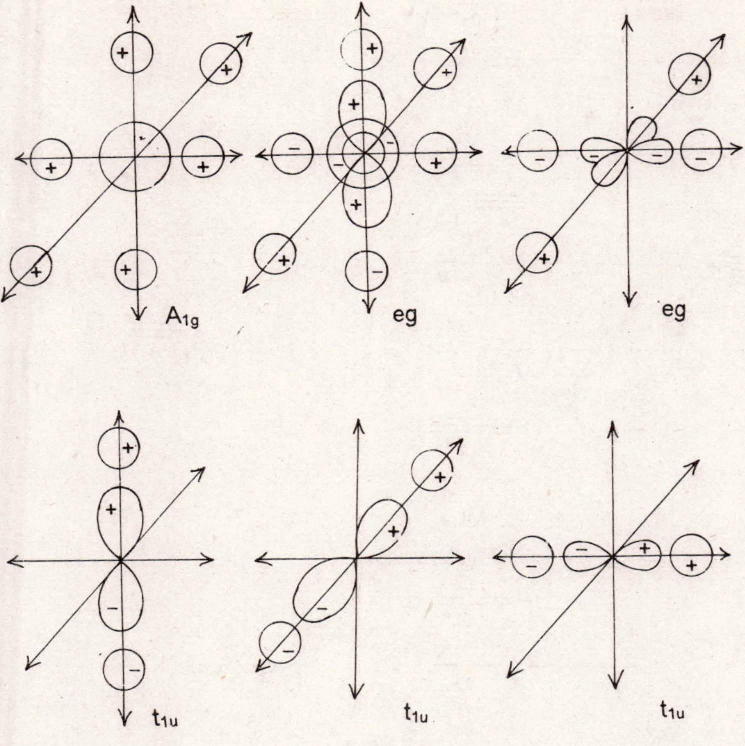

Fig. 7.14

Transition metal or heavier main group element compounds with other geometries can be explained in the same way as for AB_6 system. Four co-ordinate transition metal complexes are common with either tetrahedral (T_d) or square planar (D_{4h}) geometry. Using the same procedure as outlined for the O_h complexes the molecular orbital diagram for T_d and D_{4h} complexes can be obtained. In both the cases, mixing between metal atomic orbitals of same symmetry (*e.g.*, p & d-orbitals for T_2 symmetry for T_d point group) is neglected. In case of tetrahedral complexes, the metal complexes are largely distributed among E(non-bonding) and T_2^* (anti-bonding) orbitals with an energy gap conveniently described as Δ_t in comparison to it in O_h complexes metal electrons are distributed between a lower energy t_{2g} and higher energy e_g set of orbitals with an energy gap of Δ_o. The M.O. diagrams for T_d and D_{4h} metal complexes are given in Fig. 7.15 and Fig. 7.16.

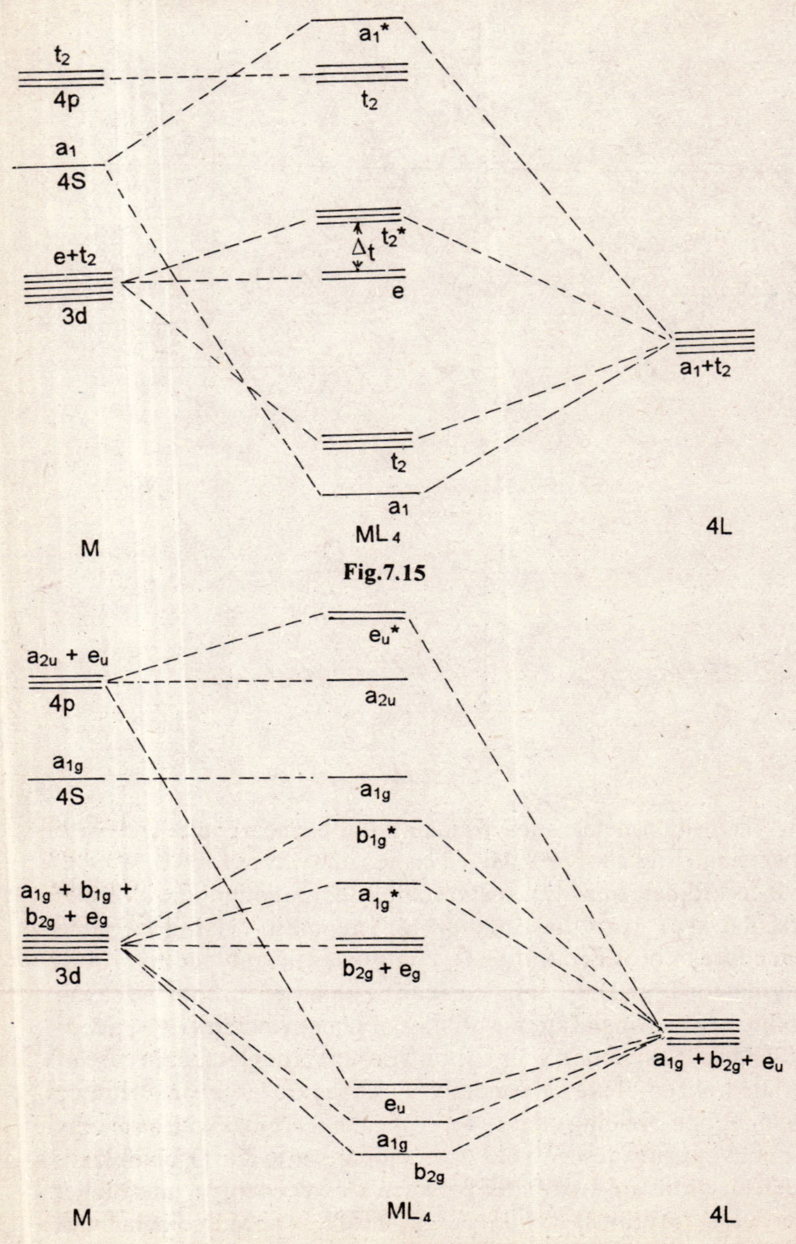

Fig.7.15

Fig. 7.16

π-bonding in AB_n molecules

Now let us consider the representative molecules which have significant π-bonding.

Example 1: Nitrite ion: According to valance bond theory following resonance structure for NO_2^- ion can be explained.

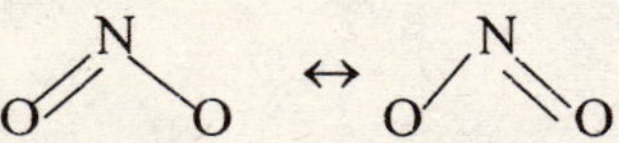

In M.O. description of bonding, the single σ-bonds are constructed exactly as in the case of H_2O. Now, let us consider the π_1 and π_2 as bases, we can generate the following representation of C_{2v} point group.

C_{2v}	E	C_2	σ_{xz}	σ_{yz}
T_π	2	0	0	–2

On reduction $T_\pi = A_2 \oplus B_1$. Using π_1 as generating function the projection operation method gives us

C_{2v}	E	C_2	σ_{xz}	σ_{yz}
π_1	π_1	$-\pi_2$	π_2	$-\pi_1$
A_2	1	1	–1	–1
B_1	1	–1	1	–1

So, the corresponding SALC's are

$$\Psi_{A2} = 1/\sqrt{2}\,(\pi_1 - \pi_2)$$
$$\Psi_{B1} = 1/\sqrt{2}\,(\pi_1 + \pi_2)$$

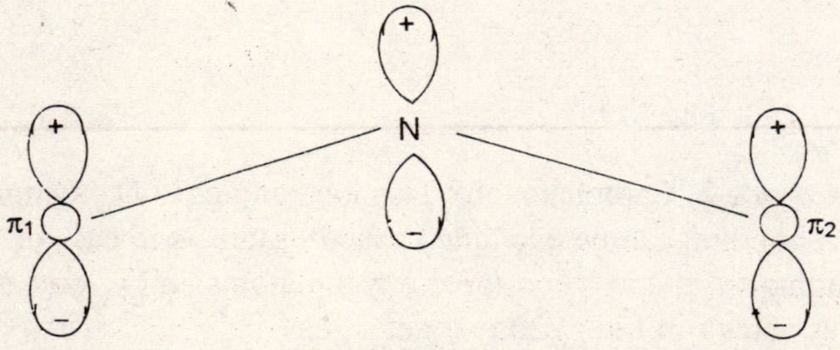

Fig. 7.17

As p_x orbitals of central N-atom has B_1 symmetry, so, it is clear that the B_1 combination can form bonding and anti-bonding molecular orbitals and A_2 forms non-bonding orbitals (Fig. 7.18).

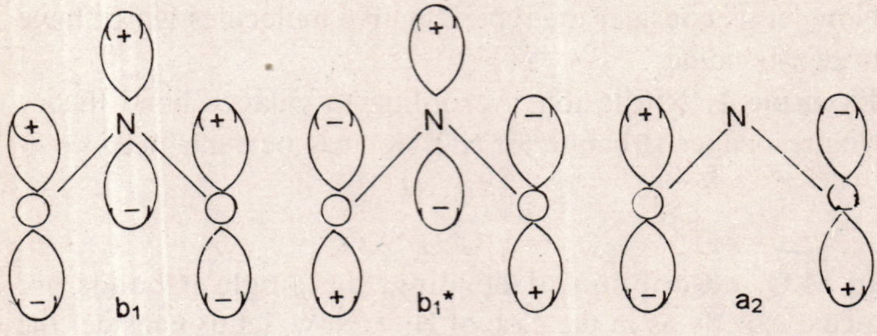

Fig.7.18

The corresponding M.O. diagram is given in Fig. 7.19.

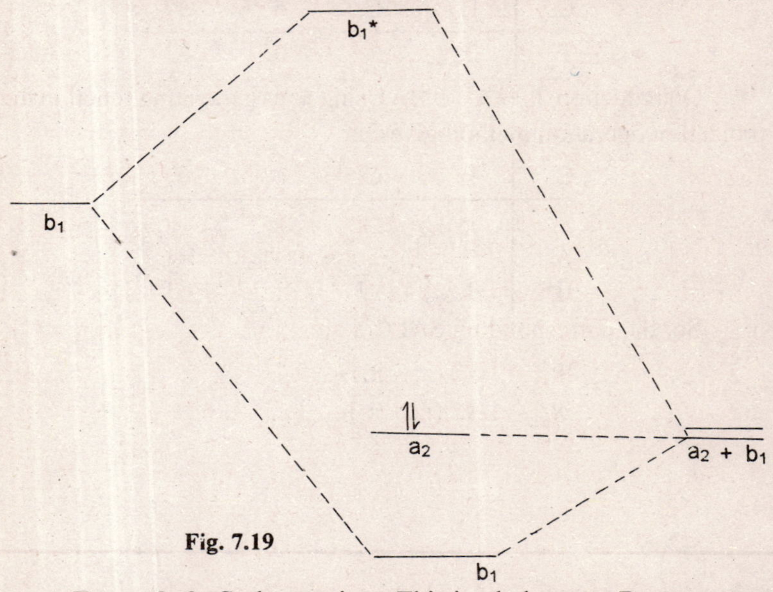

Fig. 7.19

Example 2: Carbonate ion : This ion belongs to D_{3h} symmetry and its σ-bonding can be explained exactly same as in case of BF_3. Considering π_1, π_2 and π_3 on three oxygen atoms and p_x on central carbon as shown in Fig. 7.20, we get T_π as

D_{3h}	E	$2C_3$	$3C_2$	σ_h	$2S_3$	$3\sigma_y$
T_π	3	0	−1	−3	0	1

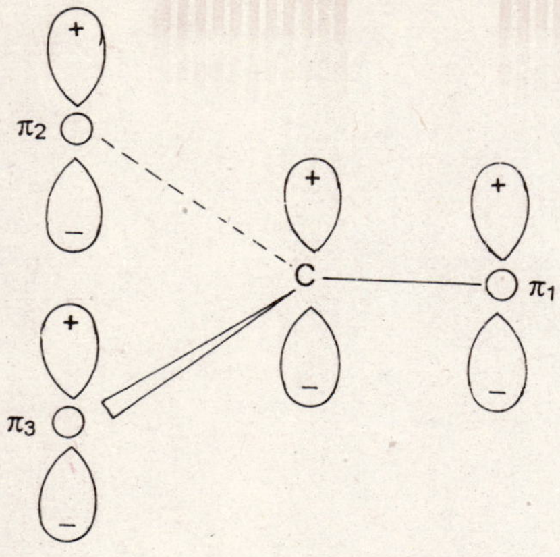

Fig. 7.20

Which on reduction give $T_\pi = A_2'' \oplus E''$ using projection operators SALC's can be otained as

$$\Psi_{A2}'' = 1/\sqrt{3}\,(\pi_1 + \pi_2 + \pi_3)$$
$$\Psi_{E}'' = 1/\sqrt{6}\,(2\pi_1 - \pi_2 - \pi_3)$$
$$\Psi_{A2}'' = 1/\sqrt{2}\,(\pi_2 - \pi_3)$$

The π-molecular orbitals for CO_3^{3-} and its corresponding M.O. given in Fig. 7.21 and 7.22.

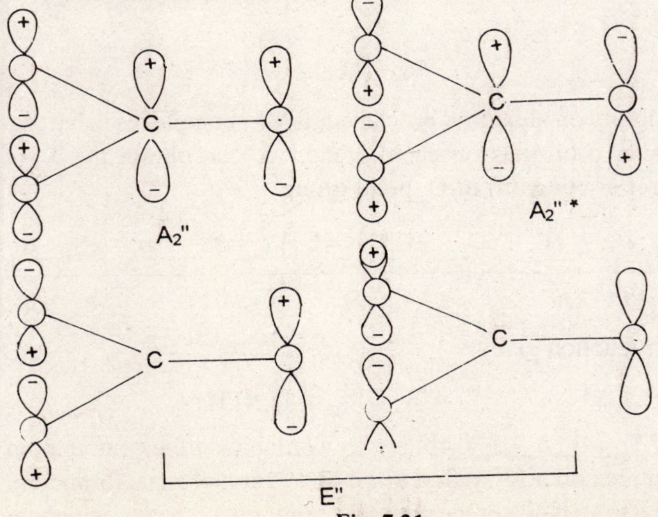

Fig. 7.21

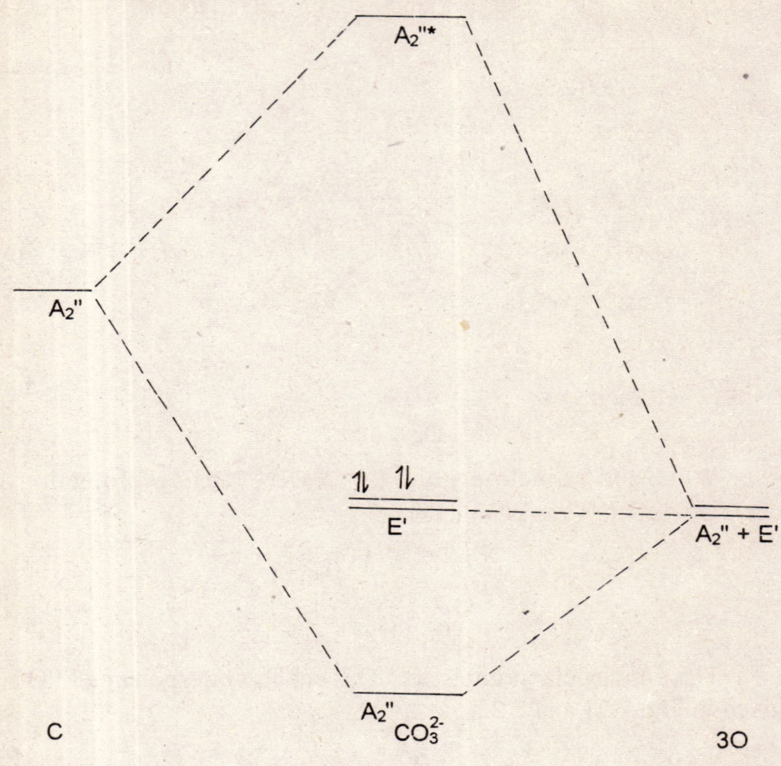

Fig. 7.22

Possibility of $_\pi$-bonding in octahedral ML_6 complexes using two perpendicular p-orbitals on each ligand, we can obtain the T_π (a reducible representation) of O_h point group.

O_h	E	$8C_3$	$6C_2$	$6C_4$	$3C_2''$	$(=C_4^2)$	i	$8S_6$	$3\sigma_h$	$6S_4$	6_{σ_d}
T_π	12	0	0	0	–4	4	0	0	0	0	0

Which on reduction gives

$$T_\pi = T_{1g} \oplus T_{2g} \oplus T_{1u} \oplus T_{2u}$$

Only T_{2g} (d_{xy}, d_{yz}, d_{xz}) orbitals set available on the central atom in ML_6 complex after formation of σ-bonds, for π orbitals formation. The π interaction of one of the metal T_{2g} orbitals (d_{yz}) with component of T_{2g} of ligand orbital set is shown in Fig. 7.24.

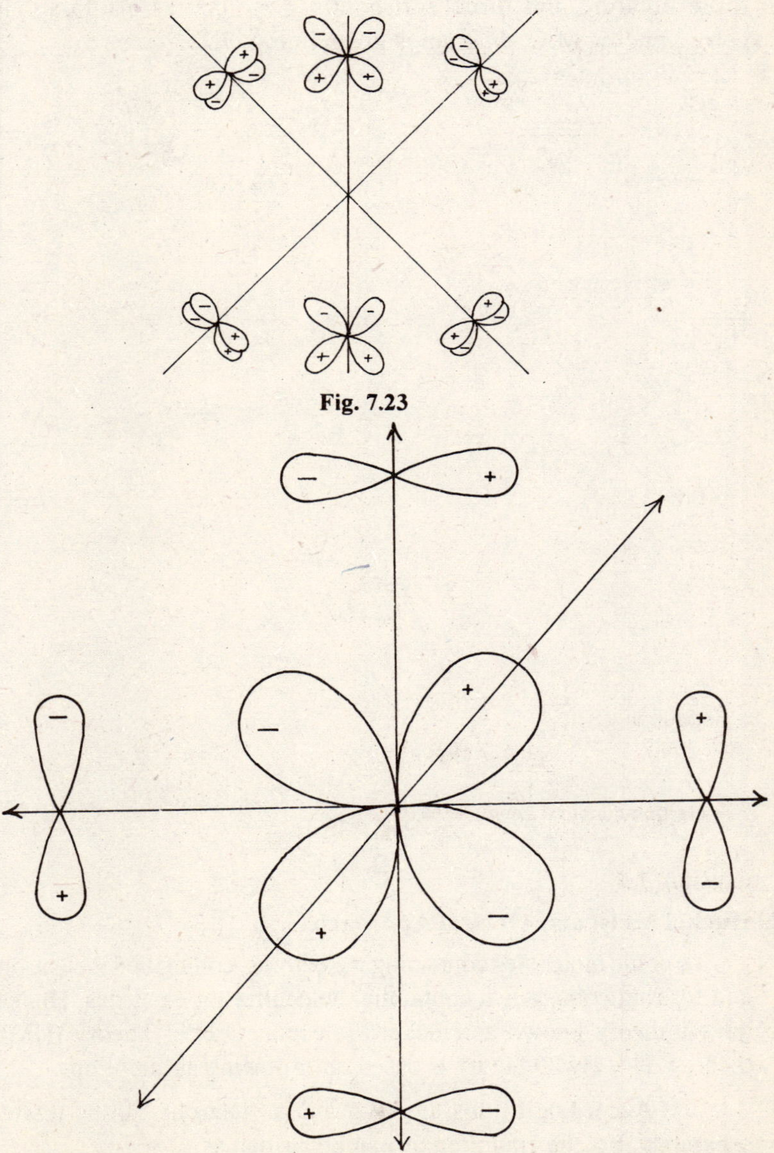

Fig. 7.23

Fig. 7.24

There are equivalent interactions in xy and xz planes giving three bonding and three anti-bonding molecular orbitals. The corresponding M.O. diagram is given in Fig. 7.25.

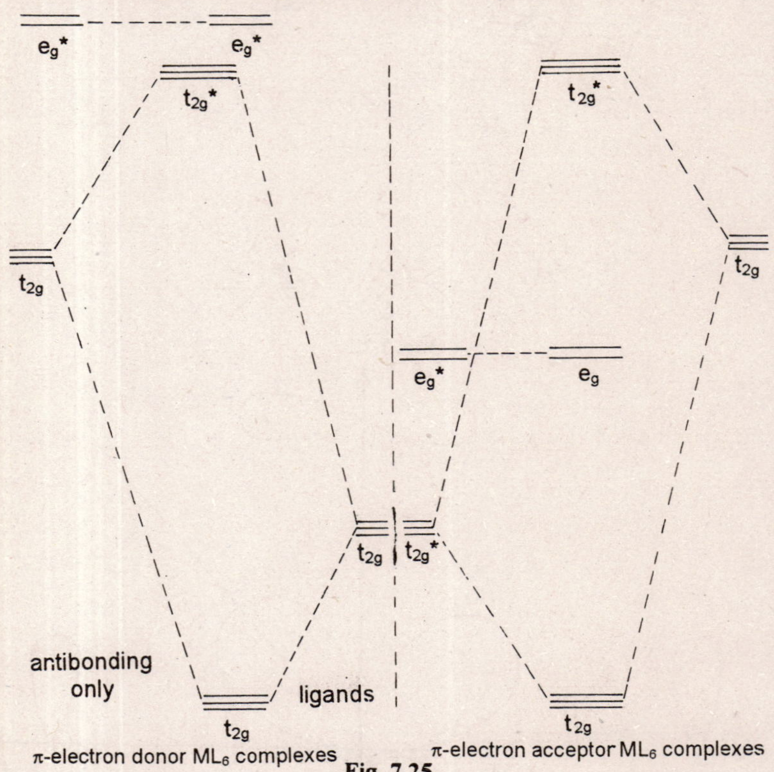

Fig. 7.25

Section 7.4

Huckel Molecular Orbital Approach

For the molecules containing π electrons, conjugated π electrons and for ring compounds containing delocalized π electrons, Huckel gave a theory known as Huckel Molecular Orbital Theory (HMO theory). This HMO theory is based on following assumption:

(i) According to this thoery σ and π electrons can be treated separately. So, the Hamiltonian can be written as

$$\hat{H} = \hat{H}_{\sigma} + \hat{H}_{\pi}$$

Similarly the wave function and energy can be written as

$$\Psi = \Psi_\sigma \Psi_\pi \text{ and}$$
$$E = E_\sigma + E_\pi$$

(ii) As π electrons are higher in energy, so, they can't interact with nucleus as they are perpendicular to it. π electrons are more energetic, therefore, they are more involved in reactions. As distance between σ and π electrons is so, large that the term $1/r_{\sigma\pi}$ in the Hamiltonian (*i.e.*, the interaction term of σ and π electrons) can be neglected. Therefore we don't consider E_σ but only E_Π.

(iii) Each π electron can be treated separately and they can't interact with each other.

So, $\hat{H} = \hat{H}_{\pi 1} + \hat{H}_{\pi 2} + \ldots\ldots$

And $\Psi_\pi = \Psi_{\pi 1} \Psi_{\pi 2} \ldots\ldots\ldots$

and $E_\pi = E_{\pi 1} + E_{\pi 2} + \ldots\ldots$

Let us now apply this theory to some suitable examples.

Example 1: Allyl system: Consider only P electron on each C-atom

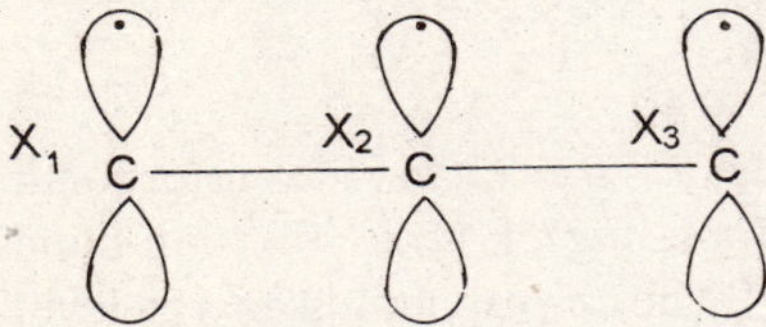

Let X_1, X_2, and X_3 be π orbitals on each carbon atom. The total M.O. can be written as

$$\Psi = C_1X_1 + C_2X_2 + C_3X_3$$

Where C_1, C_2 and C_3 are the respective coefficients. For this system the secular determinant can be written as (considering only π electrons):

$$\begin{vmatrix} H_{11} - ES_{11} & H_{12} - ES_{12} & H_{13} - ES_{13} \\ H_{21} - ES_{21} & H_{22} - ES_{22} & H_{23} - ES_{23} \\ H_{31} - ES_{31} & H_{32} - ES_{32} & H_{33} - ES_{33} \end{vmatrix} = 0$$

Where $H_{12} = H_{21} = \int X_1 |\hat{H}| X_2 \, d\tau$ and $H_{11} = \int X_1 |\hat{H}| X_1 \, d\tau$ or in general all $H_{ii} = \alpha$ (coloumb integrals) and all $H_{ij} = \beta$ (resonance integrals) and $S_{ii} = \int X_1 X_1 \, d\tau = 1$.

X_1, X_2, X_3 are to be normalized and $S_{ij} = \int X_1 X_2 \, d\tau = 0$

As X_1, X_2 and X_3 are to be orthogonal to each other. Therefore, this determinant becomes

$$\begin{vmatrix} \alpha - E & \beta & 0 \\ \beta & \alpha - E & \beta \\ 0 & \beta & \alpha - E \end{vmatrix} = 0$$

or

$$\begin{vmatrix} (\alpha - E)/\beta & 1 & 0 \\ 1 & (\alpha - E)/\beta & 1 \\ 0 & 1 & (\alpha - E)/\beta \end{vmatrix} = 0$$

let us consider α–E/β=x $\Rightarrow$ E=α–βx, such that if x = +ve the energy is higher and if x = –ve, energy is lower. So, the determinant becomes

$$\begin{vmatrix} x & 1 & 0 \\ 1 & x & 1 \\ 0 & 1 & x \end{vmatrix} = 0$$

which has the values x = 0, ±√2 and the corresponding energy values are E = $\alpha + \sqrt{2}\beta$ (bonding); E = $\alpha + 0\beta$ (non-bonding) and E = $\alpha - \sqrt{2}\beta$ (anti-bonding). The corresponding energy level diagram is given in Fig. 7.26.

—— $\alpha - \sqrt{2}\beta$	antibonding
—— α	nonbonding
—— $\alpha + \sqrt{2}\beta$	bonding

Fig.7.26

Substituting the values of x we can get the values coefficients as C_1, C_2 and C_3 and the corresponding wave functions for molecular orbitals are

$$\Psi_1 = 1/2X_1 + 1/\sqrt{2}X_2 + 1/2X_3$$
$$\Psi_2 = 1/\sqrt{2}X_1 - 1/\sqrt{2}X_3 \text{ and}$$
$$\Psi_3 = 1/2X_1 - 1/\sqrt{2}X_2 + 1/2X_3$$

Example 2: Ethylene molecule: The simplest case containing Π electrons, is ethylene molecule. Applying HMO theory to ethylene molecule the determinant can be written as

$$\begin{vmatrix} x & 1 \\ 1 & x \end{vmatrix} = 0$$

which has solution x = ±1, on substituting it we get two molecular orbitals that are $E_{bonding} = \alpha + \beta$ and $E_{antibonding} = \alpha - \beta$ and the corresponding $\Psi_{MO} = C_1X_1 + C_2X_2$ which on determination of coefficients C_1 and C_2 gives two orbitals as $\Psi_{bonding} = [1/\sqrt{2}(1-S)][X_1 + X_2]$ and $\Psi_{antibonding} = [1/\sqrt{2}(1-S)][X_1 - X_2]$.

Example 3: Butadiene : this HMO approach can also be applied to conjuagted dienes such as 1,3-Butadiene. The HMO secular determinant can be written as

$$\begin{vmatrix} x & 1 & 0 & 0 \\ 1 & x & 1 & 0 \\ 0 & -1 & x & 1 \\ 0 & 0 & 1 & x \end{vmatrix} = 0$$

which can be solved for the following values of x;
x = ± 1.718 and x = ±0.718. The corresponding energy values are $E = \alpha + 1.718\ \beta$; $E = \alpha + 0.718\ \beta$; $E = \alpha - 0.718\beta$ and $E = \alpha - 1.718\beta$ and the MO energy level diagram is given in the fig. 7.27.

____ $\alpha - 1.718\ \beta$

____ $\alpha - 0.718\ \beta$

____ $\alpha + 0.718\ \beta$

____ $\alpha + 1.718\ \beta$

Fig. 7.27

For Butadiene thus the total energy is $4\alpha + 4.472\ \beta$ corresponding energy calculations for the simplest case *i.e.*, the ethylene molecule (2 electrons case) yields the total energy value as $2\alpha + 2\beta$. So, accordingly for 4 electrons (in case of Butadiene) it should be $2(2\alpha + 2\beta)$ as Butadiene can be considered as combination of two ethylene molecules. So, there is a difference in actual and the calculated energy on this basis (*i.e.*, on the basis of ethylene molecule) *i.e.*, $\Delta E = (4\alpha + 4.472\ \beta) - (4\alpha + 4\beta) = 0.472\ \beta$. This ΔE is termed as stabilization energy that is for conjugated diene.

Example 4: Cyclobutadiene: Considering the Cyclobutadiene molecule and writing the HMO secular determinant for it and on solving it we get E = 4α + 4β for this molecule. Therefore, on comparison with the ethylene molecule (in the same way as in the above example 3) we get

$$E = (4\alpha + 4\beta) - (4\alpha + 4\beta) = 0$$

So, the stabilization energy is zero in this case. Therefore, the proposed molecule is considered to be unstable.

Example 5: Benzene : In benzene, on considering the six π-orbitals on each carbon atom and on writing T_π we get

D_{6h}	E	$2C_6$	$2C_3$	C_2	$3C_2'$	$3C_2''$	'i'	$2S_3$	$2S_6$	σ_h	$3_{\sigma v}$	$3_{\sigma v}'$
T_Π	6	0	0	0	–2	0	0	0	0	–6	2	0

Which on reduction gives

$$T_\pi = A_{2u} \oplus B_{2g} \oplus E_{1g} \oplus E_{2u}$$

On applying the projection operators SALC's can be obtained as

$$\Psi_{A2u} = 1/\sqrt{6}\,(\phi_1 + \phi_2 + \phi_3 + \phi_4 + \phi_5 + \phi_6)$$
$$\Psi_{B2u} = 1/\sqrt{6}\,(\phi_1 - \phi_2 + \phi_3 - \phi_4 + \phi_5 - \phi_6)$$
$$\Psi_{E1g} = 1/\sqrt{12}\,(2\phi_1 + \phi_2 - \phi_3 - 2\phi_4 - \phi_5 + \phi_6)$$
$$\Psi_{E1g} = 1/2\,(\phi_2 + \phi_3 - \phi_5 - \phi_6)$$
$$\Psi_{E2u} = 1/\sqrt{12}\,(2\phi_1 - \phi_2 - \phi_3 + 2\phi_4 - \phi_5 - \phi_6)$$
$$\Psi_{E2u} = 1/2\,(\phi_2 + \phi_3 - \phi_5 - \phi_6)$$

These resulting M.O.'s can be shown as in Fig. 7.28. On writing the HMO secular determinant for benzene

$$\begin{vmatrix} x & 1 & 0 & 0 & 0 & 1 \\ 1 & x & 1 & 0 & 0 & 0 \\ 0 & 1 & x & 1 & 0 & 0 \\ 0 & 0 & 1 & x & 1 & 0 \\ 0 & 0 & 0 & 1 & x & 1 \\ 1 & 0 & 0 & 0 & 1 & x \end{vmatrix} = 0$$

which on solution gives x = ±2 and ±1. Therefore, the energy level diagram for π M.O.'s can be shown as in Fig. 7.29.

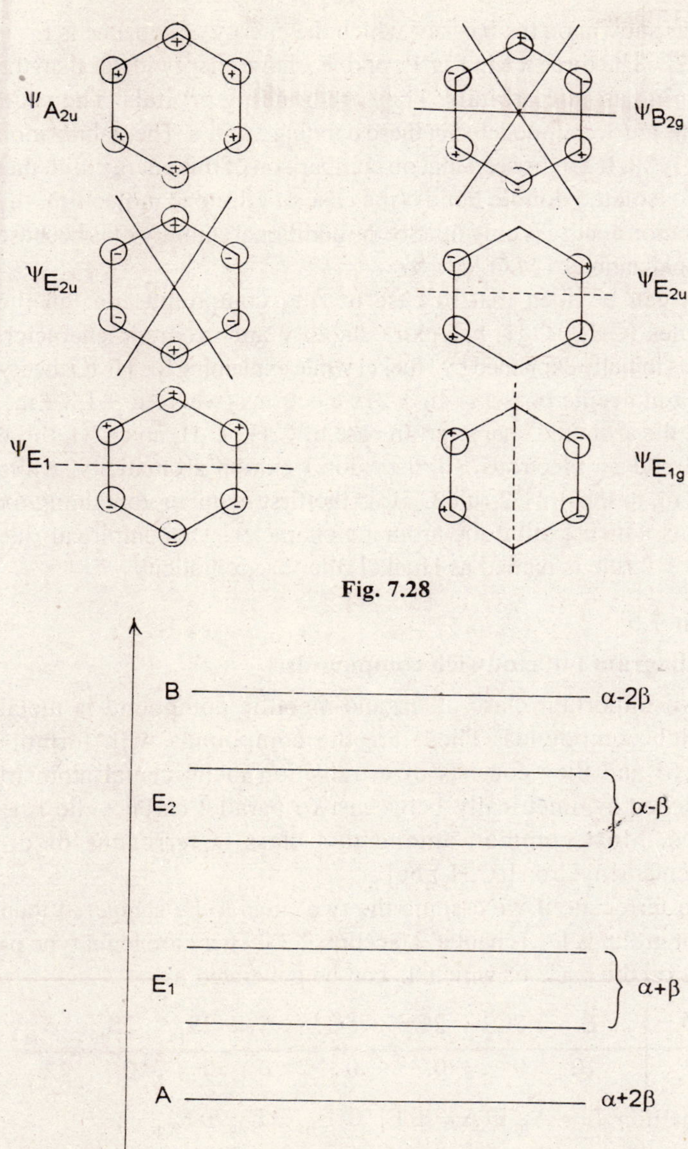

Fig. 7.28

Fig. 7.29

In this M.O. diagram the six π electrons are filled in the bonding levels as shown, on the basis of which the energy of benzene is $E_{tot} = 6\alpha + 12\beta$. The three π M.O.'s Ψ_A and Ψ_E, have energy lower than the combining atomic orbitals. They are bonding orbitals. The six π electrons are accommodated in these bonding M.O.'s. The stabilization energy is 8β. It can be seen that on comparison of this energy with that of three isolated double bonds (the case of ethylene molecule), the stabilization energy is only 6β. So, the additional stability of is because of delocalization of M.O.'s.

It can be seen that in case of ring compounds *i.e.,* in the molecules $(CH)_n$, C_6H_6 has extra stability and aromatic character. This was initially explained by Huckel while explaining the HMO theory that if a molecule possess (4n + 2)π electrons (where n = 1,2,3....), exhibit the aromatic character. In case of C_3H_3, C_4H_4 and C_5H_5 these are 3,4 and 5π electrons, so, they don't exhibit aromaticity, while according to the 4n + 2 rule C_6H_6 is the first member containing 6π electrons which exhibit the aromatic character. This empirical rule *i.e.,* 4n + 2 rule is named as Huckel rule for aromaticity.

Section 7.5

M.O. diagram for sandwich compounds

An important class of organo-metallic compound is metal sandwich compounds. These are the compounds with formula $(C_nH_n)_2M$ and they consists of a transition metal cental atom M sandwiched symmetrically between two parallel carbocyclic ring systems. Most common among this class is ferrocene or di-cyclopentadienyl iron $[(C_5H_5)_2Fe]$.

In ferrocene if we assume the two rings to be staggered than its point group is D_{5d} (chapter 2, section 2.3). There are ten π type p-orbitals on the basis of which T_π can be generated as

D_{5d}	E	$2C_5$	$2C_5^2$	$5C_2'$	'i'	$2S_{10}^3$	$2S_{10}$	$5_{\sigma d}$
T_π	10	0	0	0	0	0	0	2

On reduction $T_\pi = A_{1g} \oplus A_{2u} \oplus E_{1g} \oplus E_{1u} \oplus E_{2g} \oplus E_{2u}$

If π type p-orbitals are labeled as $\phi_1, \phi_2....\phi_5$ and $\phi_1', \phi_2'....\phi_5'$ and if their positive lobes are considered towards opposite rings, they can be represented as shown in Fig. 7.30.

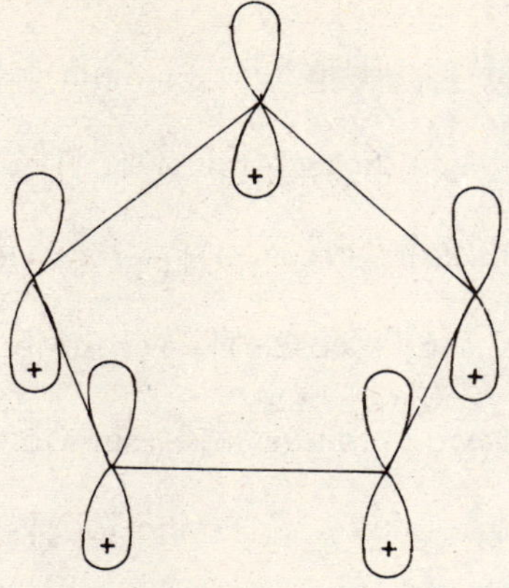

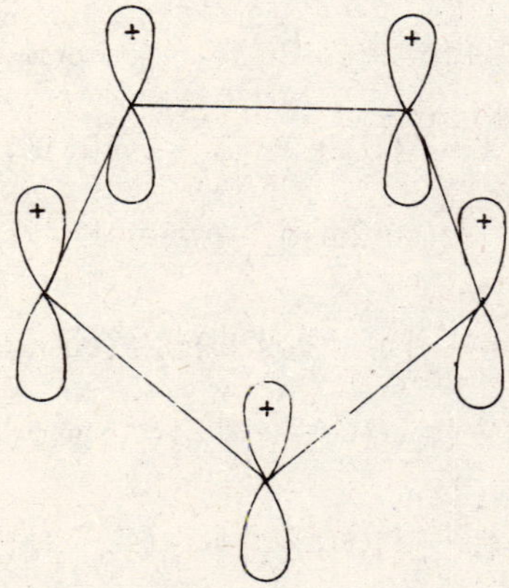

Fig.7.30

The normalized π-M.O's (SALC's) are given as

$$\Psi_{A_{1g}} = \frac{1}{\sqrt{10}}[\phi_1 + \phi_2 + \phi_3 + \phi_4 + \phi_5 + \phi_1' + \phi_2' + \phi_3' + \phi_4' + \phi_5']$$

$$\Psi_{A_{iu}} = \frac{1}{\sqrt{10}}[([\phi_1 + \phi_2 + \phi_3 + \phi_4 + \phi_5) - (\phi_1' + \phi_2' + \phi_3' + \phi_4' + \phi_5')]$$

$$\Psi_{E_{ig}(1)} = \frac{1}{\sqrt{5}}[\{\phi_1 + (\cos\omega)\phi_2 + (\cos 2\omega)\phi_3 + (\cos 2\omega)\phi_4 + (\cos\omega)\phi_5\} +$$

$$\{\phi_1' + (\cos\omega)\phi_2' + (\cos 2\omega)\phi_3' + (\cos 2\omega)\phi_4' + (\cos\omega)\phi_5'\}]$$

$$\Psi_{E_{1g}(2)} = \frac{1}{\sqrt{5}}[\{(\sin\omega)\phi_2 + (\sin 2\omega)\phi_3 - (\sin 2\omega)\phi_4 - (\sin\omega)\phi_5\} + \{(\sin\omega)\phi_2' + (\sin 2\omega)\phi_3' - (\sin 2\omega)\phi_4' - (\sin\omega)\phi_5'\}]$$

$$\Psi_{E_{1u}(1)} = \frac{1}{\sqrt{5}}[\phi_1 + (\cos\omega)\phi_2 + (\cos 2\omega)\phi_3 + (\cos 2\omega)\phi_4 + (\cos\omega)\phi_5\} - \{\phi_1' + (\cos\omega)\phi_2' + (\cos 2\omega)\phi_3' + (\cos 2\omega)\phi_4' + (\cos\omega)\phi_5'\}]$$

$$\Psi_{E_{1u}(2)} = \frac{1}{\sqrt{5}}[\{(\sin\omega)\phi_2 + (\sin 2\omega)\phi_3 - (\sin 2\omega)\phi_4 - (\sin\omega)\phi_5\} - \{(\sin\omega)\phi_2' + (\sin 2\omega)\phi_3' - (\sin 2\omega)\phi_4' - (\sin\omega)\phi_5'\}]$$

$$\Psi_{E_{2g}(1)} = \frac{1}{\sqrt{5}}[\phi_1 + (\cos 2\omega)\phi_2 + (\cos\omega)\phi_3 + (\cos\omega)\phi_4 + (\cos 2\omega)\phi_5\} + \{\phi_1' + (\cos 2\omega)\phi_2' + (\cos\omega)\phi_3' + (\cos\omega)\phi_4' + (\cos 2\omega)\phi_5'\}]$$

$$\Psi_{E_{2g}(2)} = \frac{1}{\sqrt{5}}[\{(\sin 2\omega)\phi_2 + (\sin\omega)\phi_3 - (\sin\omega)\phi_4 + (\sin 2\omega)\phi_5\} + \{(\sin 2\omega)\phi_2' + (\sin\omega)\phi_3' - (\sin\omega)\phi_4' - (\sin 2\omega)\phi_5'\}]$$

$$\Psi_{E_{2u}(1)} = \frac{1}{\sqrt{5}}[\phi_1 + (\cos 2\omega)\phi_2 + (\cos\omega)\phi_3 + (\cos\omega)\phi_4 + (\cos 2\omega)\phi_5\} - \{\phi_1' + (\cos 2\omega)\phi_2' + (\cos\omega)\phi_3' + (\cos\omega)\phi_4' + (\cos 2\omega)\phi_5'\}]$$

$$\Psi_{E_{2u}(2)} = \frac{1}{\sqrt{5}}[\{(\sin 2\omega)\phi_2 - (\sin\omega)\phi_3 + (\sin\omega)\phi_4 - (\sin 2\omega)\phi_5\} - \{(\sin 2\omega)\phi_2' + (\sin\omega)\phi_3' + (\sin\omega)\phi_4' - (\sin 2\omega)\phi_5'\}]$$

For the metal ion, the valance orbitals are five 3d-orbitals, 4s and three 4p-orbitals. The corresponding irreducible representations from D_{5d} character table for orbitals are

4s, $3d_{z^2}$	–	A_{1g}
$3d_{xz}$, $3d_{yz}$	–	E_{1g}
$3d_{xy}$, $3d_{x2-y2}$	–	A_{2g}
$4p_z$	–	A_{2u}
$4p_x$, $4p_y$	–	E_{1u}

The corresponding M.O. energy diagram is given in Fig. 7.31.

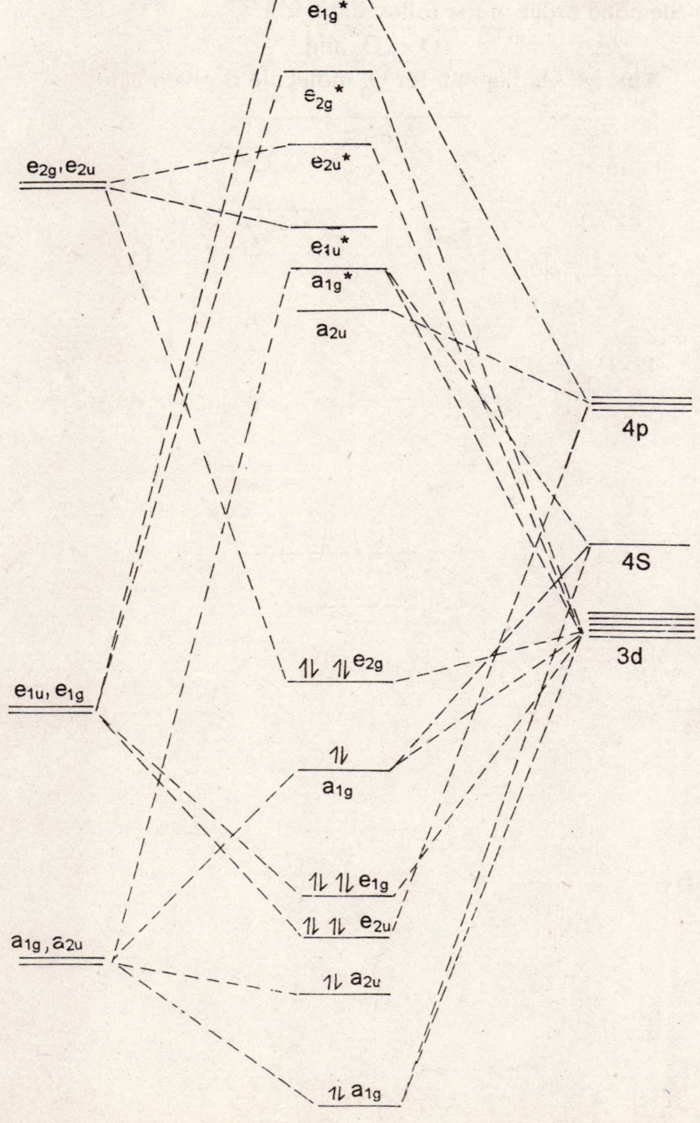

Fig. 7.31

Problems

Q. 1. Draw the M.O. diagram for O_2 molecule and work out for the bond order of the following:

O_2^+, O_2 and O_2^-.

Ans. M. O. diagram for O_2 molecule is given below:

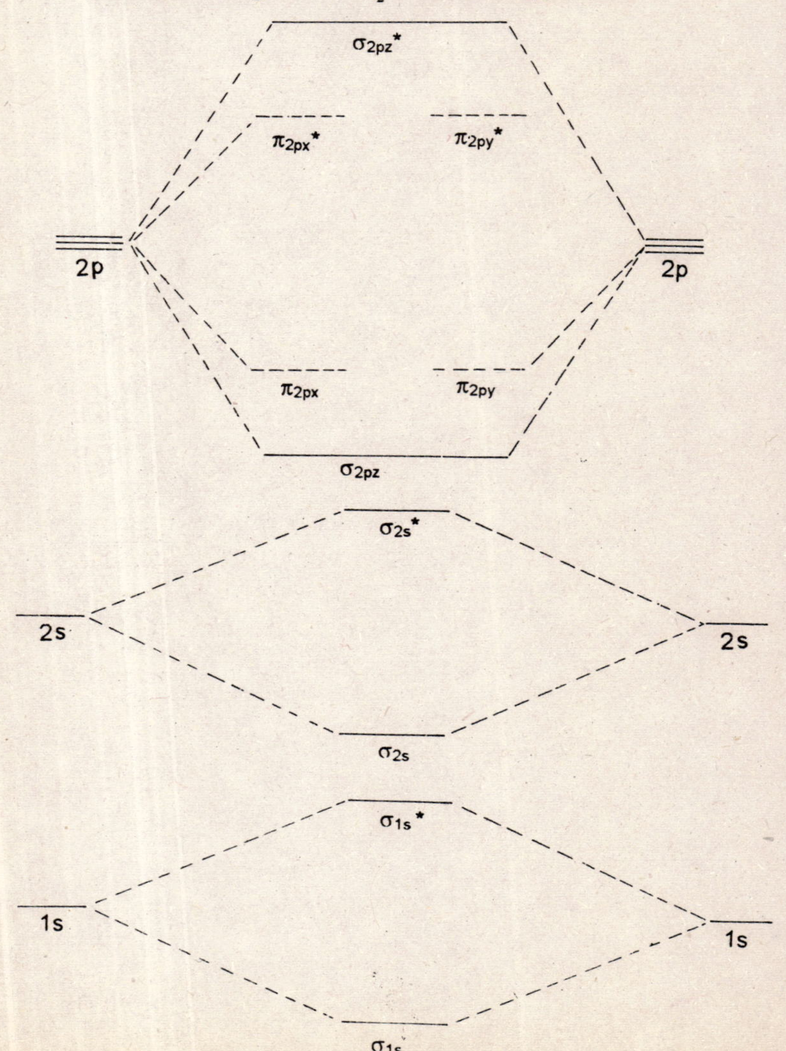

For O_2 molecule the M.O. configuration is

$$\sigma_{1s}^{2}, \sigma_{1s}^{*2}, \sigma_{2s}^{2}, \sigma_{2px}^{2}, \pi_{2pz}^{2}, \pi_{2py}^{*1}, \pi_{2pz}^{*1}$$

So, its bond order is B.O. = ½($N_b - N_a$) = ½(10 – 6) = 2

Where N_b = number of electrons in bonding M.O.'s & N_a = number of electrons in anti-bonding M.O.'s.

For O_2+ molecule the M.O. configuration is

$$\sigma_{1s}^{2}, \sigma_{1s}^{*2}, \sigma_{2s}^{2}, \sigma_{2s}^{*2}, \sigma_{2px}^{2}, \pi_{2py}^{2}, \pi_{2pz}^{2}, \pi_{2py}^{*1}, \pi_{2pz}^{*1}$$

So, B.O. = ½(10 – 5) =2.5 and

For O_2- molecular the M.O. configuration is

$$\sigma_{1s}^{2}, \sigma_{1s}^{*2}, \sigma_{2s}^{2}, \sigma_{2s}^{*2}, \sigma_{2px}^{2}, \pi_{2py}^{2}, \pi_{2pz}^{2}, \pi_{2py}^{*1}, \pi_{2pz}^{*1}$$

So, B.O. = ½(10 – 7) = 1.5.

Q. 2. On the basis of HMO approach, find out T_π for cyclopropenyl ion and reduce it. Fing its energy.

Ans. The cyclopropenyl ion is shown in Fig. It belongs to D_{3h}. Considering π-orbitals as basis its T_π can be written as

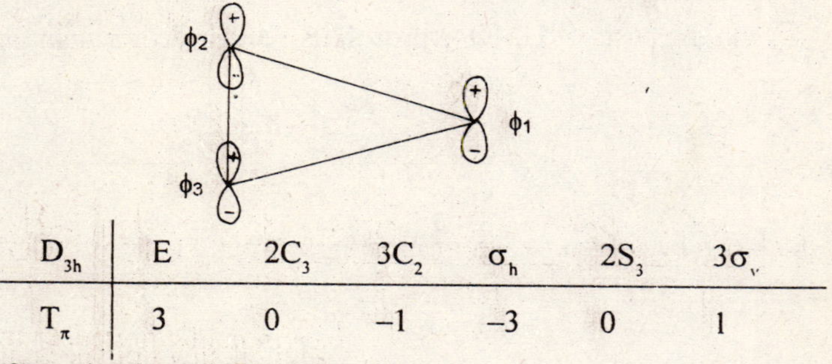

D_{3h}	E	$2C_3$	$3C_2$	σ_h	$2S_3$	$3\sigma_v$
T_π	3	0	–1	–3	0	1

On reduction which gives

$$T_\pi = A_2'' \oplus E''$$

Using projection operators, the SALC's for π-M.O.'s are

$$\Psi_{A2}'' = 1/\sqrt{3}(\phi_1 + \phi_2 + \phi_3)$$

$$\Psi_{E}'' = 1/\sqrt{6}(2\phi_1 - \phi_2 - \phi_3)$$

$$\Psi_{E}'' = 1/\sqrt{2}(\phi_2 - \phi_3)$$

Which can be shown as

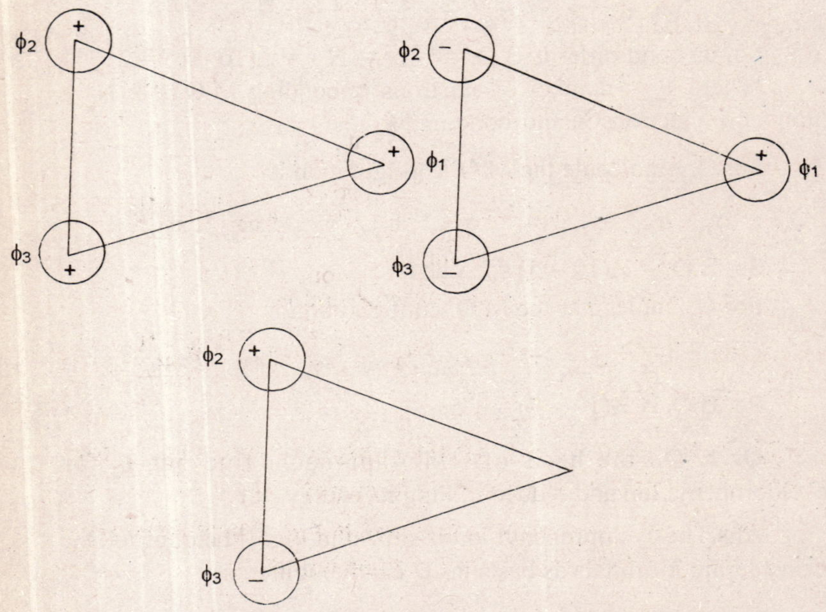

On the basis of HMO approach its secular determinant is

$$\begin{vmatrix} x & 1 & 1 \\ 1 & x & 1 \\ 1 & 1 & x \end{vmatrix}$$

which can be solved to form following π-M.O. diagram

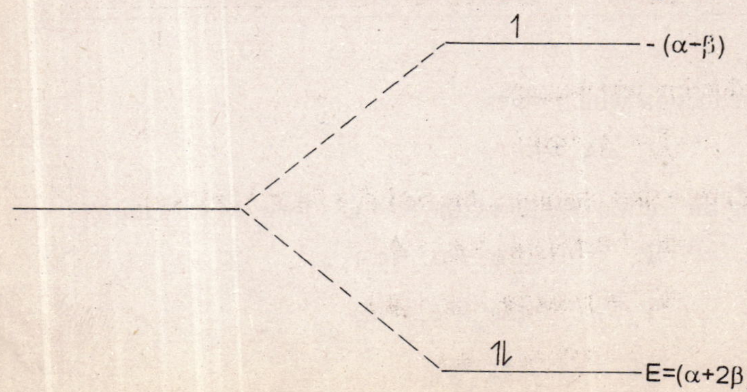

So, its energy is $(3\alpha + 3\beta)$.

Q. 3. For cyclopentadienyl ion ($C_5H_5^-$) obtain T_π and reduce it.

Ans. Cyclopentadienyl ion belongs to D_{5h} point group. Considering π p-orbitals on each carbon we can get T_π for this as

D_{5h}	E	$2C_5$	$5C_5^2$	$5C_2$	σ_h	$2S_5$	$5\sigma_v$
T_π	5	0	0	–1	–5	0	1

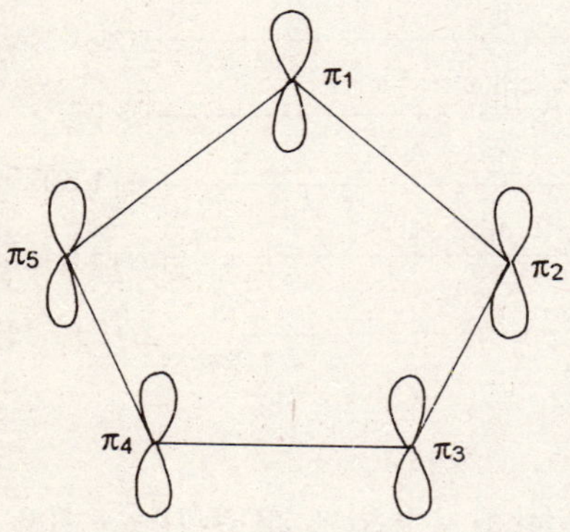

T_π on reduction reduced to the following irreducible representations

$$T_\pi = A_2'' \oplus E' \oplus E_2''.$$

Q.4. A Huckel calculation for naphathlene gives πM.O. energy as $E_i = \alpha + m_i\beta$ where ten m_i are 2.3028, 1.6180, 1.3029, 1.000, 0.6180, –0.6180, –1.000, –1.3029, –1.6180 and –2.308. Calculate the delocalization energy for naphathlene.

Ans. Ten energy levels for naphathlene are given in the fig. There are ten π electrons in naphathlene which are filled upto the bonding levels as shown in the fig. So, the total energy can be calculated on the basis of π M.O. diagram and electron distribution in them as

$$E_{\pi\ naph.} = 2(\alpha + 2.3028\beta) + 2(\alpha + 1.6180\beta) + 2(\alpha + 1.3029\beta) + 2(\alpha + \beta) + 2(\alpha + 0.618\beta)$$

$$E_{\pi\ naph.} = 10\alpha + 13.6834\beta$$

If we consider it as combination of five double bonds (compare it with ethylene) than its energy will be

$\alpha - 2.3028\beta$

$\alpha - 1.6180\beta$

$\alpha - 1.3029\beta$

$\alpha - \beta$

$\alpha - 0.6180\beta$

⇅ $\alpha + 0.6180\beta$

⇅ $\alpha + \beta$

⇅ $\alpha + 1.3029\beta$

⇅ $\alpha + 1.6180\beta$

⇅ $\alpha + 2.0328\beta$

E (for 5 ethylene) = $5(2\alpha + 2\beta) = 10\alpha + 10\beta$

The difference of these two energies *i.e.,*

$E = (10\alpha + 13.6834\beta) - (10\alpha + 10\beta)$

$E = 3.6834\beta$ which is the delocalization energy for this molecule.

Q.5. Construct π molecular orbital energy level diagram for tris (methylene) methane *i.e.,* $C(CH_2)_3$.

Ans. Tris (methylene) methane *i.e.,* $C(CH_2)_3$ has molecular point group D_{3h}. It is shown in fig.

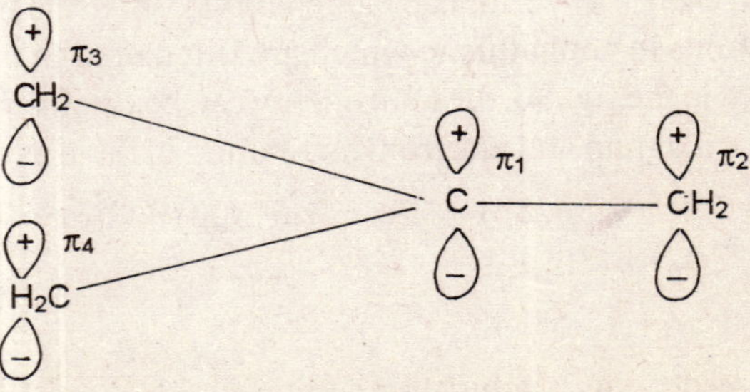

Let us consider π_1, π_2, π_3, π_4 orbitals as shown. The reducible representation for this can be generated studying the effect of various operations of D_{3h} on these π orbitals as

D_{3h}	E	$2C_3$	$3C_2$	σ_h	$2S_3$	$3\sigma_v$
Central C-atom	1	1	−1	−1	−1	1
T_π terminal carbon	3	0	−1	−3	0	1

On reduction of T_π result is

$$T_\pi = A_2'' \oplus E''$$

So, using the projection operators SALC's for the M.O.'s are

$$\Psi_{A2}'' = 1/\sqrt{3}\,(\pi_1 + \pi_2 + \pi_3)$$
$$\Psi_E'' = 1/\sqrt{6}\,(2\pi_2 - \pi_3 - \pi_1)$$
$$\Psi_E'' = 1/\sqrt{2}\,(\pi_2 - \pi_3)$$

The π M.O. diagram is shown in the fig.

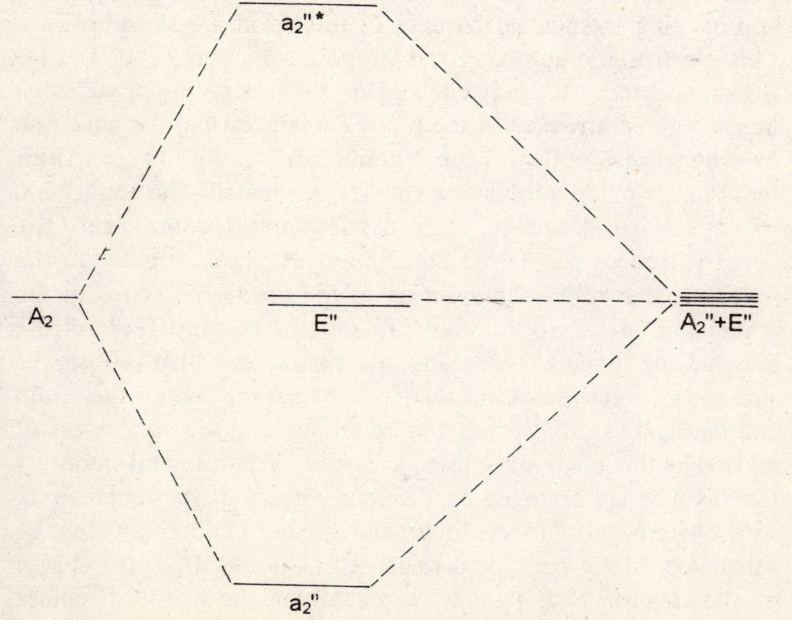

8

Group Theory and Concepts of Bonding – III

Another theory that deals with formation of transition metal complexes is the crystal field theory. This theory is based upon the concept of electrostatic attraction between metal ion and the approaching ligands in the case of transition metal complexes. This electrostatic approach was proposed by Van Arkel and De Boer. According to their concept of electrostatic approach that ligands are so arranged in the metal complexes that the attraction between the metal and the ligands are maximum and their repulsion is minimum. One of the key elements in the application of crystal field theory to transition metal complexes, the construction of so, called correlation diagrams, these are the diagrams that relate the splitting of the electronic states of the transition metal ion under the perturbation effect of the approaching ligands. This leads to a parameter which reflects the strength of electrostatic interaction between the central metal ion and the ligands. Six ligands can be arranged in the six corners of an octahedral complex while in case of a tetrahedral complex four ligands are arranged on the four corners of the tetrahedron. The electrostatic force of attraction between them can be calculated on the basis of (a) charge of the metal ion (b) charge on the approaching ligands (c) metal and the ligand distances and lastly (d) their mutual polarization.

This theory has basic assumption that ligands get arranged around the central metal ion due to electrostatic force of attraction between the metal ion and ligands resulting in the structure with least energy. Due to this electrostatic force of attraction the complex system attains stability and this system liberates energy which contributes in significant manner in the formation of a complex.

Section 8.1

Crystal Field Splitting

According to crystal field theory ligand ions create an electrical field around the central metal ion as result of which the energies of central metal orbitals get perturbed. There is a case that the electrical field generated with effect of ligands is of spherical symmetry, in such a case all s, p and d orbitals experience the uniform rise in their energies. In the case of octahedral complexes ligands approach through the six axes *i.e.,* from the corners of the octahedron, in such a case they still effect the p_x, p_y and p_z orbitals equally but in this case the d-orbitals get effected or in other words the degeneracy of p-orbitals remains same while the degeneracy of d-orbitals get resolved in case of octahedral complexes as all the d-orbitals are oriented in different ways as d_{x2-y2} and d_{z^2} are along the axes while d_{xy}, d_{yz} and d_{xz} are in between the axes system (see shapes of d-orbitals, ref. 4.3). Therefore, d-orbital's degeneracy is resolved and some splitting occurs in the d-orbitals set as result of the approach of the ligands from the corners of an octahedron to the central metal ion. This splitting can be studied with the help of group theory. Octahedral complexes belong to O_h molecular point group, as d-orbitals are symmetrical w.r.t. inversion operation, therefore, on excluding the inversion operation from the O_h point group we are left with O subgroup which contains identity and rotation operations such as C_4,C_3 and C_2. We can perform the related operations to the O subgroup on the d-orbitals and on the basis of d-orbitals we can find out the $T_{reducible}$ *i.e.*, a reducible representation assigning +1 to the orbital with no change during any operation and –1 to the orbital for any change during any operation. In identity operation on d-orbitals we get total character equal to five as all the five d-orbitals remain unchanged during E operation. During C_4 operation d-orbitals change as follows

d_{z^2}	–	+1
$d_{x^2-y^2}$	–	–1
d_{xy}	–	–1
d_{xz}	–	0
d_{yz}	–	0
Total	=	–1

So, the total character during C_4 operation is –1. It is cumbersome to imagine the changes in the d-orbitals on other rotation operations. So, we can frame a matrix for such a rotation operation. While defining any orbital as a wave function quantum mechanically we can express the wave function as a function of polar co-ordinates *i.e.*, r_1, θ and ϕ (section 4.3). As radial part is non-directional, so, it remains unchanged during any symmetry operation and as angle θ is defined relative to the principal axis (*i.e.*, the rotation axis), so, it also remains unchanged during any rotation and can also be ignored. So, it is only ϕ angle which will change and this function has its mathematical form as $e^{im\phi}$ and for d-orbitals $l = 2$ and m = –2, –1, 0, +1, +2. So, the effect of rotation by α on $e^{im\phi}$ is as follows:

$$\begin{bmatrix} e^{2i\phi} \\ e^{i\phi} \\ e^{0} \\ e^{-i\phi} \\ e^{-2i\phi} \end{bmatrix} \qquad \begin{bmatrix} e^{2i(\phi+\alpha)} \\ e^{i(\phi+\alpha)} \\ e^{0} \\ e^{-i(\phi+\alpha)} \\ e^{-2i(\phi+\alpha)} \end{bmatrix}$$

which can be expressed in the matrix as

$$\begin{bmatrix} e^{2i\alpha} & 0 & 0 & 0 & 0 \\ 0 & e^{i\alpha} & 0 & 0 & 0 \\ 0 & 0 & e^{0} & 0 & 0 \\ 0 & 0 & 0 & e^{-i\alpha} & 0 \\ 0 & 0 & 0 & 0 & e^{-2i\alpha} \end{bmatrix}$$

A general form of this rotation matrix of any set of orbitals is

$$\begin{bmatrix} e^{\ell i\alpha} & 0.......................0 & 0 \\ 0 & e^{(\ell-1)\alpha}..................0 & 0 \\ . & . \quad\quad\quad\quad . & . \\ . & . \quad\quad\quad\quad . & . \\ 0 & 0.......................e^{(1-l)\alpha} & 0 \\ 0 & 0.......................0 & e^{-\ell i\alpha} \end{bmatrix}$$

When $\alpha = 0$ each diagonal element is equal to one. The total character of this rotation matrix is the sum of diagonal elements and it is given by

$$X(\alpha) = [\sin (\ell+½)\alpha]/[\sin \alpha/2]$$

As for C_2 operation $\alpha = \pi$ radian (or 180°)

And for C_3 operation $\alpha = 2\pi/3$ radian (or 120°)

Similarly for C_4 operation $\alpha = \pi/2$ radian (or 90°)

Substituting the values of we get

$$X(\alpha) \text{ for } C_2 = 1$$

$$X(\alpha) \text{ for } C_3 = -1$$

and $$X(\alpha) \text{ for } C_4 = -1$$

Therefore, the $T_{reducible}$ using the d-orbitals set as basis function is

	E	C_4	C_3	C_2
$T_{red.}$	5	–1	–1	1

This can be reduced to the following irreducible representations for O sub group

$$T_{red.} = E \oplus T_2$$

On summarizing the inversion operation we get the irreducible representations for main O_h molecular point group as

$$T_{red.} = E_g \oplus T_{2g}$$

The corresponding orbitals to these irreducible representations are

$$E_g = d_{z^2}, d_{x^2-y^2}$$

And $$T_{2g} = d_{xy}, d_{yz}, d_{xz}$$

So, on the basis of symmetry consideration or on the basis of group theory we can conclude that a penta-degenerate set of d-orbitals splits-up into two sets, one doubly degenerate (*i.e.*, e_g set of $d_{z^2}, d_{x^2-y^2}$) and other triply degenerate (*i.e.*, t_{2g} set of d_{xy}, d_{yz}, d_{xz}) under the electrostatic effect of the approaching ligands. By the similar symmetry analysis the result summarized in the table 8.1, can be obtained. The relative difference in the energies of t_{2g} and e_g set of orbitals can be qualitatively realized on examining at the orientations of the ligands. As ligands approach from the axes or the corners of the octahedron, so, the orbitals which are directed towards the corners of the axes or the octahedron *i.e.*, the $d_{z^2}, d_{x^2-y^2}$ orbitals or the e_g set of orbitals experienced more repulsion as compared to the other d-orbitals *i.e.*, d_{xy}, d_{yz}, d_{xz} or the t_{2g} set. Therefore, the splitting that occurs can be shown as in fig. 8.1.

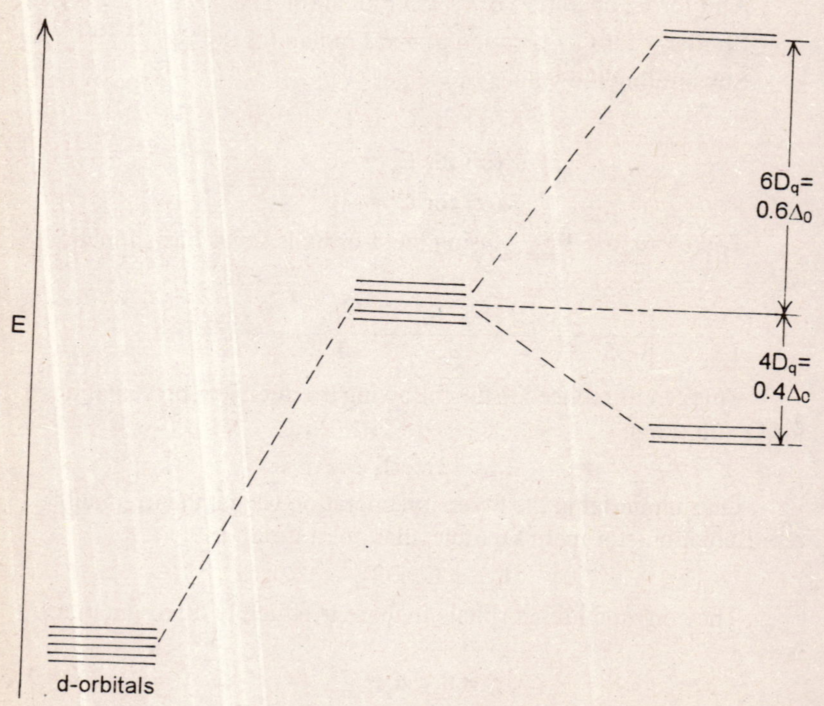

Fig. 8.1

Table 8.1: Representations for various orbitals in O_h symmetry.

Type of orbital	*ℓ value*	*Irreducible representations*
s	0	a_{1g}
p	1	t_{1u}
d	2	$e_g \oplus t_{2g}$
f	3	$a_{2u} \oplus t_{1u} \oplus t_{2u}$

The difference in the energies of e_g and t_{2g} set of orbitals is termed as Δ_0 or 10Dq as shown, where $Dq = [1/6][Ze^2 r_2^4]/[a^5]$. In this equation, e = electronic charge, Z = charge on ligand, r_2^4 = fourth power means average radius of d-orbitals and a = metal ligand distance. The rearrangement of electrons in these split d-orbitals results in lowering of the total orbital energy and which is known as crystal field stabilization energy (or CFSE).

$$T_{2g} \text{ energy decreases by } -0.4\ \Delta_0 = -4Dq$$
$$E_g \text{ energy increases by } +0.6\ \Delta_0 = +6Dq$$

Now, consider d^{p+q} orbital where p is the number of electrons in t_{2g} set of orbitals and q is the number of electrons in e_g set of orbitals. So, it is quite obvious that

Gain of the energy due to q electrons of e_g set of orbitals = $q(0.6\ \Delta_0)$;

Loss in energy due to p-electrons of the t_{2g} set of orbitals = $p(-0.4\Delta_0)$. So, total change in energy in terms of Dq is given by

$$(-4p + 6q)\ Dq \text{ or } (-0.4p + 0.6q)\ \Delta_0, \text{ which is the C.F.S.E.}$$

C.F.S.E. stabilizes the d-orbitals by lowering their energy which results from the splitting of d-orbitals into t_{2g} and e_g set of orbitals. Term P is added for each paired electron in C.F.S.E. *e.g.*, $p(-4Dq) + q(+6Dq) + nP$ where n is the number of paired electrons.

Splitting in tetrahedral field

If we apply the same procedure on the d-orbitals in the tertrahedral field *i.e.*, under the influence of T_d molecular point group, we can generate the $T_{reducible}$. On reduction of this reducible representation, we can conclude that the degeneracy of d-orbitals also get resolved in the tetrahedral field as $T_{reducible}$ get reduced to

$$T_{red.} = e \oplus t_2$$

The corresponding orbitals to these irreducible representations are

$$E = d_{z^2}, d_{x^2-y^2}$$

And $$T_2 = d_{xy}, d_{yz}, d_{xz}$$

The splitting is shown in fig. 8.2. This can be seen that there is reversal in the order of energies of the two sets as compared to octahedral field. In case of tetrahedral field the ligands approach from the corners of the tetrahedron *i.e.,* in between the axes. So, the orbitals which are located in between the axes experience more repulsion as compared to the d-orbitals which are axial or in other words d_{xy}, d_{yz}, d_{xz} (*i.e.,* t_2 set of orbitals) experience more repulsion as compare to $d_{z^2}, d_{x^2-y^2}$ (*i.e.,* set of orbitals). It is also a fact that the splitting in case tetrahedral field is approximately 4/9 of the splitting of d-orbitals in case of octahedral field *i.e.,*

$$\Delta t = 4/9\ \Delta_0$$

or $$(10Dq)_t = (4/9)\ (10Dq)_0$$

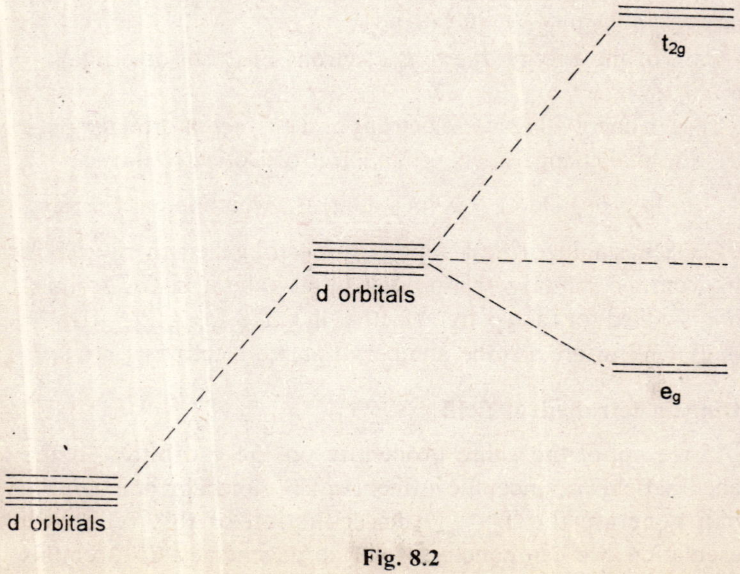

Fig. 8.2

Section 8.2

Distortion in the octahedral and tetrahedral complexes

Distortion in the octahedral complexes can be explained on the basis of John-Teller effect. In distorted octahedral field due to the effect of distortion ligands which are at axial positions in the octahedron experience comparative less or more interaction force. Distortion in the octahedral field yields the tetragonal because of which the symmetry of the ligand field is reduced to D_{4h} molecular point group. Considering the d-orbitals as basis set we can get the reducible representation which can be reduced to

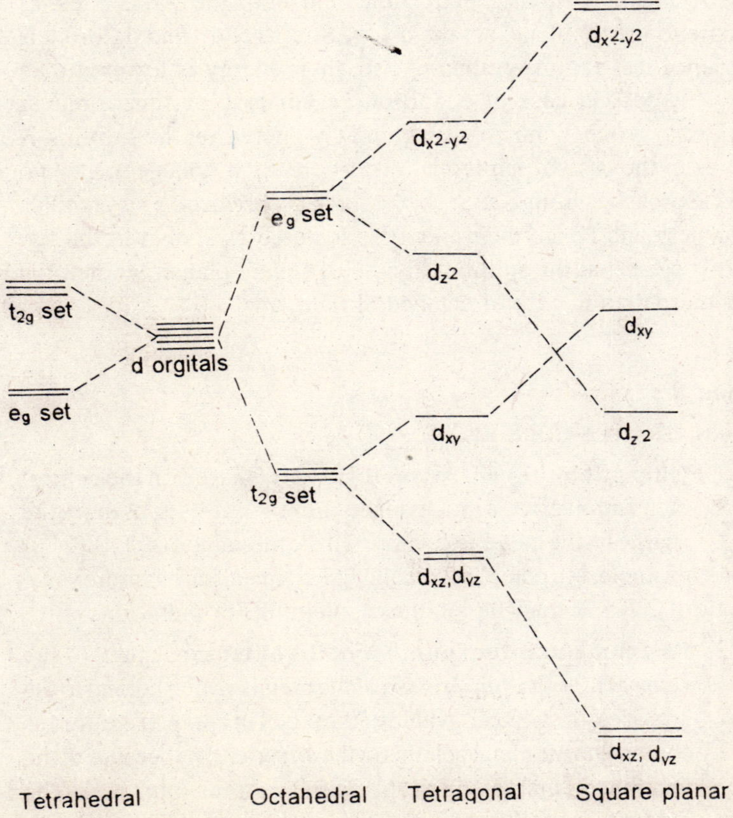

Fig. 8.3

$$T_{reducible} = a_{1g} \oplus b_{1g} \oplus b_{2g} \oplus e_g$$

The corresponding d-orbitals are

$$A_{1g} = d_{z^2}$$
$$B_{1g} = d_{x^2-y^2}$$
$$B_{2g} = d_{xy}$$
$$E_g = d_{yz}, d_{xz}$$

The doubly degenerate e_g set of orbital further split-up into d_{z^2} and $d_{x^2-y^2}$ orbitals corresponding to a_{1g} and b_{1g} symmetry and t_{2g} set of orbitals split-up into d_{xy} (b_{2g} symmetry) and d_{yz}, and d_{xz} orbitals (e_g symmetry) under the effect of distortion. In the tetragonal distortion in the octahedral field the electrostatic field along the z-axis is less as compare to the xy-plane, so, the $d_{x^2-y^2}$. Similarly d_{xz} and d_{yz} orbitals experience less repulsion than d_{xz}, so, their energy is lowered than d_{xy}. In the second case of distortion *i.e.,* in case of square planar complexes, group symmetry is again D_{4h}. Therefore, the splitting is same as in the case of tetragonal distortion. As in this geometry no ligand approaches from z-axis, so, the force experienced bv d_{z^2} orbitals is very less and hence their energy is lowered as shown in the fig. 8.3. It is clear that the splitting in case of square planar geometry is much more than in case of octahedral field, and

$$\Delta_{sp} = 1.22\Delta_0$$

Section 8.3

Factors effecting C.F.S.E.

1. **Charge on the central metal atom:** Charge on the central metal atom effects much on the value of C.F.S.E. as more the charge on the metal atom more will be the value of C.F.S.E. For example, in a complex containing Fe^{3+} metal ion has more value of C.F.S.E. than the complex containing Fe^{2+} metal ion.

2. **Principal quantum number of the outer most shell of the central metal atom:** In case of the metals which belong to the 3d series, the C.F.S.E. value of the metal in complex is more as compare to the metal belongs to the 4d series. It is because of the fact that on increasing the principal quantum number of the outer most shell there is increase in the value of the C.F.S.E. Therefore, the C.F.S.E. value in case of a platinum complex is

more as compare to the palladium metal complex.

3. **Position of the ligand in the spectrochemical series:** Ligands according to their ability to split the d-orbitals of the metal ion, are arranged in a series which is known as spectrochemical series. In this series, the ligands which are capable of producing strong field around the metal ion are placed before the ligands which are capable of producing a weak field around the central metal ion. This series is given below,
$I' < Br' < Cl' < SCN' < H_3^- < (C_2H_5O)_2^- < PS^-_2 < F' < (NH_2)_2 < CO < OH^- < C_2O_4^{2-} < H_2O < NCS^- < H^- < CN^- < NH_2CH_2CO^- < NH_3 < C_5H_5N < en < SO_3^{2-} < NH_2OH < NO_2^- < Phen < H^- < CH_3^- < CH^-CO$
Therefore, in nutshell we can say that a ligand co-ordinating through C-Atom are capable of giving more value of C.F.S.E. as compare to the Ligand co-ordinating through halogens.

4. **Size of the ligands:** The force of repulsion experienced by small ligands is Lesser from the central metal ion and hence they can approach the metal ion very closely resulting in the great splitting. Therefore, the value of Δ_0 for halogens follows the following order

$$F^- > Cl^- > Br^- > I^-$$

5. **Geometry of the complex :** It has been already discussed that with the Change of geometry of the complex the value of CFSE also changed (ref. section 8.1, 8.2). The order of splitting for different geometries is

$$\Delta_t = 0.45\ \Delta_0$$
$$\Delta_{sp} = 1.33\ \Delta_0$$

Section 8.3

Applications of C.F.S.E. value

(i) **Crystal structure of spinlas:** Mixed oxides of the formula $A^{2+} + B_2^{3+} + O_2^{4-}$ are called as spinals. There are chiefly two types of spinals (a) normal and (b) inverse spinals. In the normal spinals all the bivalent metals A^{2+} occupy one of the eight available tetrahedral holes and all B^{3+} occupy the half of the available octahedral holes. In case of inverse spinals all the A^{2+} metals and half of the B^{3+} metals are present in the octahedral holes other B^{3+} metals occupy the tetrahedral holes. The inverse spinals are represented as $B^{3+}[A^{2+} B^{3+}] O_4^{2-}$ and the normal spinals are represented as $A^{2+} [B^{3+}]O_4^{2-}$. Using the CFSE values of metal oxides we can interpret the geometry of spinals.

(ii) **Stablization of oxidation states:** CFSE value also explains that why certain oxidation states are preferably stable w.r.t. certain ligands. e.g. consider two high spin (weak field) complexes of cobalt, viz. $[Co(H_2O)_6]^{2+}$ and $[Co(H_2O)_6]^{3+}$. Experimentally it is shown that Co^{2+} is more stable than Co^{3+} because of the more CFSE value of Co^{2+} as compared to Co^{3+} in case of weak field or octahedral complexes.

(iii) **Stereochemistry of complex:** CFSE values are also used to predict the stereochemistry of a complex. e.g. Cu^{2+} ions form square planar complexes rather than tetrahedral complexes or the octahedral complexes as the CFSE of this ion in square planar geometry is more.

(iv) **Heats of hydration of bivalent ions of the first row transition series:** Heat of hydration is the heat evolved during the hydration process of metal

$$M^{2+}(g) + H_2O \rightarrow [M(H_2O)_6]^{2+}(aq)$$

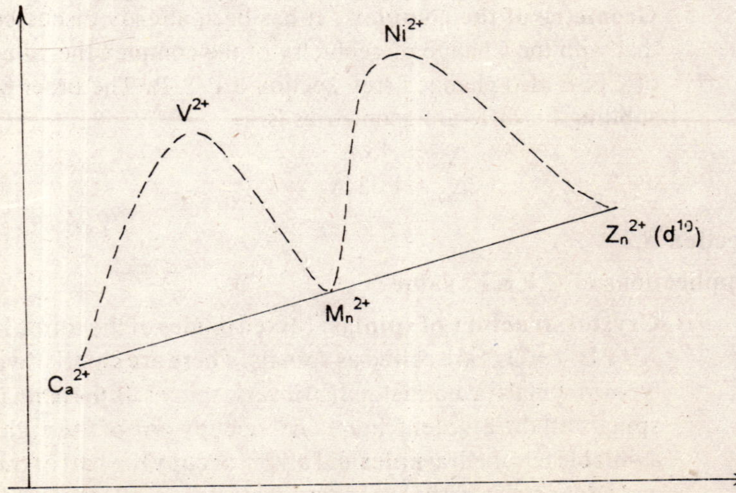

Fig. 8.4

As it is known that as the ionic radii of bivalent metal ions of the first row transition elements decrease, the experimental values of heat of hydration should regular increase from Ca^{2+} to Zn^{2+}. But its actual variation is shown in the fig. 8.4. It is due to the fact CFSE is minimum for $d^0(Ca^{2+})$, $d^5/(M_n^{2+})$ and $d^{10}(Zn^{2+})$ and maximum for V^{2+} and Ni^{2+} ions.

(v) **Magnetic properties of metal complexes:** One of the major application of CFSE is that it helps in understanding the magnetic properties of the transition metal complexes. A complex having paired electrons has no magnetic moment and such a complex is diamagnetic in nature while the complex which has one or more unpaired electrons has a definite magnetic moment and is attracted in the magnetic field. Such a substance is called as paramagnetic. Greater the number of unpaired electrons more will be the magnetic moment. As in a complex the distribution of d-electrons depends upon the crystal field splitting energy (Δ_0) and the pairing energy (P). When the splitting takes place two possibilities arise. (a) $P > \Delta_0$ and (b) $P < \Delta_0$. According to Hund's rule usually the electrons remain unpaired because of the repulsive force between the electrons prevent them from pairing but if the energy available from the splitting is more than this repulsive force pairing occurs. Hence if splitting is more electrons in d-orbitals remain unpaired *e.g.*, $[COF_6]^{3-}$ whereas when the value of Δ_0 is higher pairing occurs, *e.g.*, $[CO(NH_3)_6^{3+}]$. The spin only formula for the calculation of magnetic moment is $\mu_s = \sqrt{n(n+2)}\beta_m$ on the basis of which the magnetic moment values are given in the table 8.2.

Table 8.2: Spin only (μ_s) magnetic moment values for the ions with different numbers of unpaired electrons:

No. of unpaired electrons in the system	*Magnetic moment Spin only (μ_s)*
0	0
1	1.73
2	2.83

3	3.87
4	4.90
5	5.92

Section 8.4

Limitations of crystal field theory

(1) C.F.T considers only the d-orbitals of central metal ion and it gives no consideration at all for other *i.e.,* s, p orbitals and ligand π orbitals for which M.O. theory gives some explanation, So, to explain the properties of complexes which are formed by the involvement of π orbitals of ligands, are outside the scope of this theory.

(2) CFT is unable to account for satisfactorily for relative strengths of ligands. e.g. it gives no explanation that why H_2C is stronger ligand than OH^- in the spectrochemical series.

(3) According to CFT the bond of metal to ligand is purely ionic. It gives no account of the partly covalent nature of M-L bonds. Thus, the effects that directly dependent upon covalency can't be explained by CFT.

(4) It can't explain why four co-ordinated complexes form two structures *i.e.,* tetrahedral and the square planar.

Problems

Q. 1. Which of the following complexes amongst (i) and (ii) will have higher Δ_o and why?

(i) $[Co(NH_3)_6]^{3+}$ or $[Rh\ (NH_3)_6]^{3+}$

(ii) $[Fe\ (Ox)_3]^{4-}$ or $[Fe\ (Ox)_3]^{3-}$

Ans. (i) $[Rh\ (NH_3)_6]^{3+}$ as the principal quantum number in case of Rh is higher than as compare to Co for the last shell, so, because of increase in principal quantum number there will be approximately 33% increase in 10Dq or Δ_o.

(ii) $[Fe\ (Ox)_3]^{3-}$ will have the more 10Dq or Δ_0 value as on increasing charge on the central metal atom there is increase in the 10Dq or Δ_0.

Q. 2. Calculate the relative energies as a function of Δ_0 and P (pairing energy) for high spin and low spin octahedral complexes of Fe (II) and Co(II).

Ans. (i) High spin complex of Fe (II) (d^6) will have the configuration $t_{2g}^{\ 4}\ e_g^{\ 2}$ with the corresponding as 4(–4Dq) + 2(6Dq) + P = –4Dq + P and Fe (II) in low spin have the configuration $t_{2g}^{\ 6}\ e_g^{\ 0}$ and the corresponding Δ_o will be 6(–4Dq) + 0(6Dq) + 3P = –24Dq + 3P.

(ii) In the case of Co(II) (d^7) high spin complex the configuration of Co(II) is $t_{2g}^{\ 5}\ e_g^{\ 2}$ with equal to –8Dq + 2P and in case of low spin complex Co(II) has the configuration $t_{2g}^{\ 6}\ e_g^{\ 1}$ with value = –18Dq + 3p.

Q. 3. For Mn^{3+} ion pairing energy is 80 Kcal/mol for $[Mn(H_2O)_6]^{3+}$ Δ_0 = 60 kcal/mol and for $[Mn(CN)_6]^{3-}$ Δ_0 = 110 kcal/mol. Do these complexes have high spin or low spin configuration.

Ans. Mn^{3+} ($3d^4$) system has configuration $t_{2g}^{\ 3}\ e_g^{\ 1}$ in case of high spin complex *i.e.*, $[Mn(H_2O)_6]^{3+}$ as in this case $\Delta_0 < P$. But in the case of $[Mn(CN)_6]^{3-}$ $\Delta_0 > P$, so, this complex is low spin complex with configuration $t_{2g}^{\ 4}\ e_g^{\ 0}$.

Q. 4. Using spin only formula, calculate the magnetic moment in βm for the following ions.

Cr^{3+}, Fe^{3+}, Mn^{2+}, Cu^{2+}, Ni3+, V^{4+} and Ti^{4+}

Ans. As $\mu_s = \sqrt{n(n+2)}\ \beta m$ using spin only formula where n = number of unpaired electrons in any ion, so, the magnetic moment for these ions will be

Ion	*No. of unpaired electrons*	μ_s
$Cr^{3+}(3d^3)$	3	3.87
$Fe^{3+}(3d^5)$	5	5.92
$Mn^{2+}(3d^5)$	5	5.92
$Cu^{2+}(3d^9)$	1	1.73
$Ni^{3+}(3d^7)$	3	3.87
$V^{4+}(3d^1)$	1	1.73
$Ti^{4+}(3d^0)$	0	0.00

9

Group Theory and Spectroscopy–I

In the previous chapters, concepts of bonding are explained in light of group theory and symmetry of the molecules. Group theory is also helpful in the spectroscopic studies of the molecules.

When light is absorbed by the molecule in the UV-visible region, the absorption of in the UV–visible region is linked with the excitation of electrons from the ground state to the higher energy excited states of the molecule. As electrons are filled-up in M.O.'s in increasing order of their energy, so, on absorption of radiation electronic excitations takes place from the highest occupied molecular orbital (HOMO) to the lower unoccupied molecular orbital (LUMO). Depending upon the separation of levels (HOMO–LUMO) that are involved in a UV-visible transition, energy is absorbed in UV or in visible region. Transitions in the UV-visible region are electronic transitions which are associated with the electrical dipoles of the molecules. The symbols x, y, z can be also be taken to represent the components of dipoles. M_x, M_y, M_z of the dipole moment vector M. Since these components lie along x, y and z axes they must transform in a manner similar to the translation of vectors. Therefore, group theory is again useful in the electronic spectroscopic studies of the compounds. In this chapter, we deal with the fundamental details of UV-visible spectroscopy, various transitions involved, transition moment integral, charge transfer spectra, concepts of electronic spectroscopy as related to metal complexes and construction of Orgal diagrams etc.

Section 9.1

Fundamentals of UV-visible spectroscopy

According to the Planck's quantum theory E = hυ where υ is the frequency and E is the energy associated with a radiation and h is the Planck's constant. When electronic excitation takes place, we can express

$$\Delta E = E_{higher} - E_{lower} = h\upsilon$$

If the lower energy state is the ground state than

$$E_{excited} - E_{ground} = h\upsilon$$

Usually this difference is equal to 5eV. So,

$$E_{excited} - E_{ground} = 5eV$$

and the corresponding frequency is 1.2×10^{15} cycles/sec and the wave length according to $c = \upsilon\lambda$ is 2500 Å. So, the light of wavelength approximately 2500 Å is capable to excite the electron or for any electronic transition. The UV region light starts from blue and (~ 4000Å) and goes upto (~2000Å). We can divide it into two regions

(i) Light of 2000Å to 4000Å wavelength which is known as near UV region and

(ii) Light below 2000Å wavelength (~1000-2000Å) which is far UV or vacuum UV region. This can be very well explained by the Fig. 9.1.

Therefore, it is clear that electromagnetic radiations of UV and visible region are capable for electronic excitations in chemical species; or in other words these radiations are useful for generating electronic transitions and so electronic spectra. Therefore, this spectroscopy is known as electronic spectroscopy.

UV		Visible						
Far or Vaccuum UV	near UV	Violet	Indigo	Blue	Green	Yellow	Orange	Red

Fig. 9.1

Born-Openhiemer approximation and Frank-Condon rule

The complete internal energy of the molecule is the sum of (i) translational (E_{trans}) (ii) rotational (E_{rot}) (iii) vibration (E_{vib}) and (iv) electronic (E_{elec}) energies which can be expressed as

$$E = E_{elec} + E_{vib} + E_{rot} + E_{trans.}$$

Therefore during absorption of energy

$$\Delta E = \Delta E_{elec} + \Delta E_{vib} + \Delta E_{rot} + \Delta E_{trans.}$$

As translational distances are large, so, they can be kept out from the quantum restrictions. So, according to this approximation

$$E = E_{elec} + E_{vib} + E_{rot.}$$

Various rotational and vibrational energy levels which are associated with any electronic state are shown in Fig. 9.2. If only vibrational levels are separated associated with the electronic states, then course structure may be obtained otherwise if rotational levels are also separated in well defined manner a fine structure may be obtained.

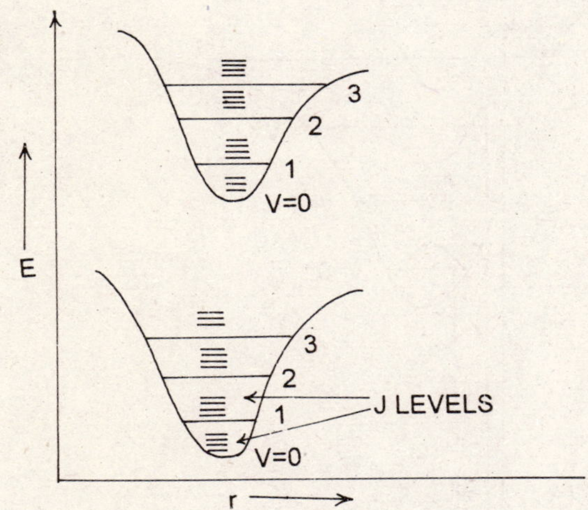

Fig. 9.2

So, for fine structure

$$\Delta E = \Delta E_{elec} + \Delta E_{vib} + \Delta E_{rot}$$

And for the course structure

$$\Delta E = \Delta E_{elec} + \Delta E_{vib}$$

There is no quantum restriction or selection rule for transitions associated with change in vibrational quantum number during an electronic transition. In the ground state the most populous level is the lowest quantum level. Therefore, transition starts from this level and is possible upto any vibrational level associated with higher electronic state. Experimentally, it is known that transitions are not equally intense. This fact is based upon a principle called Frank-Condon rule according to which "Electronic transitions are so fast that during any transition internuclear distribution are not effected", means that electronic excitations are so fast that in comparison to them atoms present in the molecule do not get sufficient time to establish them-selves in the vibrations and internuclear distributions are not possible during any electronic transition. The rule can be explained by the transitions as shown in Fig. 9.3.

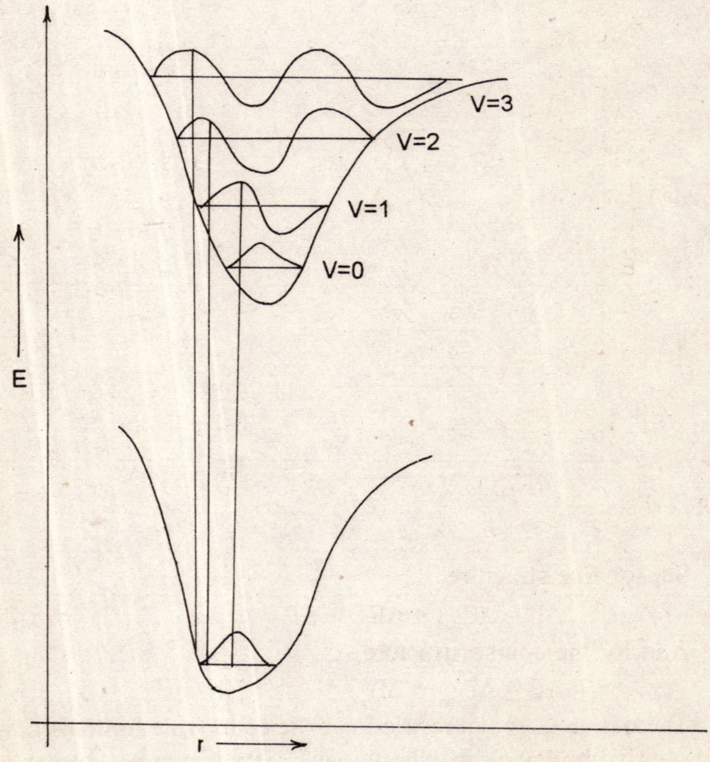

Fig. 9.3

Various Posible Transitions

As seen in the case of M.O. diagrams of molecules, the lower filled M.O.'s may be σ_b π_b or non-bonding orbitals. So, electronic transitions can take place from σ_b or π_b or n to σ_a* or π_a*. Therefore, the possible transitions are $\sigma \rightarrow \sigma^*$, $\pi \rightarrow \pi^*$, $n \rightarrow \pi^*$ and $n \rightarrow \sigma^*$, in a molecule as shown in figure 9.4. Each and every molecule is characterized by a specific number of spectral bands. As in case of saturated hydrocarbons HOMO is $\sigma_{bonding}$ and the corresponding transition is $\sigma \rightarrow \sigma^*$ which is high energy tansition, so, the absorption for this transition takes place in far UV or vacuum UV region and in case

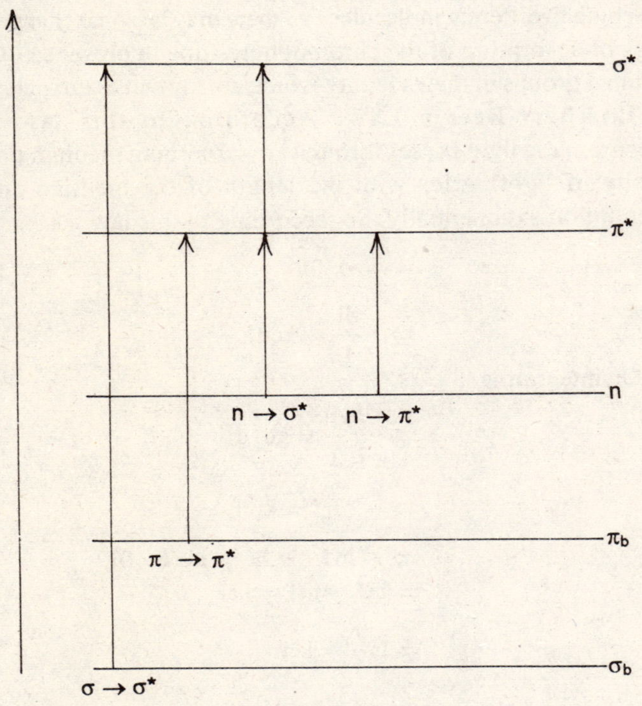

Fig. 9.4

of propane it is at λ_{max} = 135 nm. In the compounds like CH_3Cl, CH_3OH or CH_3NH_2 which have non-bonding (n) electrons, $n \rightarrow \sigma^*$ transitions are possible apart from $\sigma \rightarrow \sigma^*$ that too in far UV region *e.g.*, λ_{max} ($n \rightarrow \sigma^*$) for CH_3OH is at 183 nm, for CH_3Cl is at 173 nm and for $(CH_3)_2$ NH at 227 nm. In case of double or triple bonded compounds the highest filled M.O. is π-bonding orbital. So, the important transition in these compounds is $\pi \rightarrow \pi^*$ which appears approximately at $\lambda_{max} = 171$ nm.. In case of aldehydes and ketones $n \rightarrow \pi^*$ transition can take place in >C=O group. e.g. it appears approximately at 270-285 nm in case of carbonyl compounds.

The functional groups which show characteristic absorption in the low energy range are called as chromophores. The same functional group in different molecules does not absorb at exactly the same wavelength. The absorption a group may occur over a range of wavelength in different molecules *i.e.,* there may be some change in the energy of absorption of the chromophores due to presence of other functional groups in their vicinity which are known as auxochromes.

Lambert-Beer's Law: According to this law when monochromatic light passes through any absorbing medium then the intensity of light varies with the length of the medium and the concentration exponentially. So, according to this law

$$-\frac{dI}{I} \propto c\,dl$$

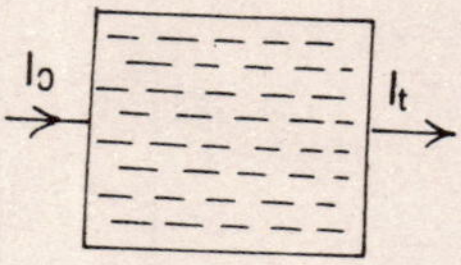

or

$$\Rightarrow -\frac{dI}{I} = kc\,dl$$

On integrating it

$$\Rightarrow -\int_{I_0}^{I_t} \frac{dI}{I} = kc\int_0^l dl$$

$$\Rightarrow -\ln I\Big|_{I_0}^{I_t} = kcl\Big|_0^l$$

$$\Rightarrow -(\ln I_t - l_n I_0) = kc(l - 0)$$

$$\Rightarrow \ln I_0 - \ln I_t = kcl$$

$$\Rightarrow \ln \frac{I_0}{I_t} = kcl$$

or

$$\log_{10} \frac{I_0}{I_t} = \frac{k}{2.303} cl$$

or

$$A = \varepsilon cl$$

Where A is the absorbance or optical density expressed as $\log_{10} \frac{I_0}{I_t}$ and k/2.303 is molar extiction coefficient and can be expressed as ε, its unit is $mole^{-1}\,dm^3 mm^{-1}$. Usually this law can be applied to explain the intensity of the spectral lines or any transition. In actual practice usually absorbance, transmittance or are plotted against wavelength or frequency. From this law

$$\in = \frac{A}{cl}$$

So, molar extinction coefficient or ε can be defined as the absorbance of one molar solution of the compound under investigation of path length 1 cm. Therefore, it is also known as molar absorption.

Section 9.2

Spectral transition probabilities

The intensity of any transition depends upon the probability of its occurring and this probability is in turn depends upon the integral known as transition moment integral which is

$$\int_{-\infty}^{\infty} \Psi_g \hat{\mu} \Psi_h d\tau$$

where Ψ_g is the wave function corresponding to the ground state and Ψ_h is the wave function of the higher state. This integral can also be written as

$$\int_{-\infty}^{\infty} \Psi_i \hat{\mu} \Psi_j d\tau$$

As $\hat{\mu}$ is the dipole moment operator here, so, a transition must involve the change in dipole moment. Mathematiclly,

$$\hat{\mu} = \hat{\mu}_x + \hat{\mu}_y + \hat{\mu}_z = e_i \Sigma x_i + e_i \Sigma y_i + e_i \Sigma z_i$$

So, I can be written as

$$\Rightarrow I \ \alpha \int \Psi_i \ \hat{\mu}_x \Psi_j d\tau$$

$$\Rightarrow I \ \alpha \int \Psi_i \ \hat{\mu}_y \Psi_j d\tau$$

and

$$I \ \alpha \int \Psi_i \ \hat{\mu}_z \Psi_j d\tau$$

For an allowed transition these integrals should be non-vanishing or should be non-zero. Here it is necessary to introduce the concept of vanishing integral.

Vansihing Integrals

Let us consider a simple integrals, of two functions, independently over the whole space

I^{st} $$\int_{-1}^{1} x^2dx$$ x^2 is even function

$$= \frac{x^3}{3}\Bigg|_{-1}^{1}$$

$$= \frac{1^3}{3} - \frac{(-1)^3}{3} = \frac{1}{3} + \frac{1}{3} \neq 0$$

IInd $$\int_{-1}^{1} xdx$$ x is odd function

$$= \frac{x^2}{2}\Bigg|_{-1}^{1}$$

$$= \frac{1^2}{2} - \frac{(-1)^2}{2}$$

$$= \frac{1}{2} - \frac{1}{2} = 0$$

Similarly this can be shown that integral of sin x over the whole space or $\int_{\infty}^{-\infty} \sin x \, dx$ is non-zero but integral of cos x over the whole space is zero. In other words, we can say that the integral of even function is non-vanishing but integral of odd function is vanishing. Now, applying this concept to the transition moment integral, if the product of $\Psi_i \, \hat{\mu} \, \Psi_j$ is even the integral is non-vanishing otherwise $\psi_i \, \hat{\mu} \psi_j$ product is odd integral is vanishing.

This can be checked group theoretically for the transition moment of any transition. Let us suppose that Ψ_i or the wave function of the lower state of the transition belongs to a irreducible representation T_σ and Ψ_j or the wave function of the higher state of the transition belongs to T_λ and since μ is the function of x, y, z, so, the irreducible representation belonging to x, y, or z can be seen from the character table of the particular point group of the molecule under consideration. So, the direct product of $T_\sigma \mu T_\lambda$ can be obtained and in this way we get another reducible representation. If this reducible representation on reduction results in A_1 (or A_{1g}) or any symmetric irreducible representation as one of the representation the product $T_\sigma \hat{\mu} T_\lambda$ will be even and the integral in this case will not be zero.

Example 9.1: For T_d point group $T_\lambda = T_2$, $\hat{\mu} = E$. Prove that the integral will be non-zero, if and only if T_σ is T_1 or T_2.

Solution: On the direct product of E T_2 can be obtained *i.e.*, $(T_1 \otimes \mu)$

T_d	E	$8C_3$	$3C_2$	$6S_4$	6σ
$T_E \otimes T_2$	6	0	–2	0	0

Which on reduction results in

$$T_{E \otimes T_2} = T_1 \oplus T_2$$

This proves that the integral is non-zero if and only if the third irreducible representation or the second function *i.e.*, T_σ is either T_1 or T_2.

Selection rules: From the discussion it is clear that for any transition the probability of the transition depends upon the transition moment integral or $P_{ij} \propto \mu_{ij}^2$

Where P_{ij} is the probability of transition from i^{th} to j^{th} levels and μ_{ij}^2 is the transition dipole moment which can be expressed as $\mu_{ij} = \int \psi_i \hat{\mu} \psi_j \, d\tau$ where ψ_i & ψ_j are the wave functions of ith & j^{th} levels associated with transition. We can express this function as

$$\Psi_{Tot} = \Psi_{elec} \Psi_{vib} \Psi_{rot} \Psi_{Nucl.}$$

According to Born-Openheimer approximation, the wave function may be defined as

$$\Psi_{Tot} = \Psi_{Elec.} \Psi_{vib.}$$

Therefore, the corresponding integral would be

$$\mu_{ij} = \int \Psi^i_{elec.} \Psi^i_{vib}\ \mu\ \Psi^j_{elec} \Psi^j_{vib} d\tau$$

As during electronic transition it may be considered that only Ψ_{elec} depends only upon $\hat{\mu}$ operator. Therefore, Ψ_{vib} may be taken as independent of $\hat{\mu}$ operator or the integral may be respresented as

$$\mu_{ij} = \int \Psi^i_{elec.} \hat{\mu} \Psi^j_{elec}\ d\tau \int \Psi^i_{vib} \Psi^j_{vib} d\tau$$

As vibrational function may be considered independent of dipole moment operator

$$\mu_{ij} = \int \Psi^i_{space} \hat{\mu} \Psi^j_{space} d\tau \int \Psi^i_{spin} \Psi^j_{spin} d\tau$$

As Ψ_{elec} may be taken as

$$\Psi_{elec} = \Psi_{space}\ \Psi_{spin}$$

Spin selection rule: This clearly shows that the spin is also important parameter which decide the fate of a transition that whether it is allowed or forbidden. If $\int \Psi^i_{spin} \Psi^i_{spin} d\tau$ is a vanishing integral the $\hat{\mu}_{ij}$ equals to zero and the transition will be forbidden. So, the transitions in which spin is conserved are allowed. Therefore, singlet to singlet and triplet to triplet transitions are allowed while singlet to triplet transition is forbidden. In conclusion

Singlet to Singlet and

Triplet to Triplet

transitions are allowed. Exception is phosphorescence in which triplet to singlet transition appears.

Orbital selection rule: For any allowed transition, the following integral should be non-zero

$$\int \Psi^i_{space} \hat{\mu} \Psi^j_{space} d\tau$$

or

$$\int \Psi^i_{orbital} \hat{\mu} \Psi^j_{orbital} d\tau \neq 0$$

Taking into account the symmetry consideration we can frame the selection rules. As $\hat{\mu}$ is the directional or vector operator with absence of centre of symmetry, so, it can be taken as ungerade *i.e.,*u. The centro-symmetric or gerade (g) function as even function is considered as even function but ungerade (u) function as odd function. So, for gerade to gerade (or g→g) transition

or $\int gug$ = odd = 0 as shown earlier that integral of odd function is zero. Similarly,

or $\int uuu$ = odd = 0, so, for u→u transition. Therefore, g→g, and u→u transition are forbidden but if we consider g→u and u→g transition, we will find that this integral is non-zero. So, the transitions will be allowed. Therefore, for g→u transition

or $\int guu$ = even = 0

and for u→g transition

or $\int uug$ = even = 0

Therefore, transitions where centro-symmetry is maintained, are forbidden. So, s→p, p→d and d→f transitions are allowed while d→d transitions are forbidden. These rules are known as Laporte selection rules. The symmetries of various M.O.'s are as follows

$\sigma_b = g;$

$\sigma^* = u;$

$\pi_b = u$ and

$\pi^* = g$

So, transitions $\sigma \rightarrow \sigma^*$ and $\pi \rightarrow \pi^*$ are allowed that appears in far UV region but in case of unsaturated hydrocarbons with conjugation, separation between π and π^* decreases and $\pi \rightarrow \pi^*$ transition can appear in the visible region. Thus long conjugated systems are coloured. Let us consider the electronic spectra of some simple compounds.

Section 9.3

Example 1: H_2O molecule: The ground state electronic M.O. configuration is $(a_1)^2 (b_2)^2 (b_1)^2 (b_2^*)^0 (a_1^*)^0$ (in terms of symmetries) as shown in the fig. 9.5.

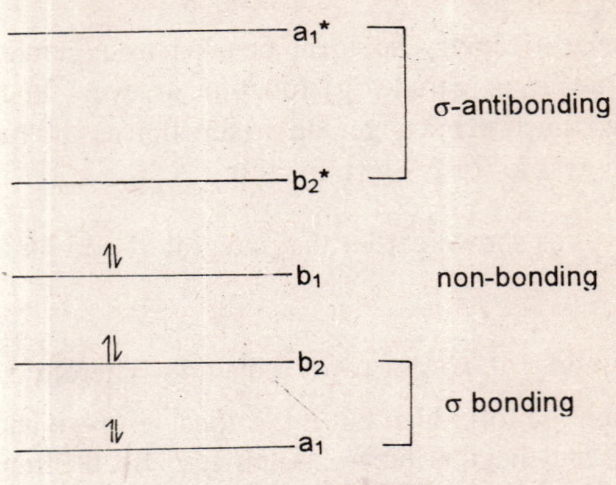

Fig. 9.5

The two lowest energy transitions that can occur will be $b_1 \rightarrow b_2^*$ and $b_1 \rightarrow a_1^*$ (which are $\sigma \rightarrow \sigma^*$ transitions). The symmetries of the excited states viz. $(a_1)^2$ $(b_2)^2$ $(b_1)^1$ $(b_2)^1$ or $(a_1)^2$ $(b_2)^2$ $(b_1)^1$ $(a_1)^1$ considering singly occupied orbitals that are A_2 and B_1 respectively, using direct product. So, the transitions in case of H_2O molecule are $A_1 \rightarrow A_2$ (electron from b_1 to b_2 M.O.'s) and $A_1 \rightarrow B_1$ (electron from b_1 to a_1). As H_2O molecule has C_{2v} molecular point group and μ belongs to A_1, B_1 and B_2 and so, the direct products are

$$(b_1 \rightarrow b_2) \qquad A_1 \times \begin{pmatrix} A_1 \\ B_1 \\ B_2 \end{pmatrix} \times A_2$$

i.e., A_2 or B_2 or B_1. So, the totally symmetric representation (*i.e.*, A_1) is absent. Therefore, this transition $(b_1 \rightarrow b_2)$ is forbidden.

IInd transition

$$(b_1 \rightarrow a_1) \qquad A_1 \times \begin{pmatrix} A_1 \\ B_1 \\ B_2 \end{pmatrix} \times B_1$$

Results of direct product for this transition are B_1 or A_1 or A_2, so, this contain A_1, the totally symmetric representation. Therefore this transition is allowed. So, in case of H_2O molecule, transition $b_1 \rightarrow a_1$ which is $n \rightarrow \sigma^*$ transition is allowed which appears in the UV region at 172 nm.

Example 2: Let us consider the example of HCHO. Its point group is again C_{2v}. The M.O. configuration for HCHO is

$1(SC)^2\ 1(SO)^2\ 2(SO)^2\ (\sigma_{CH}')^2\ (\sigma_{CH}'')^2\ (\sigma_{CO})^2\ (\pi_{CO})^2\ nb^2\ \pi_{CO}^*\ \sigma_{CO}^*\ \sigma_{CH}^*$

So, the possible transition can occur from $\sigma_{CO}^2\ \pi_{CO}^2$ and nb^2 to the higher unoccupied M.O.'s which are anti-bonding M.O.'s. So, the symmetries of the M.O.'s on >C=O have only been considered. These M.O.'s are shown in the Fig. 9.6.

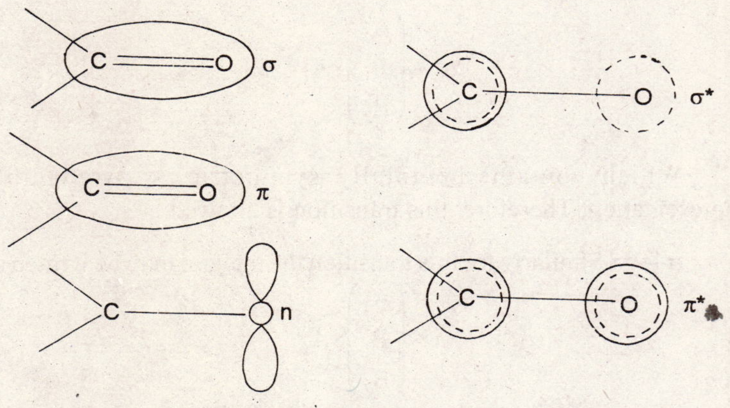

Fig. 9.6

On performing operations of C_{2v} on these M.O.'s irreducible representations can be obtained to which these M.O.'s belong

C_{2v}	E	C_2	σ_{xz}	σ_{yz}	
σ_b	1	1	1	1	a_1
σ_a^*	1	1	1	1	a_1
π_b	1	–1	1	–1	b_1
π_a^*	1	–1	1	–1	b_1
n	1	–1	–1	1	b_2

The possible transitions in this case are $n\rightarrow\pi^*$, $\pi\rightarrow\pi^*$ and $\sigma\rightarrow\sigma^*$. Let us consider them one by one

$n\rightarrow\pi^*$, the ground state has symmetry A_1. In the excited state the configuration is $n^1\pi^{*1}$, so, the corresponding symmetry by direct product is $b_1 \otimes b_2 = A_2$. So, $n \to \pi^*$ transition is considered as $A_1 \to A_2$ transition. Hence its character of the representation would be $A_1 \otimes A_2 = A_2$ (irreducible representation) which is not totally symmetric. Therefore this transition is forbidden not allowed.

$\pi \to \pi^*$ This transition which has the corresponding configuration $\pi_b^2\pi^{*0} \to \pi_b^1\pi^{*1}$. So, for the ground state the irreducible representation is $b_1 \otimes b_1 = A_1$ and for excited state it is again $b_1 \otimes b_1 = A_1$, so, this transition may be considered as

$$A_1 \times \begin{pmatrix} A_1 \\ B_1 \\ B_2 \end{pmatrix} \times A_1$$

Which contain the totally symmetric A_1 irreducible representation. Therefore, this transition is allowed.

$\sigma \to \sigma^*$ Similarly for this transition the integral may be written as

$$A_1 \times \begin{pmatrix} A_1 \\ B_1 \\ B_2 \end{pmatrix} \times A_1$$

Which contains totally symmetric irreducible representation A_1, so, the transition $\sigma \to \sigma^*$ is allowed. Similarly this can be shown that $n \to \sigma^*$ is also allowed transition.

Example 3: Let us consider the benzene molecule. This has D_{6h} molecular point group. The Huckel M.O. calculations give the results as shown in Fig. 9.7. The ground electronic state is given by $(a_{2u})^2 (e_{1g})^2 (e_{1g})^2$ which belongs to the totally symmetric with corresponding irreducible representation A_{1g}. The lowest energy transition possible is $(a_{2u})^2 (e_{1g})^2 (e_{1g})^2 (e_{2u})^1$. As the product of $(a_{2u})^2 (e_{1g})^2$ is totally symmetric, so, we have to consider only the product of $(e_{1g})^1 (e_{2u})^1$ which gives symmetry species B_{1u}, B_{2u} and E_{1u} for the excited states and E_{1u} is doubly degenerate state. Both singlet and triplet states are possible in the excited state. Fig. 9.8 show the term level diagram for benzene and the corresponding electronic spectra is given in Fig. 9.9.

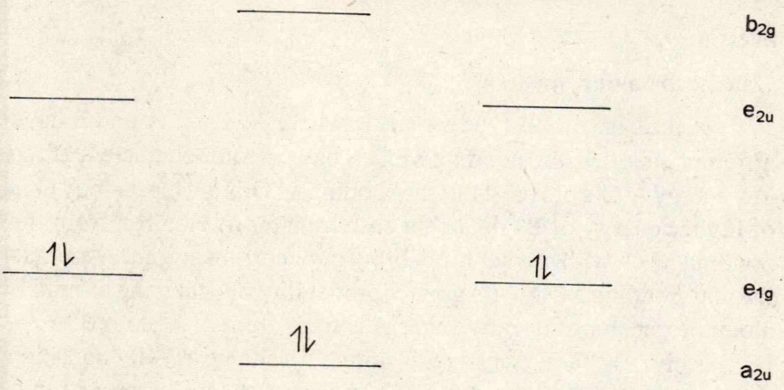

b_{2g}

e_{2u}

e_{1g}

a_{2u}

Fig. 9.7

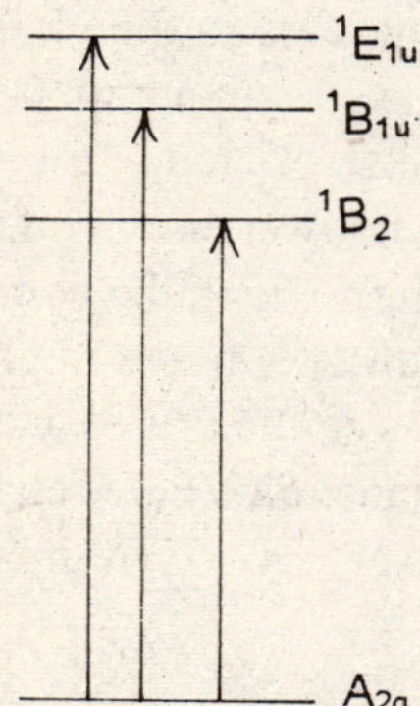

Fig. 9.8

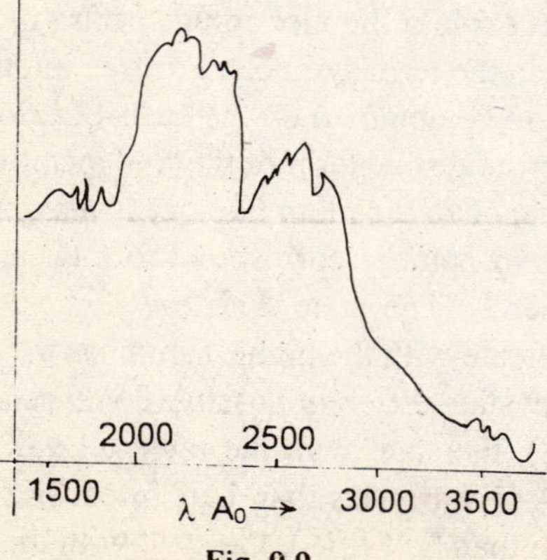

Fig. 9.9

Section 9.4

Charge transfer spectra

In the case of AB type molecule where A.O's of A and B have different energies, the bonding M.O.'s have greater characters of the low energy A.O's of A and the antibonding M.O.'s are closer to A.O.'s of higher energy of B. In such a case transfer of electron from the bonding M.O. with greater probability of occurring around A atom to the anti-bonding M.O. with greater probability of occurring around B atom, or transfer of negative charge is from atom A to atom B or A–B. This gives a charge transfer transition. In a charge transfer transition there is always change in the dipole moment, so, such a transition is allowed and occur with high intensity. In most of the cases it is a low energy transition and appears in the visible region. As a result of charge transfer transition certain compounds show coloured *e.g.,* in $M_nO_4^-$ and $Cr_2O_7^{2-}$ because of charge transfer from O^{2-} to Mn^{7+} and O^{2-} to Cr^{6+} which occur at lower energy, this transition absorbs in the visible region as a result of which the compound becomes coloured. Such a phenomenon can take place in case of metal halides also. Charge transfer from less acidic I' is easy, so, some metal iodides are coloured. In case of organic molecules charge transfer band can also be seen from I' to H_3C–N (N-methyl pyridinium iodide).

Section 9.5

Term Symbols

In order to explain the electronic spectra of transition metals, let us first explain the term symbols. For free ion the ground state or free ion terms can be obtained on the basis of LS coupling. In order to find out the ground state term of the free metal ion, Hund's rule is used. (a) The magnetic quantum number is maximized. The highest magnetic quantum number correspond to total angular momentum quantum number L. The state with the highest L has the lowest energy. (b) The state with the highest spin has the lowest energy. (c) For the multiplet state different possible J values are considered (J = L + S to L – S), the state with the lowest J value is the lowest in energy, if the A.O's are less than half filled. In the case where the A.O.'s are more than half filled, the state with the highest value of J has lowest energy. Therefore, the term symbol for the free ion *i.e.,* the term may be obtained on the basis of L, S and LS coupling. The term may be written as 2S+1

X

L+S to L – S

Where X is the term symbol that depends upon the highest angular orbital momentum quantum number *i.e.,*

$L = \Sigma m_l$

If $L = 0$ term is S

1 term is P

2 term is D

3 term is F

4 term is G etc.

2S+1 is the spin degeneracy of the state, where S is the total spin quantum number *i.e.,* $S = \Sigma s$, and and L+S to L–S represents the J-values to the state under consideration.

Let us consider d^1 case for example

+2	+1	0	–1	–2
↑				

The $m_l = 2$ hence $\Sigma m_l = L = 2$

And it is only one electron, so, its S will be ½, so, the ground term is D corresponding to L = 2 with spin degeneracy 2(1/2) + 1 = 2 and LS coupling or J values are L + S to L – S *i.e.,* 5/2, 3/2 etc. So, the term symbols for d^1 configuration state are ${}^2D_{5/2}$, ${}^2D_{3/2}$

Similarly let us consider d^2 case, for example

+2	+1	0	–1	–2
↑	↑			

The L value is 3 and S is 1. So, the term symbols are ${}^3F_4$, ${}^3F_3$, ${}^3F_2$ etc.

Section 9.6

Electronic Spectra Of Transition Metal Complexes

In case of transition metal complexes the d-d transitions are Laporte forbidden and these transitions are weak or low intense transitions, as in case of d-d transition which may be considered as (t_{2g}, e_g) *i.e.,* g → g transition, so, $\mu_{ij} = 0$ or $P_{ij} = |\mu_{ij}|^2 = 0$. Therefore, d-d transition are Laporte forbidden (Section 9.2). These transitions become sometimes allowed because of following reasons:

(i) Vibronic coupling *i.e.*, coupling of electronic and vibrational wave function is there and in vibrations centre of symmetry is lost and

(ii) In complexes, there is mixing of d and p orbitals, so, because of this mixing these transitions are no longer d-d transitions but dp-dp transitions.

Spectral terms symbols have already been explained in section 9.5. In O_h field these spectral states of free ion split up in the following manner

$$S \rightarrow A_{1g}$$
$$P \rightarrow T_{1g}$$
$$D \rightarrow E_g, T_{2g}$$
$$F \rightarrow A_{1g}, T_{1g}, T_{2g}$$
$$G \rightarrow E_g, T_2, T_1, A_1$$

As the combination of the given irreducible representations represent the A.O.'s corresponding to the above spectral states. Now, let us consider the different d^n cases one by one.

d^1 case: The free ion term for d^1 configuration is ^2D. Group theory tell us that this state splits up into T_{2g} and E_g. But group theory does not tell us about the energies of these T_{2g} and E_g state that is which of these will be of lower energy and which will be of higher energy. In this very case this transition is actually from triply degenerate $(t_{2g})^1$ $(e_g)^0$ to doubly degenerate $(t_{2g})^0$ $(e_g)^1$. The corresponding symmetries for these electronic arrangements are: for $(t_{2g})^1$ $(e_g)^0$ it is T_{2g} and for $(t_{2g})^0$ $(e_g)^1$ it is E_g. So, the splitting can be explained as it is shown in the fig. 9.10.

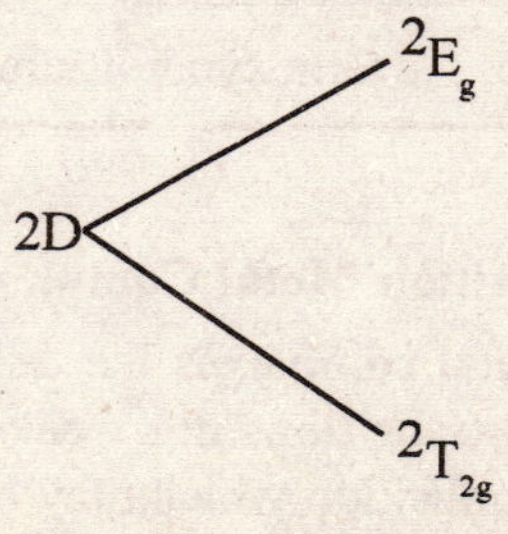

Fig. 9.10

d^9 case: In this case L = 2 and S = ½ again, so, the free ion term is ^2D. As in O_h field this term splits up into T_{2g} and E_g. The ground state electronic arrangement is shown by $(t_{2g})^6 (e_g)^3$ with unpaired electron in e_g set and the corresponding symmetry is E_g and in the excited state the electronic arrangement is $(t_{2g})^5 (e_g)^4$ showing that the unpaired electron is in t_{2g} set, so, the transition is $(t_{2g})^6 (e_g)^3$ $(t_{2g})^5 (e_g)^4$ is actually $E_g \rightarrow T_{2g}$ on the basis of symmetry consideration. Therefore, splitting pattern in d^9 is opposite to d^1 case.

d^6 case: In this case L = 2 and S = 2, so, the corresponding free ion term is $^5D_1$ which splits into E_g and T_{2g} symmetry states. The ground state electronic arrangement in this case may be represented as $(t_{2g})^4 (e_g)^2$ which corresponds to T_{2g} symmetry state and the excited state electronic arrangement is $(t_{2g})^3 (e_g)^3$ with E_g symmetry state, so, the splitting can be represented as

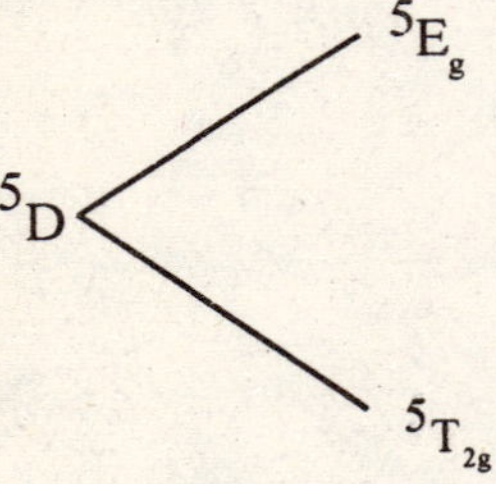

Fig. 9.11

d^4 case: In this case also L = 2 and S = 2, so, the free ion term is ^5D which splits into T_{2g} and E_g symmetry states in O_h field. The ground state electronic arrangement in this case may be represented as $(t_{2g})^3 (e_g)^1$ which correspond to E_g symmetry state and the excited state electronic arrangement is $(t_{2g})^2 (e_g)^2$ with T_{2g} symmetry state. So, the splitting can be represented as

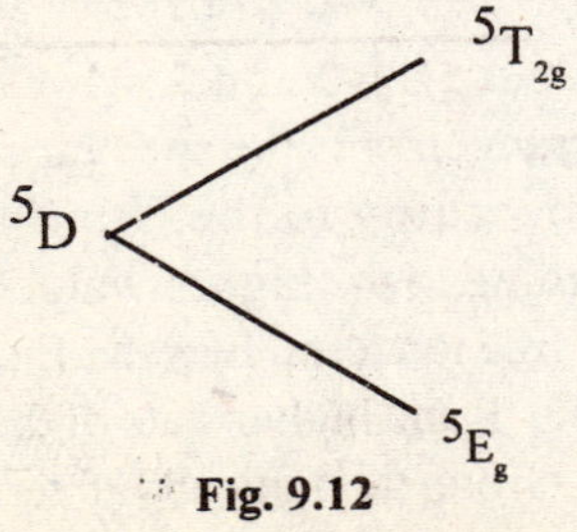

Fig. 9.12

In the case of tetrahedral field the splitting of states is just reverse that of octahedral field, but as the tetrahedral complexes has no centre of symmetry so, subscript g has not been used in the case of tetrahedral complexes.

Orgel diagrams are the compiled diagrams for these splittings for weak field cases. These are the plottings of splittings of ground states as a function of crystal field splitting. So, the diagram comparising the above discussed splittings for octahedral and tetrahedral fields is given in the Fig. 9.13.

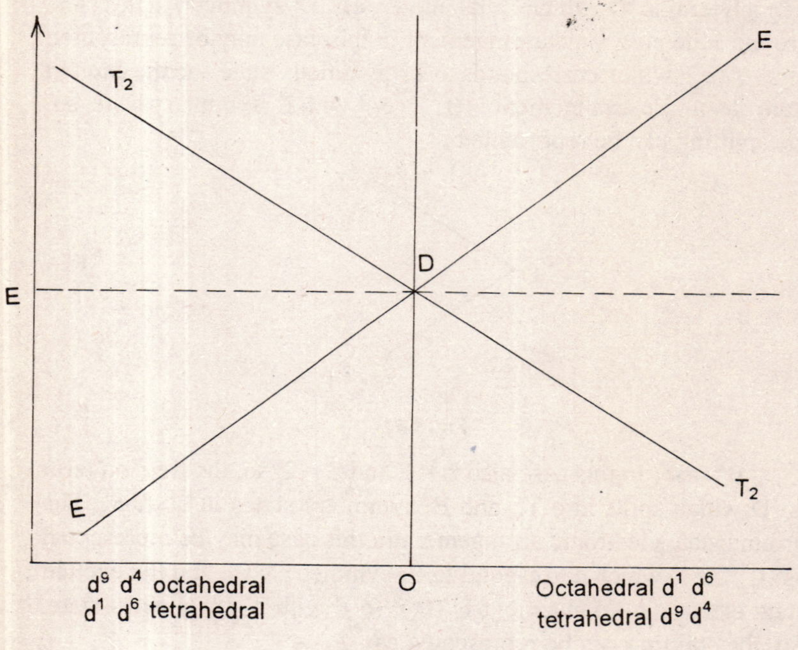

Fig. 9.13

d^2 case: In this case L = 2 and S = 1. Considering different possible microstate arrangements the spectral states possible are ^1G, ^3F, ^1D, ^3P and ^1S. According to the Hund's rule of maximum multiplicity and according to the highest value of L the ground state term is ^3F which is the free ion term. In weak field the splitting of free ion ground spectral state ^3F and higher state of the same multiplicity ^3P have to be considered. Group theoretically ^3F terms splits up into $^3A_{2g}$,

$^3T_{2g}$ and $^3T_{1g}$ and ^3P term splits into $^3T_{1g}$. On the basis of electronic arrangement the ground state arrangement is $(t_{2g})^2 (e_g)^0$ which is the lowest energy arrangement and $(t_{2g})^1 (e_g)^1$ and $(t_{2g})^0 (e_g)^2$ are the higher energy arrangements. The possible energy level diagram for the corresponding symmetry states is shown in fig. 9.14.

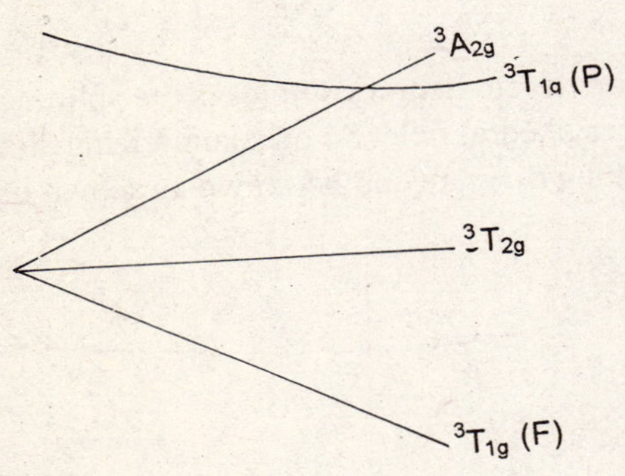

Fig. 9.14

In case of d^8 configuration splitting is represented in the fig. 9.15 with the possible transitions

$$^3T_{2g}(F) \leftarrow {}^3A_{2g}$$
$$^3T_{2g}(P) \leftarrow {}^3A_{2g}$$

for the electronic arrangement

$$(t_{2g})^6 (e_g)^2 \rightarrow (t_{2g})^5 (e_g)^3$$

and $^3T_{1g}(F) \leftarrow {}^3A_{2g}$ for the electronic arrangement $(t_{2g})^6 (e_g)^2 \rightarrow (t_{2g}) (e_g)^4$.

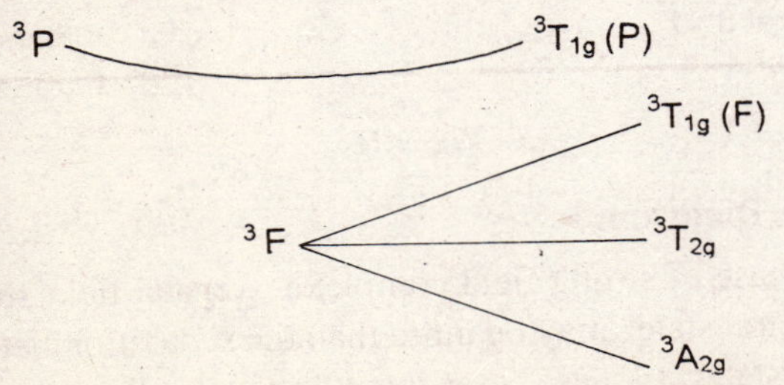

Fig. 9.15

Similarly we can consider the d^7 and d^3 case with L = 3 and S = 3/2 with the free ion term ^4F and the term with the same spin multiplicity *i.e.*, ^4P. The transitions are

d^7 case

$^4T_{1g}$ (F) → $^4T_{2g}$ (F)

$^4T_{1g}$ (F) → $^4T_{1g}$ (P)

$^4T_{1g}$ (F) → $^4A_{2g}$ (F)

d^3case

$^4A_{2g}$ (F) → $^4T_{2g}$ (F)

$^4A_{2g}$ (F) → $^4T_{1g}$ (F)

$^4A_{2g}$ (F) → $^4T_{1g}$ (P)

In the case of tetrahedral complexes the splitting is reverse as in the case of orctahedral field. So on summarizing all these splitting patterns a splitting diagram can be drawn as shown in Fig. 9.16.

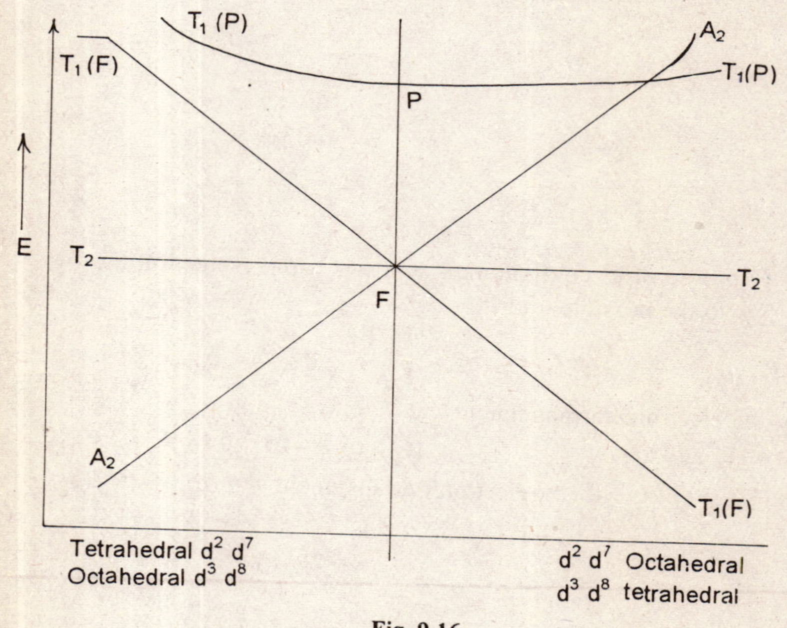

Fig. 9.16

Correlation Diagram

In the case of strong field complexes, crystal field splitting of free ion spectral states may be more than the separation between the two free ion states. In these cases the individual splittings of the free ion states are not considered. Let us consider d^2 case. The free ion states are ^3F, ^3P, ^3D, ^1G and ^1S. The possible electronic arrangements are $(t_{2g})^2 (e_g)^0$, $(t_{2g})^1 (e_g)^1$ and $(t_{2g})^0 (e_g)^2$. The electrons in the different configurations undergo orbital and spin interactions resulting in microstates. For $(t_{2g})^2$ case the number of microstates are 15 similarly for $(t_{2g})^1 (e_g)^1$ it is 24 and for $(t_{2g})^0 (e_g)^2$ it is 6. The symmetry of these states can be obtained by multiplying characters of the two electrons and then reducing the total character for irreducible representations. In the $(t_{2g})^2 (e_g)^0$ configuration both the electrons are in t_{2g}, so, the total character can be obtained by multiplying t_{2g} with $T_{red.} = A_{1g} \oplus E_g \oplus T_{1g} \oplus T_{2g}$.

As the total number of states of this symmetry are 15. We can consider the degeneracies of the states and can find out values of multiplicities as a.1 + b.2 + c.3 + d.3 = 15

As A_{1g} is one dimensional, E_g is two dimensional and T_{1g} and T_{2g} are three dimensional states and a,b,c and d are the multiplicities.

Probable values of the multiplicites [a,b,c,d] may be [1,1,3,1] or [1,1,1,3] or [3,3,1,1]. Similarly for the $(t_{2g})^1 (e_g)^1$ state the total character for reducible representation $T_{(t_{2g})^1 (eg)^1}$ is

O_h	E	$8C_3$	$6C_2$	$6C_4$	$3C_2(=C_4^2)$	'i'	$6S_4$	$8S_6$	$3\sigma_h$	$6\sigma_d$
$T_{t_{2g}^1 eg^1}$	6	0	0	0	–2	6	0	0	–1	0

Which on reducing give $T_{1g} \oplus T_{2g}$

Total number of microstates are 24. So,

$$a.3 + b.3 + c.3 + d.3 = 24$$

Therefore, [a,b,c,d] should be [1,3,1,3]. Similarly for $(t_{2g})^0 (e_g)^2$ total character is

O_h	E	$8C_3$	$6C_2$	$6C_4$	$3C_2(=C_4^2)$	'i'	$6S_4$	$8S_6$	$3\sigma_h$	$6\sigma_d$
$T_{t_{2g}^0 eg^2}$	4	1	0	0	4	4	0	1	4	0

Which on reduction results

$$T_{t_{2g}^0 eg^2} = A_{1g} \oplus A_{2g} \oplus E_g$$

Since the total number of microstates are 6. The two possible sets for a,b,c values are [3,1,1] or [1,3,1]. As in this case

$$a.1 + b.1 + c.2 = 6$$

Fig. 9.17 shows the correlation diagram for O_h field which is based upon the following facts:

(a) Weak field states can be correlated to strong field states and in doing this the symmetry properties of the system does not change. So, the number of spectral states, their symmetries and the multiplicities remain same in weak as well as in strong field. There should be one to one correspondence between weak field and strong field states.

(b) The states with same symmetry and multiplicity can't cross in the diagram which means that

(i) Number of energy states do not change from L.H.S. weak field side to R.H.S. the strong field side.

(ii) Types of energy states should also remain same.

(iii) The terms of the same multiplicity should not cross each other.

(iv) There should be 1:1 correspondence between energy states both the sides.

Hole Formalism

This is a relationship that makes many of the problems for determining the states of an electronic configuration and considering its splitting as shown in orgel and correlation diagrams, easier to some extent. According to hole formalism d^{10-n} configuration behaves in the same way at all points of the axis of the energy level as the corresponding d^n configuration except that all energies interaction with the environment will have the opposite sign. The n holes in the d shell may be treated.

The splitting patterns (either in octahedral or in tetrahedral fields) for d^n configuration is just inverted in the case of d^{10-n} configuration. This relation is also termed as $d^n - d^{10-n}$ relation.

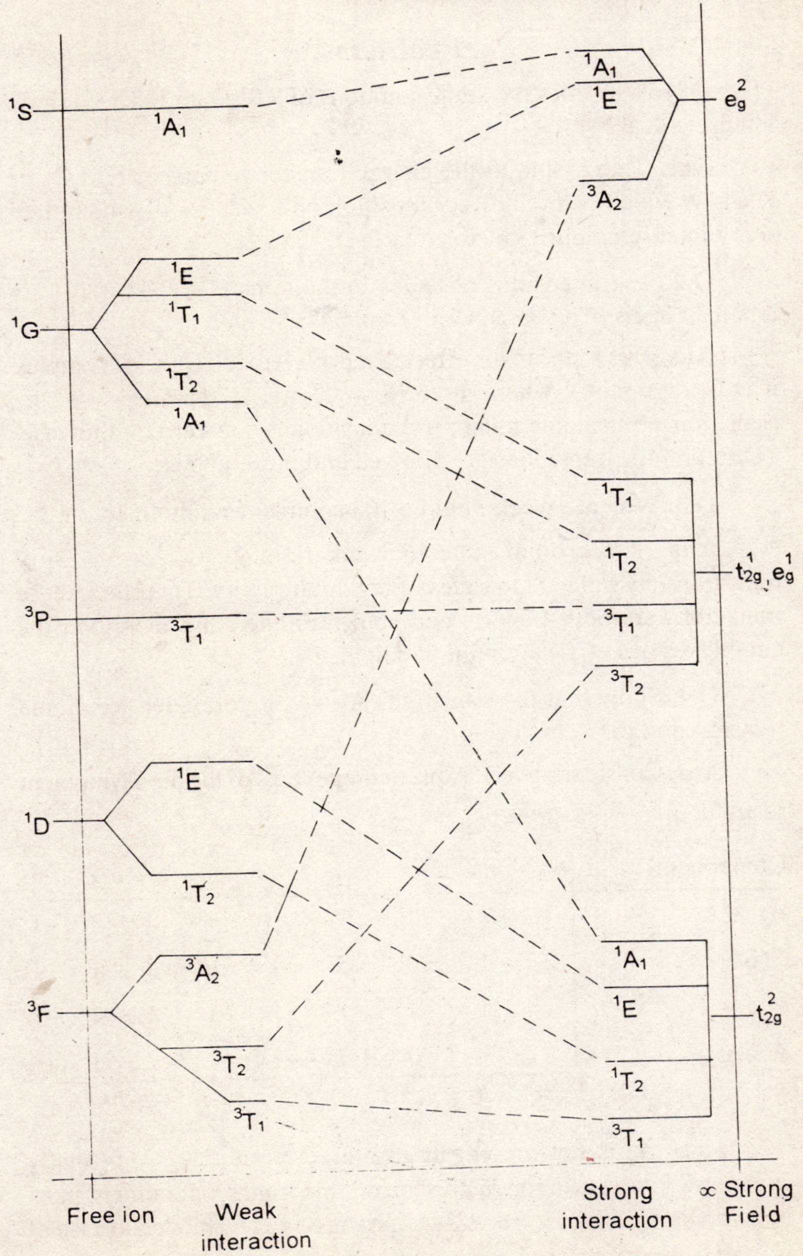

Fig. 9.17

Problems

Q. 1. Why does the dilute solutions of KM_nO_4 or $K_2Cr_2O_7$ show intense colour?

Ans. This is due to the charge transfer in case of KM_nO_4 or $K_2Cr_2O_7$ which is an allowed transition and because of which they show intense colour (Ref. section 9.4).

Q.2. The intensities of bands in the trans $[Co\ (NH_3)_4\ Cl_2]^+$ is less than in cis $[Co\ (NH_3)_4\ Cl_2]^+$. Explain?

Ans. This is due to the effect that in cis $[Co\ (NH_3)_4\ Cl_2]^+$ complex it is C_{2v} symmetry where there is no centre of symmetry, so, the transition moment integral for d-d transition is non-zero for this case. Therefore, the transitions are allowed and more intense.

Q.3. Why are weak field Fe(III) complexes light in colour?

Ans. The ground state in weak field Fe(III) is 6_D. Only transitions possible are to states of less multiplicity. Therefore, these transitions are both Laporte and spin forbidden and therefore, the complexes of Fe(III) are light in colour.

Q.4. Show that the transitions $A_1 \rightarrow A_2$ is forbidden for dipole transmission in C_{3v} molecule.

Ans. Considering all the three components of the dipole moment operator $\hat{\mu}$

Components of $\hat{\mu}$	X			Y			Z		
A_1	1	1	1	1	1	1	1	1	1
$T(\hat{\mu})$	2	−1	0	2	−1	0	1	1	1
A_2	1	1	−1	1	1	−1	1	1	−1
$A_1T(\hat{\mu})A_2$	2	−1	0	2	−1	0	1	1	−1
	E			E			A_2		

Since A_1 is not present in any direct product *i.e.,* no totally symmetric representation is present, so, the transition moment must be a vanishing integral. So, $A_1 \rightarrow A_2$ transition is forbidden in case of C_{3v} point group.

Q.5. Is $p_x \rightarrow p_y$ an allowed transition in tetrahedral molecule.

Ans. The tetrahedral molecule possess T_d symmetry. So, refer to the character table of T_d point group, we have to check whether $p_x\ \hat{\mu}\ p_y$ integral non-zero or not.

T_d	E	$8C_3$	$3C_2$	$6\sigma_d$	$6S_4$	
p_x	3	0	–1	–1	1	T_2
$\hat{\mu}$	3	0	–1	–1	1	T_2
p_y	3	0	–1	–1	1	T_2
	27	0	–1	–1	1	T_{red}.

Which on reduction shows that it contains A_1 (totally symmetric) irreducible representation. So, the given integral will be non-vanishing and $p_x \rightarrow p_y$ transition will be allowed transition.

10

Group Theory and Spectroscopy–II

In the previous chapter, symmetry and group theoretical aspects of UV–visible or electronic spectroscopy are discussed. When a molecule is provided energy in the i.r. region, molecule sets-up in the vibrations and rotations. The energies corresponding to vibrations and rotations are also quantized because of which corresponding spectra can be obtained. For a given molecule, there are large number of complex vibrations possible, relatively small number of specific vibrations or normal modes of vibration can be understood. The displacement of atoms from their mean position is defined in terms of normal co-ordinates, which are functions of angles and distances. In the normal modes of vibration all the atoms move in phase with one another i.e., they pass through their mean positions at the same instant. For degenerate normal modes of vibration the symmetry operations will transform the degenerate set of vibrations into linear combination of mutually degenerate normal co-ordinates or wave functions. So, each normal mode is associated with the symmetry or the irreducible representations. Therefore, symmetry and group theoretical studies are useful to predict the normal (vibration) modes and, so, also the infrared and Raman active bands. In this chapter the introduction to the vibration-rotation and Raman spectra, selection rules, various modes of vibrations, concepts of group-theoretical analysis of various vibration modes with some suitable examples of i.r., spectra of metal complexes, are given

Section 10.1

Fundamentals of vibration-rotation spectroscopy

Let us consider a diatomic molecule in the vibration-rotation region that lies between 2.5 μ to 25 μ vibration levels are quantized as shown in Fig. 10.1. The energy of molecule may be defined by the formula

$$E_{tot} = E_{vib} + E_{rot}$$

Where $E_{rot} = E_J = [h^2/8\pi^2 I]\ [J(J + 1)]$

$$E_{rot} = E_J/hc = [h/8\pi^2 Ic][J\ (J+ 1)] = B\ J(J + 1)$$

corresponding energy of vibration (according to the concept of harmonic oscillator) is

$$E_{vib} = (v + 1/2)h\upsilon_0$$

Where v is the vibration and J is the rotation quantum number.

$$E_{vib} / hc = (v + 1/2)\overline{\upsilon}_0$$

Therefore,

$$E_{tot} = (v + 1/2)\overline{\upsilon}_0 + [h/8\pi^2 Ic][J(J+1)]$$

Where v = ±1 with corresponding J values as Δ J = ±1, ±2, ±3, ... and υ_0 can be expressed as

$$\upsilon_0 = \frac{1}{2\pi}\sqrt{\frac{k}{\mu}}$$

Where k is a constant known as force constant and μ is the reduced mass. As molecules are not the hard compact structures of atoms, we can consider the molecule with the combination of balls of different masses and bonds as spring. Therefore, on absorption of energy in the i.r. region vibration are set-up in the molecule, for which frequency of vibrations are given above along with energy of vibration.

Because of quantization of vibration and rotation levels as shown in Fig. 10.1, transitions from (v, J) to (v', J') are possible. Basically two types of vibrations (i) stretching and (ii) bending modes are possible in the molecule.

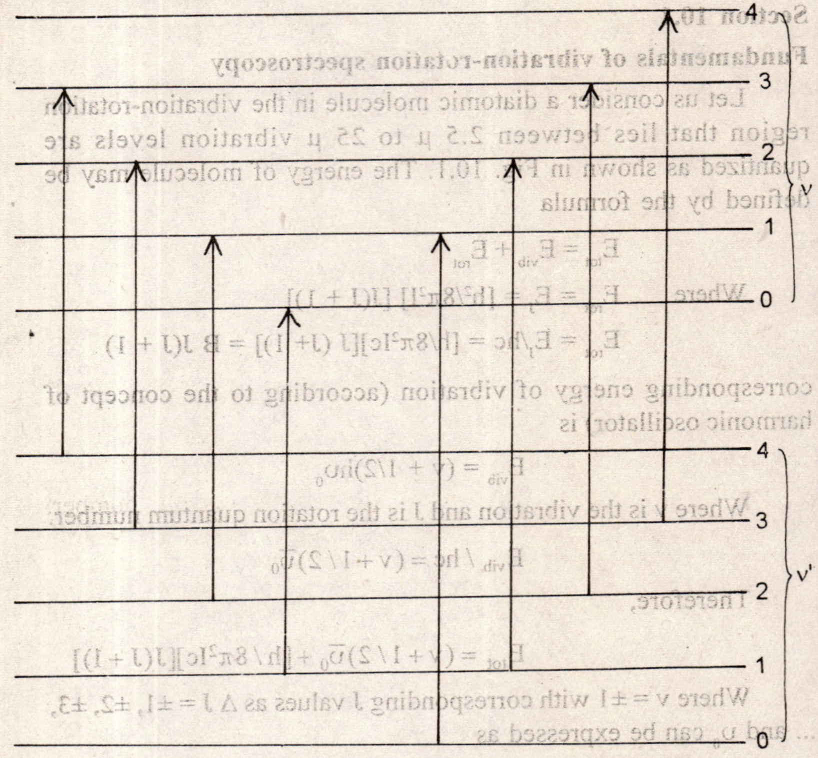

Fig. 10.1

Section 10.2

Modes of interactions

It is expected that change in dipole moment during vibration of diatomic molecule will occur if and only if there exists a permanent dipole moment within the molecule, because an oscillating electrical field from the molecule is generated when there is change in the dipole moment during vibration of molecule. Hence, homo-nuclear diatomic molecules are infra-red inactive molecules but hetero-nuclear diatomic molecules are infer-red active.

Chiefly there are two types of modes of vibrations(i) stretching in which the distances between the atoms increase and decrease along the axes. Stretching vibration modes may be of two types *i.e.*, (a) symmetric stretching and (b) anti-symmetric stretching. These modes are shown in fig. 10.2. Second type of modes of vibrations are (ii) bending modes which may be in plane or out of plane bending. Scissoring and rocking modes in plane bending modes and twisting and wagging are out of plane bending modes.

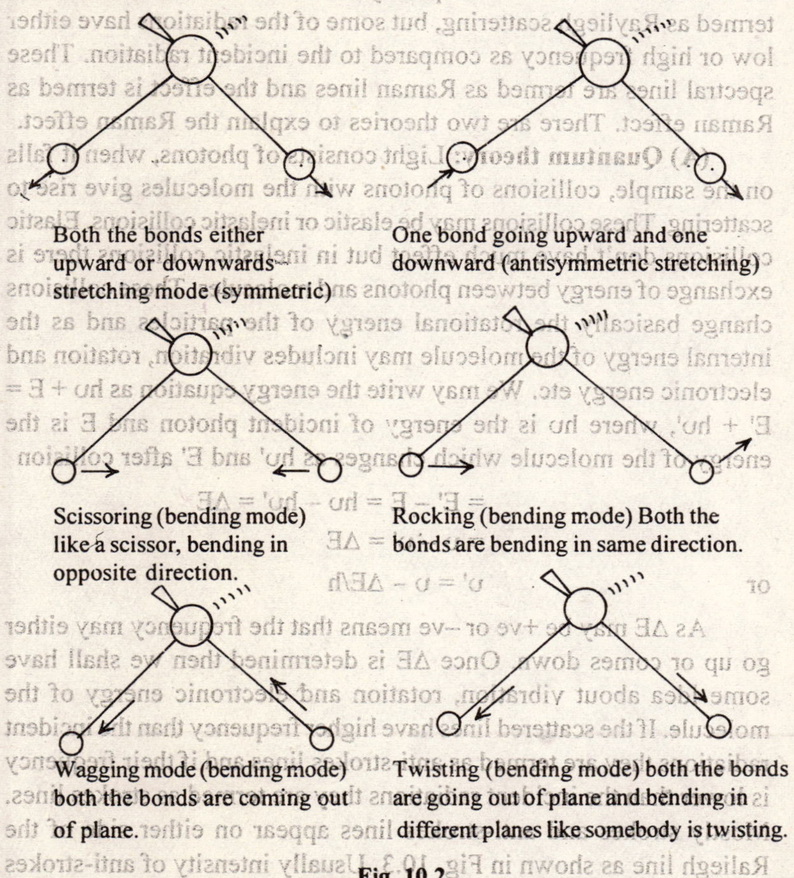

Fig. 10.2.

Selection Rules: $\Delta v = \pm 1$, transitions with this change in the quantum number are allowed and all other are forbidden because all other values do not commute with the Hamiltonian operator and their solutions are not acceptable.

Section 10.3

Fundamentals of Raman Spectroscopy

It was observed by Prof. C.V. Raman that when radiation is passed through a solution, the scattered rays consist of wave length different from the incident radiation. Among these radiation most of the radiation has the same frequency as the incident radiation. This is termed as Rayliegh scattering, but some of the radiations have either low or high frequency as compared to the incident radiation. These spectral lines are termed as Raman lines and the effect is termed as Raman effect. There are two theories to explain the Raman effect.

(A) Quantum theory: Light consists of photons, when it falls on the sample, collisions of photons with the molecules give rise to scattering. These collisions may be elastic or inelastic collisions. Elastic collisions don't have much effect but in inelastic collisions there is exchange of energy between photons and molecules. These collisions change basically the rotational energy of the particles and as the internal energy of the molecule may includes vibration, rotation and electronic energy etc. We may write the energy equation as $h\upsilon + E = E' + h\upsilon'$, where $h\upsilon$ is the energy of incident photon and E is the energy of the molecule which changes as $h\upsilon'$ and E' after collision

$$= E' - E = h\upsilon - h\upsilon' = \Delta E$$

$$= \upsilon - \upsilon' = \Delta E$$

or $$\upsilon' = \upsilon - \Delta E/h$$

As ΔE may be +ve or –ve means that the frequency may either go up or comes down. Once ΔE is determined then we shall have some idea about vibration, rotation and electronic energy of the molecule. If the scattered lines have higher frequency than the incident radiations they are termed as anti-strokes lines and if their frequency is lower than the incident radiations they are termed as strokes lines. Mostly strokes and anti-strokes lines appear on either side of the Raliegh line as shown in Fig. 10.3. Usually intensity of anti-strokes lines is very small because of less population or v = 1 level (Ref. Figure 10.3).

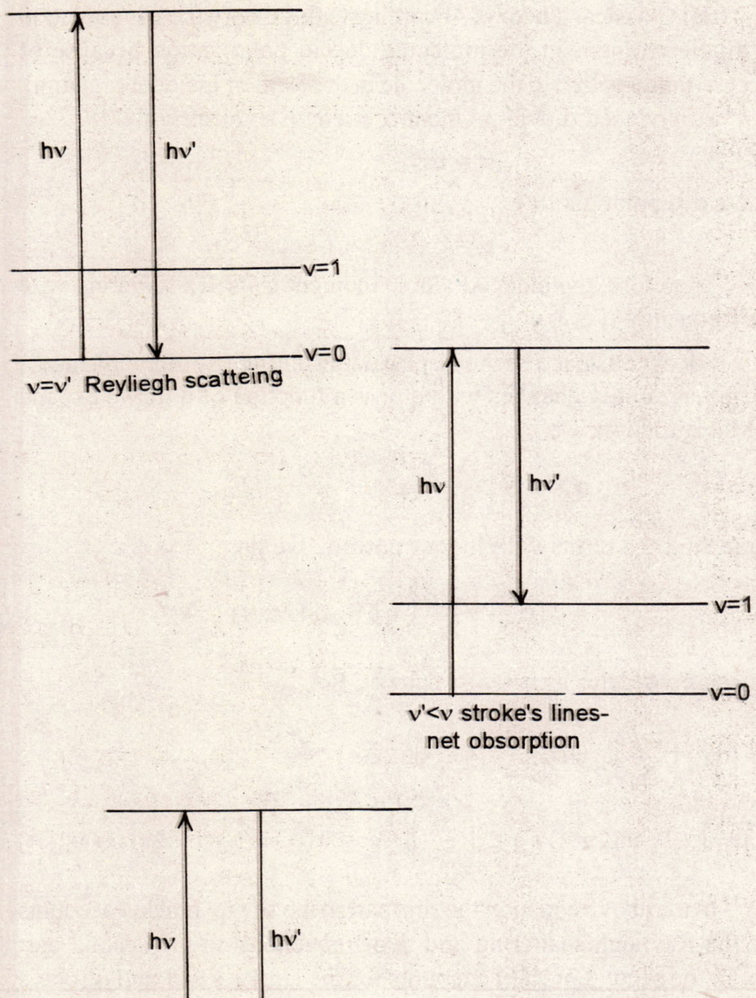
hν
hν'
v=1
v=0
ν=ν' Reyliegh scatteing
hν
hν'
v=1
v=0
ν'<ν stroke's lines-
net obsorption
hν
hν'
v=1
v=0
ν'>ν anti stroke's lines-
net loss in energy

Fig. 10.3

(B) **Classical Theory:** According to this theory there is induction of dipole moment in the molecule due to polarization because of electric field applied to the molecule in the form of incident radiation.

So, induced dipole moment α strength of electric field.

or $$\mu = \alpha E$$

where α is polarizability.

as $$E = E_0 \sin 2\pi \upsilon t, \text{ so,}$$

Therefore, the induced dipole moment is itself oscillating with the frequency (i.e. υ).

Let us consider q be the displacement during any vibration mode. So, polarizability changes with q or is a function of q. If we expand in Maclaurian series.

$$\alpha = \alpha_e + \left(\frac{\partial \alpha}{\partial q}\right)_q q + \frac{1}{2}\left(\frac{\partial^2 \alpha}{\partial q^2}\right) q^2 + ...$$

neglecting the terms with higher powers, we get

$$\mu = [\alpha_e + \left(\frac{\partial \alpha}{\partial q}\right)_q q][E_0 \sin 2\pi \upsilon t]$$

we can consider q as $q = q_0 \sin 2\pi \upsilon_{vib} t$. So,

$$\mu = [\alpha_e + \left(\frac{\partial \alpha}{\partial q}\right) q_0 \sin 2\pi \upsilon_{vib} t][E_0 \sin 2\pi \upsilon t]$$

$$\Rightarrow \mu = [\alpha_e E_0 \sin 2\pi \upsilon t] + q_0 E_0 \left(\frac{\partial \alpha}{\partial q}\right)[\frac{1}{2} \cos 2\pi(\upsilon - \upsilon_{vib}) t \cos 2\pi(\upsilon + \upsilon_{vib}) t]$$

In the above equation the first part in the square bracket accounts for the Rayliegh scattering and two frequencies in the second part *i.e.,* in the second bracket accounts for the stroke's and anti-stroke's lines.

From the above equation, it is clear that when $(\partial \alpha / \partial q) = 0$ we should not observe any Raman effect but only Rayliegh scattering is observed. So, for a mode to be Raman active there should be $(\partial \alpha / \partial q) \neq 0$ *i.e.,* change in polarizability with the displacement should not be zero or in conclusion there should be change in the polarizability. This is the selection rule for Raman spectroscopy but in contrast for a mode to be i.r. active there should be change in dipole moment. With the help of group theory we can predict the normal modes of a molecule and their i.r. and Raman activity. Some examples are discussed in section 10.4.

Section 10.4

Normal Modes

Let us consider a molecule having N atoms. If we specify the position of each atom in space with the co-ordinates x, y, z, then there will be 3N co-ordinates for the entire molecular system to specify it. Since the molecule has translational, rotational, and vibrational motion, these 3N co-ordinates can be assigned to each of the motions as expressed in table 10.1.

Table 10.1

Motion	*Degrees of freedom required to describe the motion*
Translation	3
Rotation (about the centre of mass)	2 (linear) 3(non-linear)
Vibration	3N – 5(linear) 3N – 6 (Non-linear)

Group theoretical analysis of normal modes

Group theoretical analysis of normal modes for poly-atomic molecules can be done on the basis of following steps:

(i) Find out the molecular point group for the molecule for which normal modes are to determined.

(ii) Obtained the reducible representation involving the co-ordinates at each and every atom *i.e.*, T_{3N} which can be obtained by considering axes system at every atom and finding the characters of the corresponding matrices of the symmetry operations of that point group. T_{3N} can also be obtained by an alternate method *i.e.*, consider only one x, y, z co-ordinates axis frame of the molecule and find out the reducible representation and multiply the characters of this reducible representation with the number of un-shifted atoms in various symmetry operations.

(iii) Reduce T_{3N} to get the corresponding irreducible representations.

(iv) Subtract the irreducible representations corresponding to translational mode (*i.e.*, irreducible representations corresponding to x, y, z basis functions as translation of a molecule as a whole can be expressed with reference to x, y, z reference frame of axes). Now, we are left with the irreducible representations corresponding to rotational and vibration or normal modes.

(v) Now, subtract the rotation modes (*i.e.*, irreducible representations with basis functions, the rotation vectors *i.e.*, R_x, R_y and R_z).

(vi) Remaining are the vibration modes or the normal modes of the molecule. In order to check their i.r. activity or Raman activity, we have to consider the transition moment integral *i.e.*, μ_{ij} (ref. section 9.2) *i.e.*, for a particular vibration mode whether it is zero or not. For this we have to consider Ψ_i *i.e.*, the state function as the totally symmetric irreducible representation (*i.e.*, A_1 or A_{1g} etc.) and μ is dipole moment operator which is a function of x, y, z, and the irreducible representations corresponding to x, y, z functions may be considered as a representation for μ. Ψ_j is the functions corresponding to the irreducible representation which express the normal modes. If for a particular transition ($\Psi_i \rightarrow \Psi_j$) using direct product of representation this integral is zero, the transition is forbidden otherwise it is allowed (ref. section 9.3). In this way the i.r. activity for the various normal modes can be checked.

(vii) In order to check the Raman activity the same integral with polarizability operator *i.e.*, $\int \Psi_i \, \alpha \, \Psi_j \, d\tau$ can be checked for the transition in the same way. α may be taken as function of x^2, y^2, z^2, xy, yz, xz, $x^2 - y^2$ etc.

Now, let us try to find out the normal modes for some simple molecules.

Example 1: Let us consider H_2O molecule. Its molecular point group is C_{2v}. Considering x, y, z axes, T_{xyz} can be obatined as

C_{2v}	E	C_2	σ_{xz}	σ_{yz}
T_{xyz}	3	–1	1	1
No. of unshifted atoms	3	1	1	3
T_{3N}	9	–1	1	3

T_{3N} can be obtained by multiplying the number of un-shifted atoms in various symmetry operations with T_{xyz} as shown above. T_{3N} can also be obtained considering x, y, z reference frame of axes on each and every atom and by obtain in the corresponding matrices for various symmetry operations.

E operation. In E operation x1, y1, z1, x2, y2, z2, x3, y3, and z3 remains as such and the correspondingly 9×9 matrix represent the operation whose character is 9.

$$\begin{bmatrix} x1' \\ y1' \\ z1' \\ x2' \\ y2' \\ z2' \\ x3' \\ y3' \\ z3' \end{bmatrix} = \begin{bmatrix} 1 & 0 & 0 & 0 & 0 & 0 & 0 & 0 & 0 \\ 0 & 1 & 0 & 0 & 0 & 0 & 0 & 0 & 0 \\ 0 & 0 & 1 & 0 & 0 & 0 & 0 & 0 & 0 \\ 0 & 0 & 0 & 1 & 0 & 0 & 0 & 0 & 0 \\ 0 & 0 & 0 & 0 & 1 & 0 & 0 & 0 & 0 \\ 0 & 0 & 0 & 0 & 0 & 1 & 0 & 0 & 0 \\ 0 & 0 & 0 & 0 & 0 & 0 & 1 & 0 & 0 \\ 0 & 0 & 0 & 0 & 0 & 0 & 0 & 1 & 0 \\ 0 & 0 & 0 & 0 & 0 & 0 & 0 & 0 & 1 \end{bmatrix} \begin{bmatrix} x1 \\ y1 \\ z1 \\ x2 \\ y2 \\ z2 \\ x3 \\ y3 \\ z3 \end{bmatrix}$$

So, the character for E operation is 9.

C_2 operation. In C_2 operation the corresponding 9×9 matrix is respresented as

$$\begin{bmatrix} x1' \\ y1' \\ z1' \\ x2' \\ y2' \\ z2' \\ x3' \\ y3' \\ z3' \end{bmatrix} = \begin{bmatrix} 0 & 0 & 0 & -1 & 0 & 0 & 0 & 0 & 0 \\ 0 & 0 & 0 & 0 & -1 & 0 & 0 & 0 & 0 \\ 0 & 0 & 0 & 0 & 0 & 1 & 0 & 0 & 0 \\ -1 & 0 & 0 & 0 & 0 & 0 & 0 & 0 & 0 \\ 0 & -1 & 0 & 0 & 0 & 0 & 0 & 0 & 0 \\ 0 & 0 & 1 & 0 & 0 & 0 & 0 & 0 & 0 \\ 0 & 0 & 0 & 0 & 0 & 0 & -1 & 0 & 0 \\ 0 & 0 & 0 & 0 & 0 & 0 & 0 & -1 & 0 \\ 0 & 0 & 0 & 0 & 0 & 0 & 0 & 0 & 1 \end{bmatrix} \begin{bmatrix} x1 \\ y1 \\ z1 \\ x2 \\ y2 \\ z2 \\ x3 \\ y3 \\ z3 \end{bmatrix}$$

The character for C_2 operation is -1.

σ_{xz} operation. In this operation the following 9×9 matrix can be obtained.

$$\begin{bmatrix} x1' \\ y1' \\ z1' \\ x2' \\ y2' \\ z2' \\ x3' \\ y3' \\ z3' \end{bmatrix} = \begin{bmatrix} 0 & 0 & 0 & -1 & 0 & 0 & 0 & 0 & 0 \\ 0 & 0 & 0 & 0 & -1 & 0 & 0 & 0 & 0 \\ 0 & 0 & 0 & 0 & 0 & 1 & 0 & 0 & 0 \\ -1 & 0 & 0 & 0 & 0 & 0 & 0 & 0 & 0 \\ 0 & -1 & 0 & 0 & 0 & 0 & 0 & 0 & 0 \\ 0 & 0 & 1 & 0 & 0 & 0 & 0 & 0 & 0 \\ 0 & 0 & 0 & 0 & 0 & 0 & 1 & 0 & 0 \\ 0 & 0 & 0 & 0 & 0 & 0 & 0 & -1 & 0 \\ 0 & 0 & 0 & 0 & 0 & 0 & 0 & 0 & 1 \end{bmatrix} \begin{bmatrix} x1 \\ y1 \\ z1 \\ x2 \\ y2 \\ z2 \\ x3 \\ y3 \\ z3 \end{bmatrix}$$

The character for this 9×9 matrix is 1.

σ_{yz} operation. In this operation the following 9×9 matrix can be obtained.

$$\begin{bmatrix} x1' \\ y1' \\ z1' \\ x2' \\ y2' \\ z2' \\ x3' \\ y3' \\ z3' \end{bmatrix} = \begin{bmatrix} -1 & 0 & 0 & 0 & 0 & 0 & 0 & 0 & 0 \\ 0 & 1 & 0 & 0 & 0 & 0 & 0 & 0 & 0 \\ 0 & 0 & 1 & 0 & 0 & 0 & 0 & 0 & 0 \\ 0 & 0 & 0 & -1 & 0 & 0 & 0 & 0 & 0 \\ 0 & 0 & 0 & 0 & 1 & 0 & 0 & 0 & 0 \\ 0 & 0 & 0 & 0 & 0 & 1 & 0 & 0 & 0 \\ 0 & 0 & 0 & 0 & 0 & 0 & -1 & 0 & 0 \\ 0 & 0 & 0 & 0 & 0 & 0 & 0 & 1 & 0 \\ 0 & 0 & 0 & 0 & 0 & 0 & 0 & 0 & 1 \end{bmatrix} \begin{bmatrix} x1 \\ y1 \\ z1 \\ x2 \\ y2 \\ z2 \\ x3 \\ y3 \\ z3 \end{bmatrix}$$

Therefore, the character of this 9×9 matrix is 3. The corresponding operations are shown in Fig. 10.4.

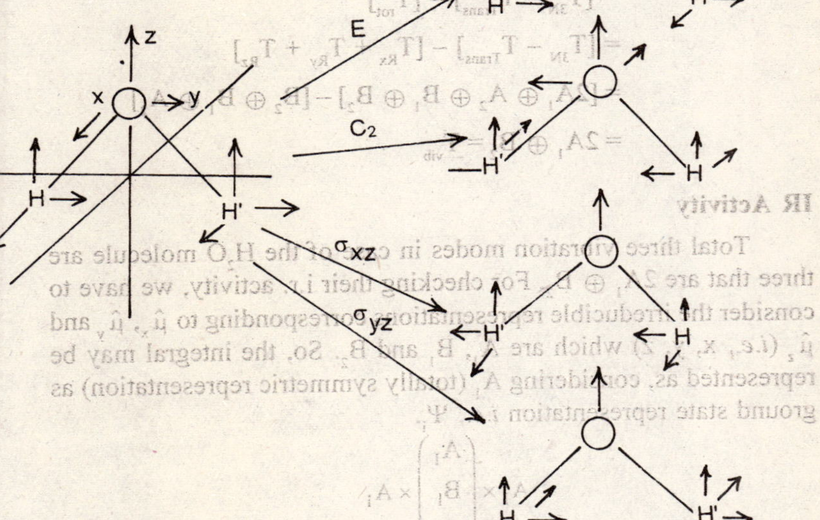

Fig. 10.4

So, T_{3N} is

C_{2v}	E	C_2	σ_{xz}	σ_{yz}
T_{3N}	9	–1	1	3

On reduction, T_{3N} results the following irreducible representations

$$T_{3N} = 3A_1 \oplus A_2 \oplus 2B_1 \oplus 2B_2.$$

The irreducible representation corresponding to translation mode *i.e.,* corresponding to the x, y, z basis functions are B_1, B_2, and A_1. So, on subtracting the translation modes, we are left with

$$\begin{aligned} T_{3N} - T_{Trans} &= T_{3N} - (T_x + T_y + T_z) \\ &= (3A_1 \oplus A_2 \oplus 2B_1 \oplus 2B_2) - (B_1 \oplus B_2 \oplus A_1) \\ &= 2A_1 \oplus A_2 \oplus B_1 \oplus B_2 \end{aligned}$$

The irreducible representations corresponding to the rotation modes *i.e.*, corresponding to the R_x, R_y and R_z basis functions are B_2, B_1 and A_2. So, on subtracting these modes also we are left with the normal or the vibration modes for the H_2O molecule *i.e.*,

$$[T_{3N} - T_{Trans}] - [T_{rot}]$$
$$= [T_{3N} - T_{Trans}] - [T_{Rx} + T_{Ry} + T_{Rz}]$$
$$= [2A_1 \oplus A_2 \oplus B_1 \oplus B_2] - [B_2 \oplus B_1 \oplus A_2]$$
$$= 2A_1 \oplus B_2 = T_{vib}$$

IR Activity

Total three vibration modes in case of the H_2O molecule are three that are $2A_1 \oplus B_2$. For checking their i.r. activity, we have to consider the irreducible representations corresponding to $\hat{\mu}_x$, $\hat{\mu}_y$ and $\hat{\mu}_z$ (*i.e.*, x, y, z) which are A_1, B_1 and B_2. So, the integral may be represented as, considering A_1 (totally symmetric representation) as ground state representation *i.e.*, Ψ_i.

$$A_1 \times \begin{pmatrix} A_1 \\ B_1 \\ B_2 \end{pmatrix} \times A_1$$

In the results of this direct product we will also get one representation as A_1 (*i.e.*, totally symmetric representation), so, the corresponding integral for $A_1(\Psi_i) \rightarrow A_1(\Psi_j)$ transition will be non-zero and the transition will be allowed or we can say that the A_1 mode is i.r. active. Second vibration mode *i.e.*, again $A_1(\Psi_j)$ may be considered if we take the direct products of $A_1B_1A_1$ or $A_1B_2A_1$. So, out of $2A_1$ normal modes only one is i.r. active. For B_2 normal mode the corresponding integral may be represented as

$$A_1 \times \begin{pmatrix} A_1 \\ B_1 \\ B_2 \end{pmatrix} \times B_2$$

In the above direct products one direct product *i e.*, $A_1B_2B_2$ yield the totally symmetric A_1 irreducible representation. Therefore the transition $A_1(\Psi_1) \rightarrow B_2(\Psi_1)$ is allowed and B_2 vibration is *i.r.*, active.

Raman activity: In order to check the Raman activity we

have to consider polarizability operator *i.e.,* α which is the function of x^2, y^2, z^2, xy, xz, yz, or $x^2 - y^2$ and in this case the corresponding representations are A_1, A_2, B_1 and B_2. On considering the corresponding integrals in the same way we can find out that all the three modes in case of water molecule are Raman active. As the integrals corresponding to these normal modes are

$$A_1 \times \begin{pmatrix} A_1 \\ A_2 \\ B_1 \\ B_2 \end{pmatrix} \times A_1$$

And

$$A_1 \times \begin{pmatrix} A_1 \\ A_2 \\ B_1 \\ B_2 \end{pmatrix} \times B_2$$

Results of above direct products yield the totally symmetric irreducible representations. So, all the three modes in case of water molecule are Raman active. The same results are true for SO_2 which also belongs to C_{2v} point group.

According to 3N–6 also there are 3(3) – 6 = 3 *i.e.,* three modes possible for the water molecule.

Example 2: Let us consider NH_3 molecule. Its point group is C_{3v}. Considering x, y, z axes on the molecule T_{xyz} can be represented as

C_{3v}	E	$2C_3$	$3\sigma_v$
T_{xyz}	3	0	1
No. of Unshifted Atoms	4	1	2
T_{3N}	12	0	2

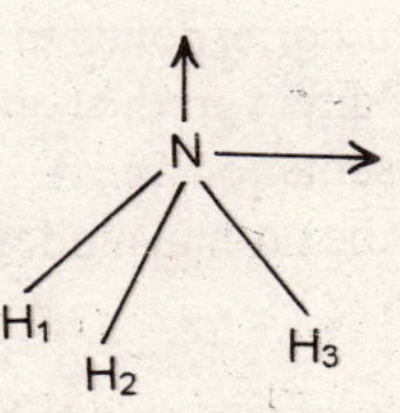

T_{3N} can be obtained by the multiplication of number of un-shifted atoms with T_{xyz} as shown above. On reduction T_{3N} we get $3A_1 \oplus A_2 \oplus 4E$.

So, $T_{3N} = 3A_1 \oplus A_2 \oplus 4E$

On subtracting the irreducible representations corresponding to translation modes *i.e.*, the irreducible representations whose basis functions are x, y, z, we get

$$T_{3N} - T_{trans} = (3A_1 \oplus A_2 \oplus 4E) - (A_1 \oplus E)$$
$$= 2A_1 \oplus A_2 + 3E$$

Now, on subtracting the irreducible representation corresponding to the rotation modes (*i.e.*, corresponding to the rotational vectors R_x, R_y, R_z), we get

$$=[T_{3N} - T_{trans}] - [T_{rot}]$$
$$=[2A_1 \oplus A_2 \oplus 3E] - [A_2 \oplus E]$$
$$=2A_1 \oplus 2E$$

So, the normal modes in case of NH_3 are $2A_1 + 2E$. According to 3N—6 the normal modes are $3(4) - 6 = 6$ *i.e.*, two of A_1 and two of E (doubly degenerate and hence 4 vibrations) symmetry species. So, total six modes are there which are represented as

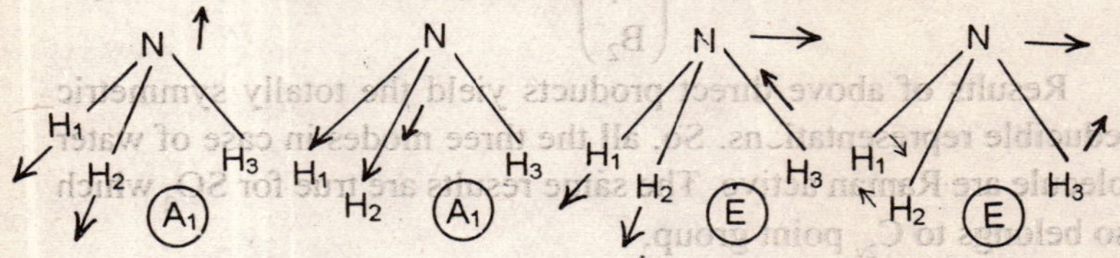

IR activity: For checking the i.r. activity of these modes, let us consider the following integrals

$$A_1 \times \begin{pmatrix} A_1 \\ E \end{pmatrix} \times A_1$$

As μ corresponds to the A_1 to E irreducible representations. In the above direct product one is $A_1A_1A_1$, whose result is totally symmetric representation *i.e.*, A_1, so, the result of this direct product is non-zero. So, the normal modes A_1 are i.r. active. Similarly for 2E modes.

$$A_1 \times \begin{pmatrix} A_1 \\ E \end{pmatrix} \times A_1$$

C_{3v}	E	$2C_3$	$3\sigma_v$
E^2	4	1	0

This representation shown above can be reduced to $A_1 \oplus A_2 \oplus$

E as it contains A_1 irreducible representation, so, the modes corresponding 2E are also i.r. active. In conclusion we can say that for NH_3 all the modes are i.r. active.

Raman activity : As is the function of x^2, y^2, z^2, xy, xz, yz, or $x^2 - y^2$. From the character table of C_{3v} point group it is clear that irreducible representations for are A_1 and E. Therefore, similarly as shown above, this can be shown that all the modes *i.e.*, $2A_1 + 2E$ in case of NH_3 are Raman active also.

Example 3: Let us consider the example of CO_2. Its point group is $D_{\infty h}$. Considering x, y, z co-ordinates T_{xyz} can be obtained as

$D_{\infty h}$	E	$2C_\infty(\phi).....\infty\sigma_v$	'i'	$2S_\infty(\phi).....\infty C_2'$
T_{xyz}	3	$1+2\cos\phi....1$	–3	$-1+2\cos\phi.....-1$
No. of Un-shifted Atoms	3	3..............3	1	1...................1
T_{3N}	9	$3+6\cos\Phi....3$	–3	$-1+2\cos\Phi.....-1$

On subtracting the irreducible representations for translation modes (*i.e.*, representations for x, y, z basis functions)

$D_{\infty h}$	E	$2C_\infty(\phi).....\infty\sigma_v$	'i'	$2S_\infty(\phi).....\infty C_{2'}$	
T_{3N}	9	$3+6\cos\phi....3$	–3	$-1+2\cos\phi.....-1$	
T_{trans} (i.e.π_u^+,Σ_u^+)	3	$1+2\cos\phi....1$	–3	$-1+2\cos\phi.....-1$	xyz
$T_{3N}-T_{trans}$	6	$2+4\cos\phi...2$	0	0...................0	
T_{rot} (π_g)	2	$2\cos\phi.......0$	2	$-2\cos\phi..........0$	
T_{vib} ($T_{3N}-T_{trans}-T_{rot}$)	4	$2+2\cos\Phi...2$	–2	$2\cos\Phi..........0$	

On subtracting the T_{rot} *i.e.*, representations corresponding to rotational vectors, we get T_{vib} as shown above which on reduction gives the following normal modes: Σ_g^+, Σ_u^+, π_u

I.R. activity: As the totally symmetric state is the Σ_g^+ which is the representation for ground state and for μ whose components are μ_x μ_y μ_z.π_u and Σ_u^+ irreducible representations corresponds to x, y, z,. So in the integral the direct product combinations are

$$\Sigma_g^+ \times \begin{pmatrix} \Sigma_u^+ \\ \pi_u \end{pmatrix} \times \begin{matrix} \Sigma_g^+ \\ \Sigma_u^+ \\ \pi_u \end{matrix}$$

From the above direct product combinations it is clear that only Σ_u^+, π_u are the i.r. active modes as all g states are i.r. inactive or otherwise the integral where the combination is gug will be a vanishing integral. Since Σ_g^+ must be a totally symmetric vibration *i.e.,* there is no change in the dipole moment in this vibration. So, it is i.r. inactive. In conclusion only Σ_u^+, π_u are i.r. active modes.

Raman activity : For Raman activity we have to take into consideration the polarizability which is the function of x^2, y^2, z^2, xy, xz, yz, or $x^2 - y^2$ and the corresponding irreducible representations for these basis functions are Σ_g^+, π_g and Δ_g. Therefore, the integral can be shown in the following direct products

$$\Sigma_g^+ \times \begin{pmatrix} \Sigma_u^+ \\ \pi_u \\ \Delta_g \end{pmatrix} \times \begin{matrix} \Sigma_g^+ \\ \Sigma_u^+ \\ \pi_u \end{matrix}$$

Only the first vibration mode *i.e.,* Σ_g^+ would yield the non-zero value for the integral, so, this mode is Raman active but Σ_u^+, π_u are Raman inactive modes.

Rule of mutual exclusion: The above observation for CO_2 molecule can be framed as a rule *i.e.,* rule of mutual exclusion according to which "In case of linear molecules which possess centre of symmetry, i.r. active vibrations (or the normal modes) are Raman inactive and vice-versa". Here in this case (*i.e.,* in CO_2) Σ_g^+ is i.r. inactive but Raman active mode but Σ_u^+, π_u modes are i.r. active but Raman inactive modes. Therefore, CO_2 is the exclusive example of this rule *i.r.,* rule of mutual exclusion.

Section 10.5
Sepctra of metal complexes

IR spectroscopy: When ligands or molecules get co-ordinated with the central metal atom to form a complex, some properties of the ligand change and a M-L bond is formed in case of complex. IR is an important tool for the structure elucidation for complexes. The co-ordination of ligand molecule with the metal ion can change the symmetry of the molecule. It is normally reduced and hence the bands which are i.r. inactive in the free ion or molecule may become i.r. active on the co-ordination and the degenerate vibrations may split up in the complex due to the lowering of symmetry. The lowering of symmetry in complexes involving poly-atomic anionic ligands can be caused by a number of mechanisms. This may be due to site symmetry lowering. This effect arises because of the fact that the point symmetry selection rule, which are based on the isolated molecules, do not hold good in the crystalline state and hence degenerate vibrations become active. The magnitude of splitting caused by this method is generally very small. Let us now consider the various ions whose symmetry may change on co-ordination.

(A) Complexes containing perchlorate ions: The free ion has regular tetrahedral structure and it belongs to point group T_d, symmetry having nine vibration degrees of freedom distributed between four normal modes of vibrations viz. υ_1, the A_1Type (C*l*O) stretching (Raman active) υ_2, the E type (OC*l*O) bending (Raman active), υ_3, the T_2 type (C*l*O) stretching and the υ_3 T_2 type (C*l*O) bending (both i.r. and Raman active). Isolated ClO_4 ion does show only two absorptions in the i.r. region due to υ_3 (at **ca 1100**cm^{-1}). The causes for the lowering of symmetry are (i) covalent bonding of the perchlorate to the metals through one or two of its oxygen atoms (ii) non-uniform field due to the ligand molecules around the cation and (iii) distortion of this perchlorate tetrahedron in the crystal lattice. In covalent co-ordinated perchlorates υ_3 (~1100 cm^{-1}) and υ (~620 cm^{-1}) is split while the T_d symmetry is lowered to C_{3v}, the triple degeneracy is removed and the υ_1 (~930 cm^{-1}) and υ (~460 cm^{-1}) modes become i.r. active. The occurence of two strong bands at about 1090-1085 cm^{-1} and 650-620 cm^{-1} in the spectra of perchlorate complexes is attributed to υ_3 and υ_4 vibrations of the ionic perchlorate with T_d symmetry suggested that the perchlorate ions remain outside the co-ordination sphere. Same discussion is valid for sulfato ligand.

(B) Complexes with nitrite ions: Nitrite ions may be present in the complex in three form (i) nitrite anionic ligand (ii) NO_2^- co-ordinated to the metal atom through nitrogen atom. Such complexes are termed as nitro complexes. (iii) NO_2^- co-ordinated through one oxygen atom, nitrito complexes. Anionic NO_2^- group has C_{2v} symmetry and show the *i.r.*, bands at ~1335 cm^{-1} (υ_3), 1250 cm^{-1} (υ_1) and 850 cm^{-1} (υ_2). In N-co-ordinated nitrito ligand, C_{2v} symmetry of the ion is maintained. So, v_3, v_1, v_2 bands appear with some change in their positions. A new band appear at ~625 cm^{-1} corresponds to out of plane bending mode of NO_2^-. If the co-ordination is through oxygen the symmetry is reduced to C_s. The two bands occurs at ~1468 cm^{-1} and ~1065 cm^{-1} and υ_2 occurs at 825 cm^{-1}.

(C) Complexes containing nitrate ligands: The free nitrate ion has D_{3h} symmetry and has four fundamental vibrations viz. υ_1, the A_1 type (NO stretching), υ_2 the A_2 type (No bending), υ_3 the E type (NO stretching) and υ_4 the E type (NO bending). On lowering the symmetry from D_{3h} to C_{2v} or C_s, all the vibration modes split, but the number of infra-red active fundamental vibrations is same for C_{2v} and C_s symmetry. Uni-dentate, symmetrical bi-dentate and bridging nitrate groups are possible which may possess C_{2v} symmetry. Lever separation rule is used to distinguish between uni-dentate and bi-dentate ligands. According to this separation rule the separation ($\upsilon_1 - \upsilon_4$) for mono-dentate nitrate groups appear at 5–25 cm^{-1} and for bi-dentate groups it is 20–60 cm^{-1}. Same is the case with carbonate ligand.

In this way group theory is helpful in the study of i.r. spectra of metal complexes. It is also useful in the Raman spectral studies of metal complexes.

Raman spectroscopy: Raman spectroscopy is also helpful to study the M-L vibration, since M-L bonds are relatively weaker, the stretching and the bending skeletal virbations of the complex occur at low energy. These vibrations may occur below 400 cm^{-1} or even below 250 cm^{-1} and may be beyond the range of i.r. spectrophotometer. But Raman shifts corresponding to the low energy vibrations can be easily seen in the Raman spectrum of a complex, *e.g.*, in case of a square planar complex with amine ligands, three Raman lines are expected and only two Raman lines corresponding to M-N are observed at about 525 cm^{-1} (A_{1g}) and 505 cm^{-1} (B_{2g}) in the case of $[Pt(NH_3)_4]Cl_2$. In case of tetrahedral complexes there are four vibrations two stretching and two bending and all are Raman active. In case of $[Zn(NH_3)_4Cl_2$ these vibrations occur at 435 and 410 cm^{-1} with A_1 and T_2 symmetries.

Vibronic Coupling: It is mechanism which suggest that there is coupling of vibration and electronic wave functions. Some of the vibrations of the complex distorn the octahedron in such a way that the centre of symmetry is destroyed as the vibration takes place. The states of d^n configuration are no longer retain their g character *i.e.*, centre of symmetry. In such a condition the d-d transitions become slightly allowed This disturbance from the centre of symmetry during normal vibrations can be shown in the Fig. 10.5. The process of vibronic coupling can be explained effectively on the basis of group theory.

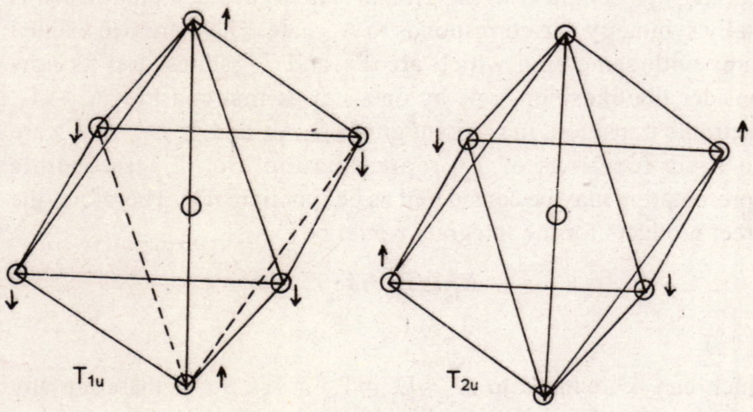

Fig. 10.5

The vibration wave function of a molecule can be written as the product of individual vibration which may be 3N–6 where N is the number of atoms in the molecule. The vibration wave function may be taken as product of 3N–6 functions each of one normal modes. Usually, we express the Ψ_v the vibration wave function belonging to ground state as totally symmetric. In the excited state the normal mades may be totally symmetric or not. The total function may be written as the product of three wave functions which are Ψ_{vib}, Ψ_{rot} and Ψ_{elc}. So,

$$\Psi = \Psi_{elec}\,\Psi_{rot}\,\Psi_{vib}$$

So,

$$H\Psi_{elec} = E_{elec}\Psi_{elec}$$
$$H\Psi_{vib} = E_{vib}\Psi_{vib}$$
$$H\Psi_{rot} = E_{rot}\Psi_{rot}$$

The total energy of the molecular system in such a case will be

$$E = E_{elec} + E_{vib} + E_{rot}$$

In the above shown total wave function the Ψ_{rot} the rotational wave function may be assumed as independent of two. So, in the transition moment integral we must consider only

$$\int \Psi_{elec}\Psi_{vib}\,\hat{\mu}\,\Psi_{vib}\Psi_{elec}d\tau$$

In order to show this concept, let us take example of $[Co(NH_3)_6]^{3+}$ complex. In the ground state of this complex which is totally symmetry can corresponds to A_{1g} state. There are two excited states with same spin which are T_{1g} and T_{2g} states. Let us now consider the transitions one by one. Let us first consider $A_{1g} \rightarrow T_{1g}$ electronic transition. In O_h point group $\hat{\mu}_x$, $\hat{\mu}_y$ and $\hat{\mu}_z$ *i.e.*, x, y, z are the basis functions of T_{1u} representation. So, T_{1u} irreducible representation may be considered as be function of μ. Therefore, the direct products for the integrals would be

$$A_{1g} \otimes T_{1u} \otimes T_{1g}$$
$$= T_{1u} \otimes T_{1g} = T_{red}$$

which can be reduced to $A_{1u} \oplus E_u \oplus T_{1u} \oplus T_{2u}$. So, if there are any normal mode whose first excited states belong to any of these above mentioned irreducible representations these will be non-zero integral, and for AB_6 octahedral molecule this can be shown that the normal modes are A_{1g}, E_g, $2T_{1u}$ and T_{2u}. Therefore, while pure electronic transition $A_{1g} \rightarrow T_{1g}$ is not allowed, all those transitions in which there is simultaneous excitation of a vibration of T_{1u} or T_{2u} are symmetry allowed due to vibronic coupling. Similarly for $A_{1g} \rightarrow T_{2g}$ by direct product method the integral is

$$A_{1g} \otimes T_{1u} \otimes T_{2g}$$
$$= T_{1u} \otimes T_{2g}$$

which may be reduced to $A_{2n} \oplus E_u \oplus T_{1u} \oplus T_{2u}$. Therefore, by the same reason $A_{1g} \rightarrow T_{2g}$ transition can occur if there is simultaneous excitation of a T_{1u} or T_{2u} vibration.

In case of tetrahedral complexes, there is no centre oi symmetry and the respective orbitals don't have 'g'or 'u' designation. The designation E and T_2 may be given to the orbitals in tetrahedral symmetry. The E orbitals are non-bonding and purely metal d atomic orbitals, so, they retain their 'g' character. The T_2 molecular orbitals are formed by mixing of d and p orbitals *i.e.*, by the mixing of orbitals of character 'g' and 'u' and this dp mixing imparts some 'u' character to the m.o., therefore the orbital selection rules can be relaxed in this case as a result of which the molar absorptivity of the tetrahedral complexes is approximately 10^2 greater than that of octahdedral complexes.

Problems

Q.1. Give the SALC's corresponding to bond stretching and in and out of plane bending for BF_3.

Ans. BF_3 has D_{3h} point group. Considering the bond vectors as basis function in each mode the correspond SALC's are

B–F stretching $(A_1') \; 1/\sqrt{3}\,(\phi_1 + \phi_2 + \phi_3)$

$(E') \; 1\sqrt{6}\,(2\phi_1 - \phi_2 - \phi_3)$

$1/\sqrt{2}\,(\phi_2 - \phi_3)$

In plane bending $(E') \; 1/\sqrt{6}\,(2\phi_1 - \phi_2 - \phi_3)$

$1/\sqrt{2}\,(\phi_2 - \phi_3)$

Out of plane bending $(A_2'') \; 1/\sqrt{3}\,(\Phi_1 + \Phi_2 + \Phi_3)$

Q. 2. For $POCl_3$ and $PtCl_4^{2-}$ find out the normal modes.

Ans. $POCl_3$ has point group C_{3v}. The reducible T_{3N} representation can be find out and reduced to

$$T_{3N} = 4A_1 \oplus A_2 \oplus 5E$$

$PtCl_4^{2-}$ has point group D_{4h} and the corresponding T_{3N} is

$$T_{3N} = A_{1g} \oplus A_{2g} \oplus B_{1g} \oplus B_{2g} \oplus E_g \oplus 2A_{2u} \oplus B_{2u} \oplus 3E_u$$

On subtracting the T_{rot} and $T_{trans.}$ We get

$$T_{vib}\,(POCl_3) = 3A_1 \oplus 3E \text{ Similarly}$$

$$T_{vib}\,(PtCl_4^{2-}) = A_{1g} \oplus B_{1g} \oplus B_{2g} \oplus A_{2u} \oplus B_{2u} \oplus 2E_u$$

Q.3. For SO_2 molecule, the following spectroscopic data are available. Predict the geometry of the molecule.

Frequency (cm^{-1})	IR	Raman
519	Active	Active
	(PQR)	(polarized)
1151	Active	Active

	(PQR)	(polarized)
1361	Active (PQR)	Active

Ans. As all the transitions include PQR branches, so, this may be concluded that molecule may not be linear *i.e.,* non-linear. All the lines are i.r. and Raman active so, there is absence of centre of symmetry in this molecule, (according to rule of mutual exclusion), As some Raman lines are polarized, which gives the indication that the oxygen atoms are placed symmetrically with respect to s-atom. Therefore, its geometry will be

Appendix–I

Chapter-wise Assignment and Problems for Revision

Chapter–1

Q.1. Define the following terms:

(1) Symmetry Element
(2) Symmetry operation
(3) Proper axis of rotation
(4) Improper axis of rotation
(5) Plane of symmetry
(6) Inversion centre
(7) Identity

with examples.

Q.2. List the symmetry operations related to following:

(1) C_5 (2) C_4 (3) S_3 (4) i^2 (5) S_6

[**Hint:** List the related symmetry operations to these elements, as it is mentioned in the Q.1 of the problem set of chapter–1].

Q.3. Which of the following molecules contain dipole moment

(1) CO_2 (2) BF_3 (3) SF_6 (4) trans-$C_2H_2Cl_2$

[**Hint:** In these molecules identify the various symmetry elements such as C_n axes, 'i' and plane etc. and accordingly look for the conditions for existence of dipole moment in a molecule].

Q.4. Which of the following molecules possess the inversion centre:

(1) BF_3 (2) $CHCl_3$ (3) trans-$C_2H_2Cl_2$ (4) cis-$C_2H_2Cl_2$
(5) PCl_5

Q.5. How is symmetry related to optical activity?
[**Ref.** Section 1.3]

Chapter–2

Q.6. Systematically find the point groups for the following molecules:

(1) $[PtCl_4]^{2-}$ (2) H_3PO_4 (3) NO (4) SO_3
(5) C_6H_5OH (6) Allene (7) CH_3Cl (8) CO_2
(9) XeF_4 (10) CH_3CHO

Q.7. Give the equivalent operations for the following

(1) $C_3^1.C_3^2$ in C_{3v} point group

(2) C_2.i in case of C_{2h} point group

(3) $\sigma_v.\sigma_\upsilon^1$ in case of C_{2v} point group

(4) $C_2.\sigma_h$ in C_{2h} point group

Q.8. What is a group multiplication table? Explain by taking example of C_{2h} point group.
[**Ref.** Section 2.5]

Q.9. Out of the following point group which has the centre of inversion as group element.

D_2, D_{2h}, C_{2h}, C_{3v}, D_{5h} and C_i

Q.10. Out of the following which point group represent the group of low symmetry.

C_2, C_i, C_s, C_{2v}, C_{3v}, C_1 and D_2

Chapter–3

Q.11. Explain the matrix representations for various symmetry operations.
[**Ref.** Section 3.2]

Q.12. Construct the group multiplication table for C_{2v} and C_{3v} point group using matrices for different symmetry operations.
[**Ref.** Section 3.3]

Q.13. What do you understand by reducible and irreducible representation?
[**Ref.** Section 3.5]

Q.14. State and explain C the great orthogonality theorem (G.O.T.). Explain the reduction formula.
[**Ref.** Section 3.6 and 3.7]

Q.15. What do you understand by character table? Explain its various portions.
[**Ref.** Section 3.8]

Q.16. State and explain the Mulikan notations for various representations.
[**Ref.** Section 3.7]

Q.17. Find the missing characters

C_{3v}	E	C_3	$3\sigma_v$
A_1	1	1	1
A_2	1	X_1	X_2
E	2	–1	0

Chapter-4

Q.18. Explain the postulates of quantum mechanics.
[**Ref.** Section 4.1]

Q.19. Give the criteria for a well behaved wave function.
[**Ref.** Section 4.2]

Q.20. Explain the shapes of various orbitals.
[**Ref.** Section 4.3]

Q.21. Explain the effect of transformation operator on the p_z, p_y and p_x orbitals in case of C_{2v} point group.
[**Ref.** Section 4.4]

Q.22. What do you understand by direct product representation?
[**Ref.** Section 4.5]

Q.25. Find the following direct products.

(1) $C_4v : B_1B_2$

(2) $C_2v : A_1A_2$

(3) $C_3v : A_1E$

(4) $C_3v : A_2E$

(5) $C_2v : A_1B_1$

Chapter–5

Q.24. Explain the concept of projection operators
[**Ref.** Section 5.1]

Q.25. Give the use of projection operators to construct SALC's.
[**Ref.** Section 5.2]

Chapter–6

Q.26. State the assumptions of valance bond theory and explain the formation of σ and π orbitals.
[**Ref.** Section 6.1]

Q.27. Explain the concept of hybridization the light of Group theory taking suitable examples.
[**Ref.** Section 6.2]

Chapter–7

Q.28. Explain the points on which the molecular orbital theory is based. Explain the L.C.A.O.–M.O. concept.
[**Ref.** Section 7.1]

Q.29. Enlist the similarities and differences in the Valance bond and molecular orbital theories.
[**Ref.** Section 7.2]

Q.30. Give and explain the M.O. diagrams for:

(1) σ-bonding in H_2O molecule

(2) σ-bonding in a planar AB_3 molecule

(3) σ-bonding in CH_4 molecule

[**Ref.** Section 7.3]

Q.31. Explain the concept of π-bonding in AB_n molecules in light of M.O. theory.

[**Ref.** Section 7.3]

Q.32. Draw the M.O. diagrams for π electron acceptor and π electron donor ML_6 metal complexes.

[**Ref.** Section 7.3]

Q.33. Give the assumptions of the Huckel Molecular Orbital Approach.

[**Ref.** Section 7.4]

Q.34. Draw and explain the M.O. diagram of Ferrocene

[**Ref.** Section 7.5]

Chapter–8

Q.35. Explain the splitting of d-orbitals in octahedral and in tetrahedral fields.

[**Ref.** Section 8.1]

Q.36. How is C.F.S.E. useful to explain the variation of heats of hydrations? Explain taking example of 3d series.

[**Ref.** Section 8.3]

Q.37. Explain the square planar and tetragonal distortions in the complexes.

[**Ref.** Section 8.1]

Q.38. Explain the limitations of crystal field theory.

[**Ref.** Section 8.4]

Chapter–9

Q.39. State and explain the Franck-Condon rule.

[**Ref.** Section 9.1]

Q.40. What are vanishing integrals? What is their significance?

Q.41. Write a note on spectral transition probabilities.
[**Ref.** Section 9.2]

Q.42. Explain the Laporte selection rules.

Q.43. Explain the concept of term symbols. Construct the Orgal diagram for d^2–configuration.
[**Ref.** Section 9.5, 9.6]

Q.44. Explain the concept of 'Hole Formalism'.
[**Ref.** Section 9.6]

Q.45. Write a note on Correlation diagram.
[**Ref.** Section 9.6]

Chapter–10

Q.46. Explain the terms:
(1) Stretching vibrations
(2) Bending virbations

Q.47. What are normal modes? Explains the step-wise procedure to find out the normal modes group theoretically.
[**Ref.** Section 10.4]

Q.48. Give the systematic normal modes analysis for:
(1) H_2O (2) NH_3 and (3) CO_2 molecules
[**Ref.** Section 10.4]

Q.49. Explain the concept of vibronic coupling taking suitable example.
[**Ref.** Section 10.5]

Appendix–II

Some Common Character Tables

C_1	E
A	1

C_s	E	σ		
A′	1	1	x, y, R_z	x^2, y^2, z^2, xy
A″	1	–1	z, R_x, R_y	yz, xz

C_i	E	'i'		
A_g	1	1	R_x, R_y, R_z	x^2, y^2, z^2, xy, yz, xz
A_u	1	–1	x, y, z	

C_2	E	C_2		
A	1	1	z, R_z	x^2, y^2, z^2, xy
B	1	–1	x, y, R_x, R_y	yz, xz

C_3	E	C_3^1	C_3^2		$\varepsilon = \exp(2\pi i/3)$
A	1	1	1	z, R_z	x^2+y^2, z^2
E	1	ε	ε*	x,y R_x, R_y	x^2-y^2, xy, yz, xz
	1	ε*	ε		

C_4	E	C_4^1	$C_4^2(C_2)$	C_4^3		
A	1	1	1	1	z, R_z	x^2+y^2, z^2
B	1	–1	1	–1		x^2-y^2, xy
E	1	i	–1	–1	x,y, R_x, R_y	yz, xz
	1	–i	–1	i		

C_5	E	C_5	C_5^2	C_5^3	C_5^4		$\varepsilon = \exp(2\Pi i/5)$
A	1	1	1	1	1	z, R_z	x^2+y^2, z^2
E_1	1	ε	ε^{2*}	ε^{2*}	ε^*	x, y, R_x, R_y	yz, xz
	1	ε^*	ε^{2*}	ε^2	ε		
E_2	1	ε^2	ε^*	ε	ε^{2*}		x^2--y^2, xy
	1	ε^{2*}	ε	ε^*	ε^2		

D_2	E	$C_2(x)$	$C_2(y)$	$C_2(z)$		
A	1	1	1	1		x^2, y^2, z^2
B_1	1	–1	1	–1	z, R_z	xy
B_2	1	–1	1	–1	y, R_y	xz
B_3	1	1	–1	–1	x, R_x	yz

D_3	E	$2C_3$	$3C_2$		
A_1	1	1	1		x^2+y^2, z^2
A_2	1	1	–1	z, Rz	
E	2	–1	0	x,y R_x, R_y	x^2-y^2, xy, xz, yz

D_4	E	$2C_4$	$C_4^2(=C_2)$	$2C_2'$	$2C_2''$		
A_1	1	1	1	1	1		x^2+y^2, z^2
A_2	1	1	1	–1	–1	z, R_z	
B_1	1	–1	1	1	–1		x^2-y^2
B_2	1	–1	1	–1	1		xy
E	2	0	–2	0	0	x,y R_x, R_y	xz, yz

D_6	E	$2C_6$	$2C_3$	C_2	$3C_2'$	$3C_2''$		
A_1	1	1	1	1	1	1		x^2+y^2, z^2
A_2	1	1	1	1	–1	–1	z, R_z	
B_1	1	–1	1	–1	1	–1		
B_2	1	–1	1	–1	–1	1		
E_1	2	1	–1	–2	0	0	x,y R_x, R_y	xz, yz
E_2	2	–1	–1	2	0	0		x^2-y^2, xy

C_{2v}	E	C_2	σ_{xz}	σ_{yz}		
A_1	1	1	1	1	z	x^2, y^2, z^2
A_2	1	1	–1	–1	R_z	xy
B_1	1	–1	1	–1	x, R_y	xz
B_2	1	–1	–1	1	y, R_x	yz

C_{3v}	E	$2C_3$	$3\sigma_v$		
A_1	1	1	1	z	x^2+y^2, z^2
A_2	1	1	–1	R_z	
E	2	–1	0	x,y R_x, R_y	x^2-y^2, xy, xz, yz

C_{2h}	E	C_2	'i'	σ_h		
A_g	1	1	1	1	R_z	x^2+y^2, z^2
B_g	1	–1	1	–1	R_x,R_y	xz,yz
A_u	1	1	–1	–1	z	
B_u	1	–1	–1	1	x,y	

C_{3h}	E	C_3	C_3^2	σ_h	S_3	S_3^5	$\varepsilon = \exp$	$(2\pi i/3)$
A'	1	1	1	1	1	1	R_z	x^2+y^2, z^2
E'	1	ε	ε*	1	ε	ε*	x,y	x^2-y^2,xy
	1	ε*	ε	1	ε*	ε		
A"	1	1	1	–1	–1	–1	z	
E"	1	ε	ε*	–1	–ε	–ε*		
	1	ε*	ε	–1	–ε*	–ε	Rx,Ry	xy, xz

C_{4h}	E	C_4	C_2	C_4^3	'i'	S_4^3	σ_h	S_4		
A_g	1	1	1	1	1	1	1	1	R_z	x^2+y^2, z^2
B_g	1	–1	1	–1	1	–1	1	–1		x^2-y^2, xy
E_g	1	i	–1	–i	1	i	–1	–i	R_x,R_y	xy,xz
	1	–i	–1	i	1	–i	–1	i		
A_u	1	1	1	1	–1	–1	–1	–1	z	
B_u	1	–1	1	–1	–1	1	–1	1		
E_u	1	i	–1	–i	–1	–i	1	i		
	1	–i	–1	i	–1	i	1	–1	x,y	

D_2h	E	$C_2(z)$	$C_2(y)$	$C_2(x)$	'i'	σ_{xy}	σ_{xz}	σ_{yz}		
A_g	1	1	1	1	1	1	1	1		x^2,y^2,z^2
B_{1g}	1	1	–1	–1	1	1	–1	–1	R_x	xy
B_{2g}	1	–1	1	–1	1	–1	1	–1	R_y	xz
B_{3g}	1	–1	–1	1	1	–1	–1	1	R_z	yz
A_u	1	1	1	1	–1	–1	–1	–1		
B_{1u}	1	1	–1	–1	–1	–1	1	1	z	
B_{2u}	1	–1	1	–1	–1	1	–1	1	y	
B_{3u}	1	–1	–1	1	–1	1	1	–1	x	

D_{3h}	E	$2C_3$	$3C_2$	σ_h	$2S_3$	$3\sigma_v$		
A_1'	1	1	1	1	1	1		x^2+y^2, z^2
A_2'	1	1	–1	1	1	–1	R_z	
E'	2	–1	0	2	–1	0	x, y	x^2-y^2,xy
A_1''	1	1	1	–1	–1	–1		
A_2''	1	1	–1	–1	–1	1	z	
E"	2	–1	0	–2	1	0	R_x,R_y	xz,yz

D_{2d}	E	$2S_4$	C_2	$2C_2'$	$2\sigma_d$		
A_1	1	1	1	1	1		x^2+y^2, z^2
A_2	1	1	1	–1	–1	R_z	
B_1	1	–1	1	1	–1		x^2-y^2
B_2	1	–1	1	–1	1	z	xy
E	2	0	–1	0	0	x,y R_x,R_y	xz,yz

D_{3d}	E	$2C_3$	$3C_2$	'i'	$2S_6$	$3\sigma_d$		
A_{1g}	1	1	1	1	1	1		x^2+y^2, z^2
A_{2g}	1	1	–1	1	1	–1	R_z	x^2-y^2, xy, xz, yz
E_g	2	–1	0	2	–1	0	R_xR_y	
A_{1u}	1	1	1	–1	–1	–1		
A_{2u}	1	1	–1	–1	–1	1	z	
E_u	2	–1	0	–2	1	0	x,y	

D_{4d}	E	$2S_8$	$2C_4$	$2S_8^3$	C_2	$4C_2'$	$4\sigma_d$		
A_1	1	1	1	1	1	1	1		x^2+y^2, z^2
$A_{\cdot}$	1	1	1	1	1	–1	–1	R_z	
B_1	1	–1	1	–1	1	1	–1		
B_2	1	–1	1	–1	1	–1	1	z	
E_1	2	2	0	–2	–2	0	0	x,y	
E_2	2	0	–2	0	2	0	0		x^2-y^2,xy
E_3	2	–2	0	2	–2	0	0	R_x,R_y	xz,yz

S_4	E	S_4	C_2	S_4^3		
A	1	1	1	1	R_z	x^2+y+z^2
B	1	–1	1	–1	z	x^2-y^2,xy
E	1	–i	–1	–i		
	1	–i	–1	I	x,y R_x, R_y	xz,yz

S_6	E	C_3	C_3^2	'i'	S_6^5	S_6		ε=exp(2πi/3)
A_g	1	1	1	1	1	1	R_z	x^2+y^2, z^2
E_g	1	ε	ε*	1	ε	ε*		x^2-y^2, xy
	1	ε*	ε	1	ε*	ε	R_x, R_y	xy, yz
A_u	1	1	1	–1	–1	–1	z	
E_u	1	ε	ε*	–1	–ε	–ε*		
	1	ε*	ε	–1	–ε*	–ε	x,y	

T	E	$4C_3$	$4C_3^2$	$3C_2$		$\varepsilon = \exp(2\pi i/3)$
A	1	1	1	1		$x^2+y^2+z^2$
E	1	ε	ε^*	1		$2z^2-x^2-y^2$, x^2-y^2
	1	ε^*	ε	1		
T	3	0	0	–1	R_x, R_y, R_z, x, y, z	xy, xz, yz

T_d	E	$8C_3$	$3C_2$	$6S_4$	$6_{\sigma d}$		
A_1	1	1	1	1	1		$x^2+y^2+z^2$
A_2	1	1	1	–1	–1		
E	2	–1	2	0	0		$2z^2-x^2-y^2$, x^2-y^2
T_1	3	0	–1	1	–1	R_x, R_y, R_z	
T_2	3	0	–1	–1	1	x, y, z	xy, xz, yz

O	E	$6C_4$	$3C_2(=C_4^2)$	$8C_3$	$6C_2$		
A_1	1	1	1	1	1		$x^2+y^2+z^2$
A_2	1	–1	1	–1	–1		
E	2	0	2	–1	0		$2z^2-x^2-y^2$, x^2-y^2
T_1	3	1	–1	0	–1	R_x, R_y, R_z x, y, z	
T_2	3	–1	–1	0	1		xy, xz, yz

O_h	E	$8C_3$	$6C_2$	$6C_4$	$3C_2(=C_4^2)$	'i'	$6S_4$	$8S_6$	$3_{\sigma h}$	$6_{\sigma d}$
A_{1g}	1	1	1	1	1	1	1	1	1	1
A_{2g}	1	1	–1	–1	1	1	–1	1	1	–1
E_g	2	–1	0	0	2	2	0	–1	2	0
T_{1g}	3	0	–1	1	–1	3	1	0	–1	–1
T_{2g}	3	0	1	–1	–1	3	–1	0	–1	1
A_{1u}	1	1	1	1	1	–1	–1	–1	–1	–1
A_{2u}	1	1	–1	–1	1	–1	1	–1	–1	1
E_u	2	–1	0	0	2	–2	0	1	–2	0
T_{1u}	3	0	–1	1	–1	–3	–1	0	1	1
T_{2u}	3	0	1	–1	–1	–3	1	0	1	–1

$C_{\infty v}$	E	$2C_\infty(\Phi)$......$\infty\sigma_v$		
$A_1 = \Sigma^+$	1	1................1	z	x^2+y^2, z^2
$A_2 = \Sigma^-$	1	1.............–1	R_z	
$E_1 = \pi$	2	2cos ϕ.........0	x,y R_x, R_y	xz, yz
$E_2 = \Delta$	2	2cos2 ϕ.......0		x^2-y^2, xy
$E_3 = \phi$	2	2cos3 ϕ.......0		

$D_{\infty h}$	E	$2C_\infty(\phi)$.....$\infty\sigma_v$	'i'	$2S_\infty(\phi)$....∞C_2		
Σ_g^+	1	1................1	1	1................1		x^2+y^2, z^2
Σ_g^-	1	2.............–1	1	1.............–1	R_z	
Π_g	2	2cosϕ.........0	2	–2cosϕ.......0	R_x,R_y	xz, yz
Δ_g	2	2cos2ϕ........0	2	2cos2ϕ.......0		x^2-y^2, xy
	...	... .	...	... .		
Σ_u^+	1	1.................1	–1	–1..............1	z	
Σ_u^-	1	1..............–1	–1	–1..............1		
Π_u	2	2cosϕ............0	–2	2cosϕ.........0	x,y	
Δ_u	2	2cos2ϕ........0	–2	–2cos2ϕ......0		
	...	... .	...	... .		

Bibliography

1. *Chemical Applications of Group Theory*–F. A. Cotton.
2. *A Textbook of Advanced Inorganic Chemistry*–F. A. Cotton and G. Wilkinson.
3. *Group Theory in Chemistry and Physics*–M. G. Arora.
4. *Group Theory and its Application to Chemistry*–K. V. Raman.
5. *Group Theory and its Chemical Applications*–P. K. Bhattacharya.
6. *Group Theory and Symmetry in Chemistry*–Gurdeep Raj, Ajay Bhagi and Vinod Jain.
7. *Introduction to Ligand Fields*–B. N. Figgis.
8. *Atomic Structure and Chemical Bond*–Manas Chanda.
9. *Introduction to Ligand Field Theory*–C. J. Ballhausen.
10. *Inorganic Electronic Spectroscopy*–A. B. P. Lever.
11. *Infra-red and Raman Spectra of Inorganic and Co-ordination Compounds*–K. Nakamoto.
12. *Valency and Molecular Structure*–E. Cartmell and G.W. Fowles.
13. *Physical Methods in Inorganic Chemistry*–R.S. Drago.
14. *An Introduction to Quantum Chemistry*–A. K. Chandra.

15. *Quantum Chemistry Including Spectroscopy*–B. K. Sen.

16. *Quantum Chemistry*–I.N. Levine.

17. *Quantum Chemistry*–R.K. Prasad.

18. *Quantum Mechanics*–G.R. Chatwal and S.R. Anand.

19. *A Textbook of Physical Chemistry*, Vol. 4–K.L. Kappor.

20. *Chemical Spectroscopy (in Hindi)*–O.P. Srivastava and Kishor Arora.

21. *Textbook of Molecular Spectrosocpy*–G.M. Barrow.

22. *Spectroscopy*–B.K. Sharma.

23. *Inorganic Chemistry, Principles of Structure and Reactivity* (Fourth Edition)–J.E. Huheey, E.A. Keiter, R.I. Kieter.

❑❑❑